ANATOMY
& MASSAGE

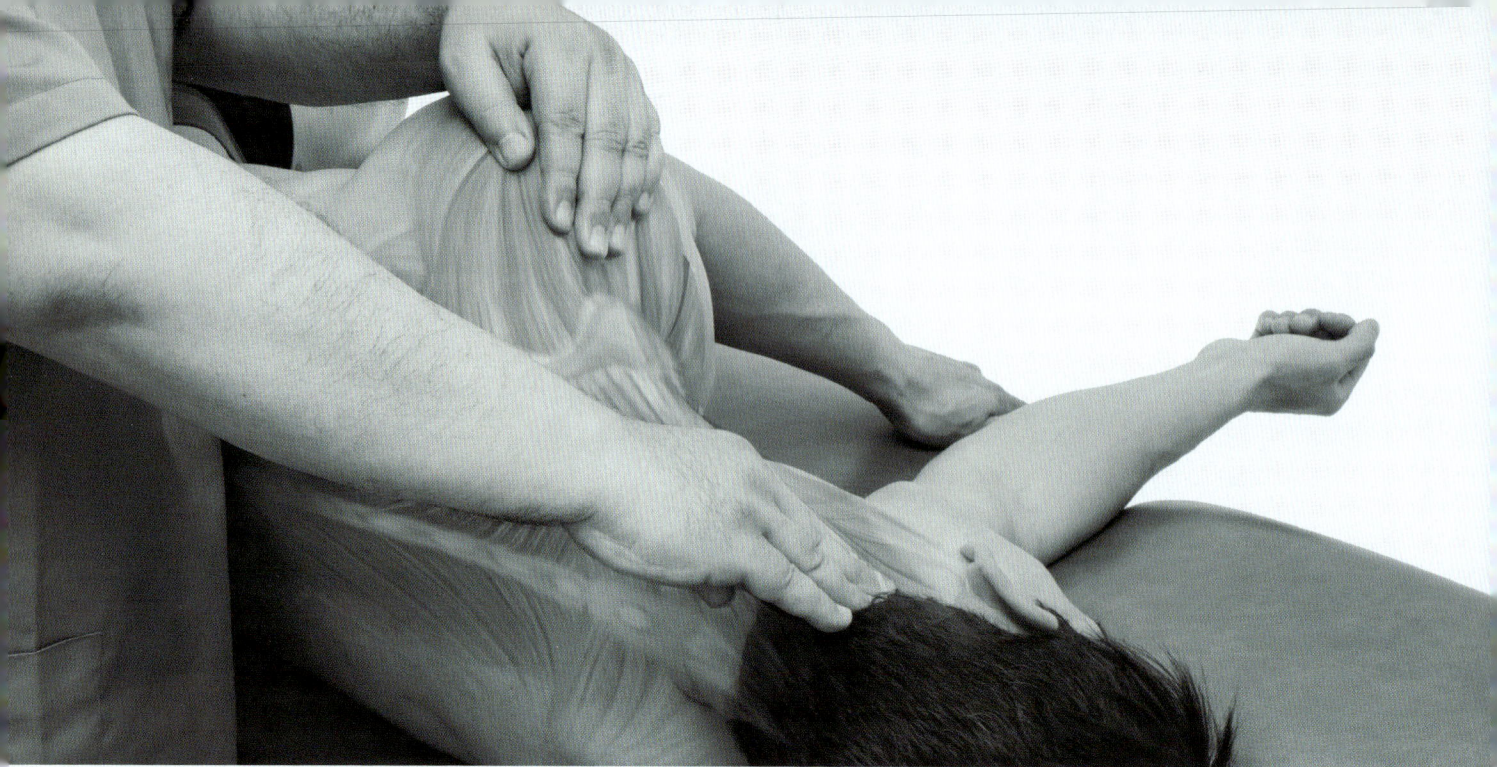

Compiled and produced by:
Editorial Paidotribo

Editorial direction: María Fernanda Canal
Scientific review: Prof. Víctor Götzens
Text: Josep Mármol Esparcia and Artur Jacomet Carrasco
Image coordination: Guillermo Seijas Albir
Corrections: Aurora Zafra, Roser Pérez
Graphic design: Toni Inglès
Illustrations: Myriam Ferrón
Photographs: Nos i Soto
Layout: Estudi Toni Inglès
Typesetting: www.satzstudio-hilger.de
Translator: Ian Hayden Jones
Pre-printing: Estudi Genís

Spanish original title:
Anatomía & Masaje Deportivo
© 2017 Editorial Paidotribo—World Rights
Published by Editorial Paidotribo, Spain

Exclusive worldwide publishing rights
ISBN: 978-1-78255-138-6
E-Mail: info@m-m-sports.com
www.m-m-sports.com
Printed in Spain

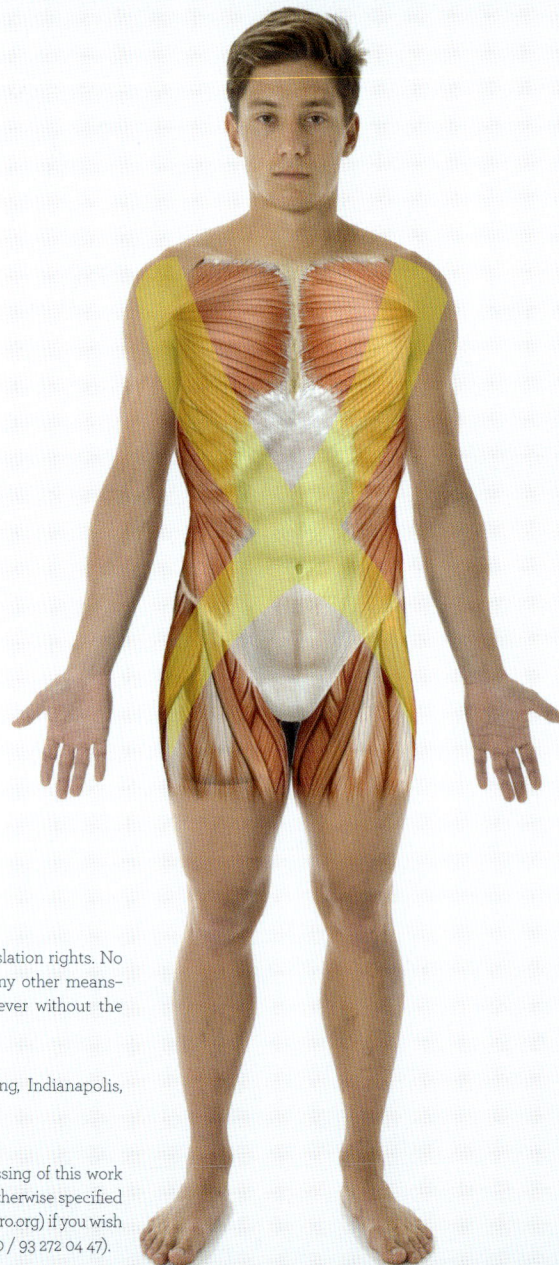

Anatomy & Massage
British Library Cataloguing in Publication Data
A catalogue record for this book is available from the British Library

Maidenhead: Meyer & Meyer Sport (UK) Ltd., 2018
ISBN: 978-1-78255-138-6

© 2018 by Meyer & Meyer Sport (UK) Ltd.
Aachen, Auckland, Beirut, Cairo, Cape Town, Dubai, Hägendorf, Hong Kong, Indianapolis, Manila, New Delhi, Singapore, Sydney, Tehran, Vienna
Member of the World Sport Publishers' Association (WSPA)

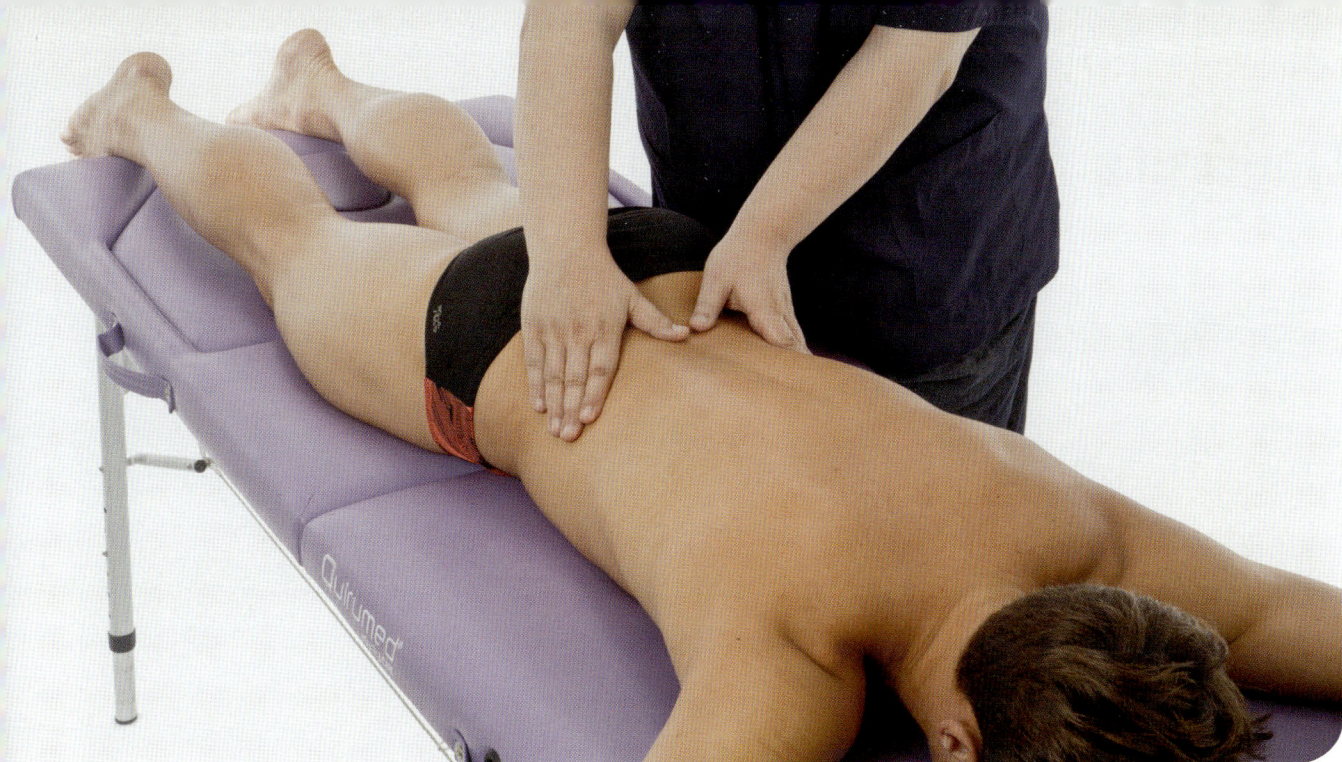

Josep Mármol, Artur Jacomet

ANATOMY & MASSAGE

Detailed illustrated techniques, including new insights
into massaging myofascial tissue

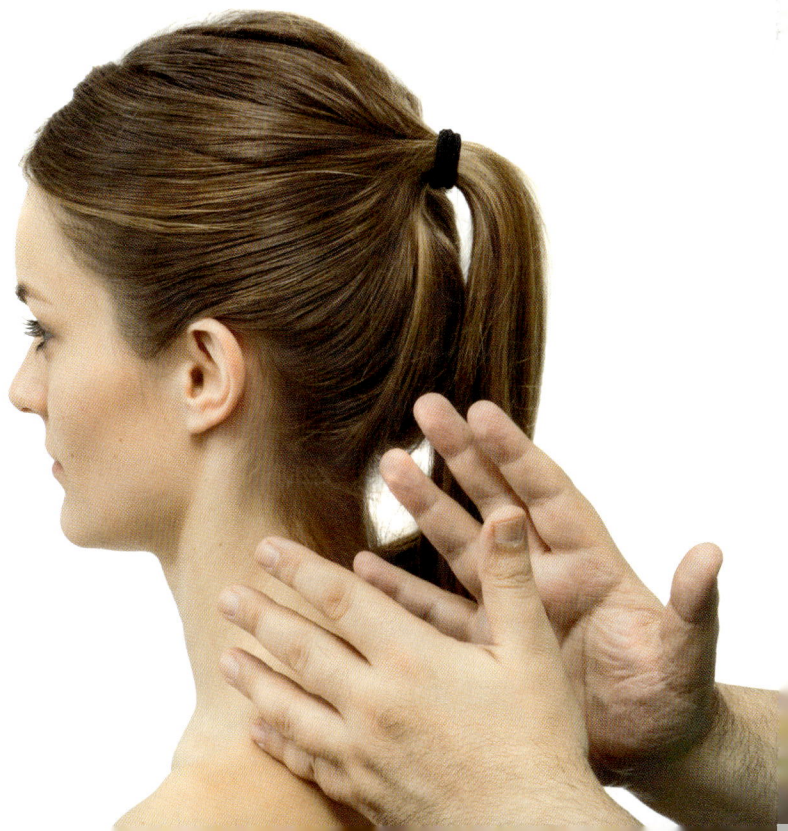

Paidotribo

Contents

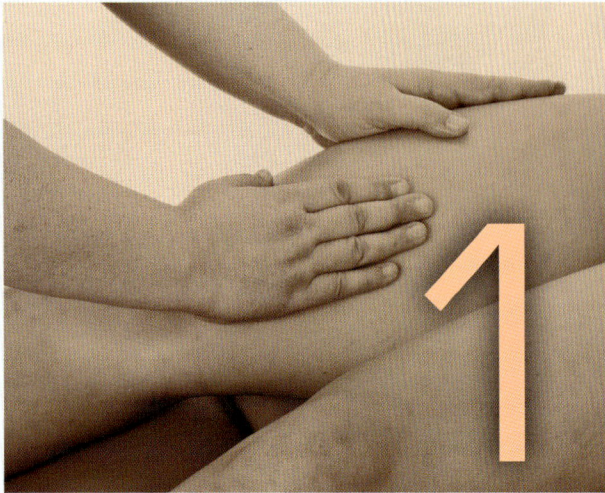

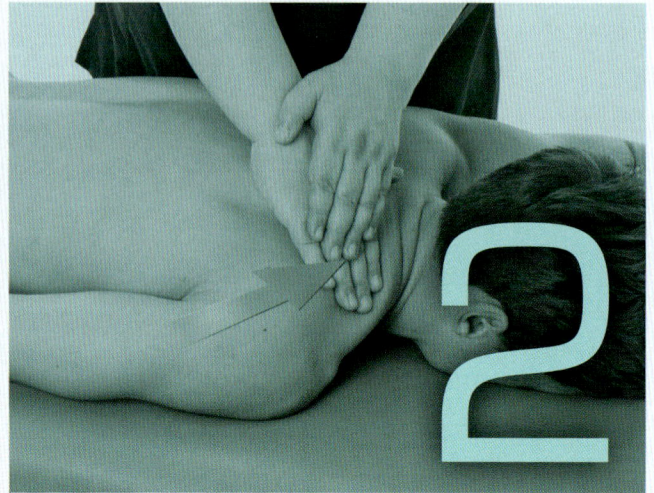

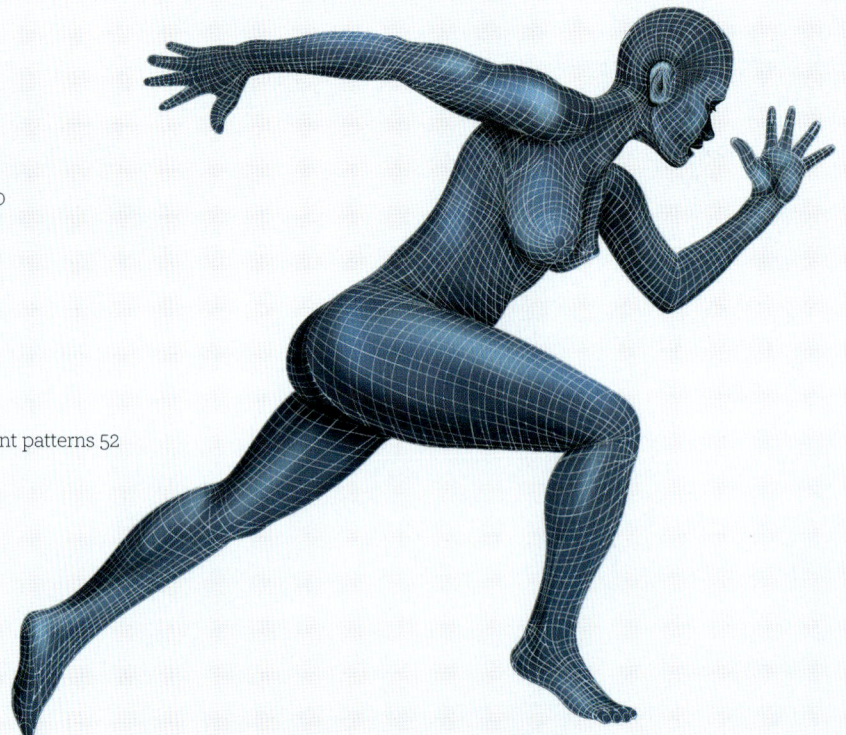

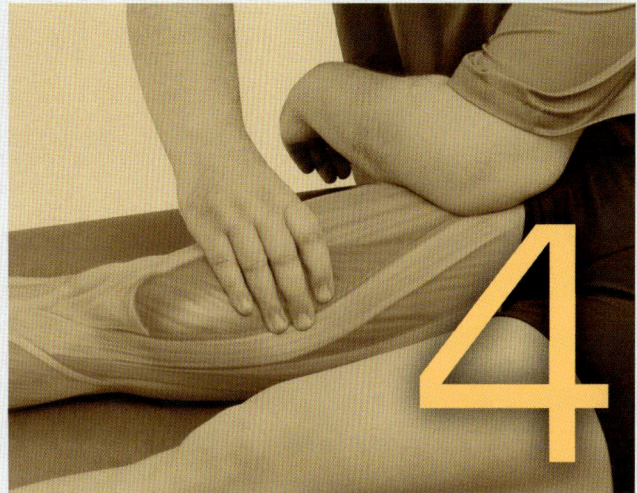

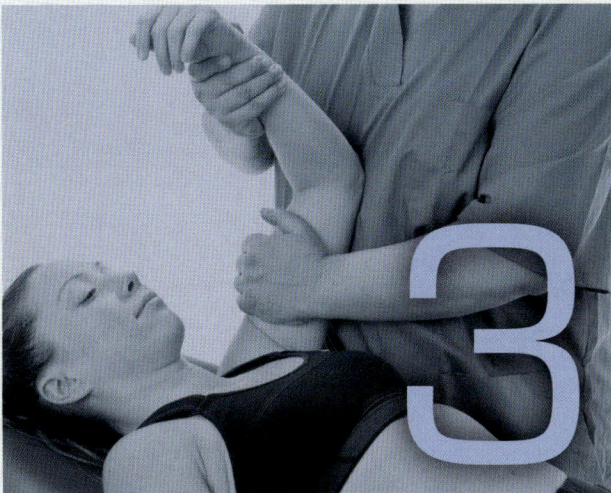

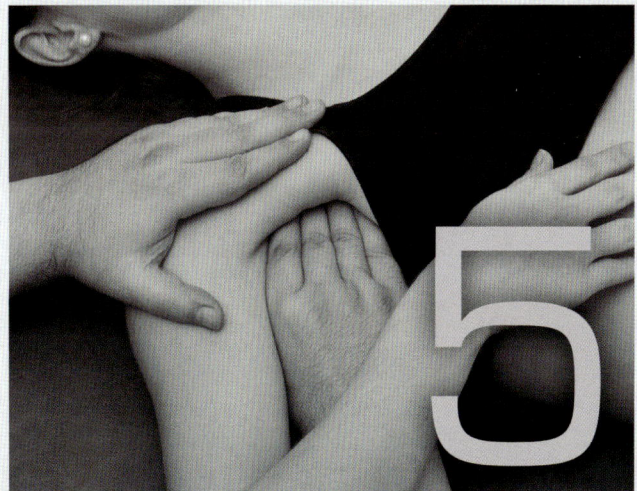

Introduction

Sports massage is a method of manual therapy integrated within conditioning programs for the professional athlete, enhancing efficiency, prevention injury and promoting health.

Sports massage for professionals and amateurs

With trained hands, massage provides continuous assessment, while seeking to promote venous return, reduce physical recovery times and avoid the risk of tissue damage.

High-performance sports teams are staffed with qualified personnel who apply massage on a regular basis, according to a training plan or given competition. But the amateur athlete may not enjoy these benefits and will only seek professional help when experiencing discomfort.

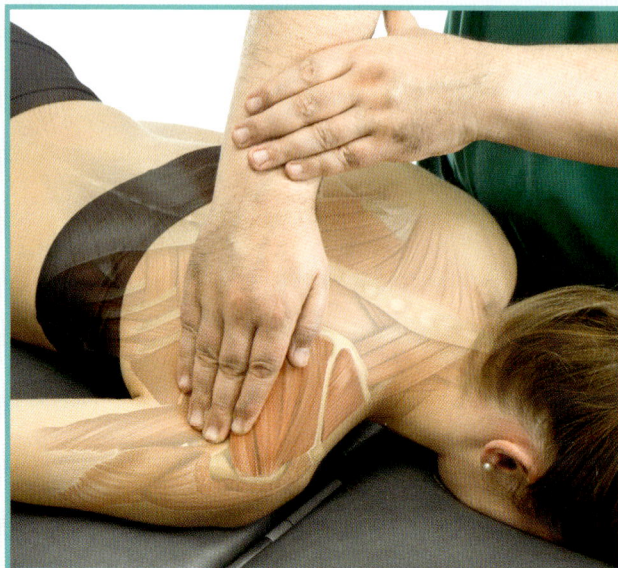

Purpose of this illustrated book

This work aims to be a useful tool for casual, amateur or high-performance athletes, as well as for all those professionals who want athletes' physical and emotional well-being to function efficiently. It is also a practical resource for the student and other curious learners who want to know more about this discipline.

This book describes in a practical way the implementation of manual massage according to the concept of "myofascial chains." This is a holistic formula for the valuation of bodily structures and the body as a whole, as well as their roles, position and movement. It will allow practitioners to note and subsequently see the results of their treatment.

The text has been combined with a large quantity of color illustrations, anatomical schematics, flow diagrams and video clips, demonstrating the different techniques required to facilitate its implementation.

We hope that *Anatomy & Massage* will enable better understanding of athlete massage and contribute to the general welfare of the population, in particular to the physically active who participate in sports.

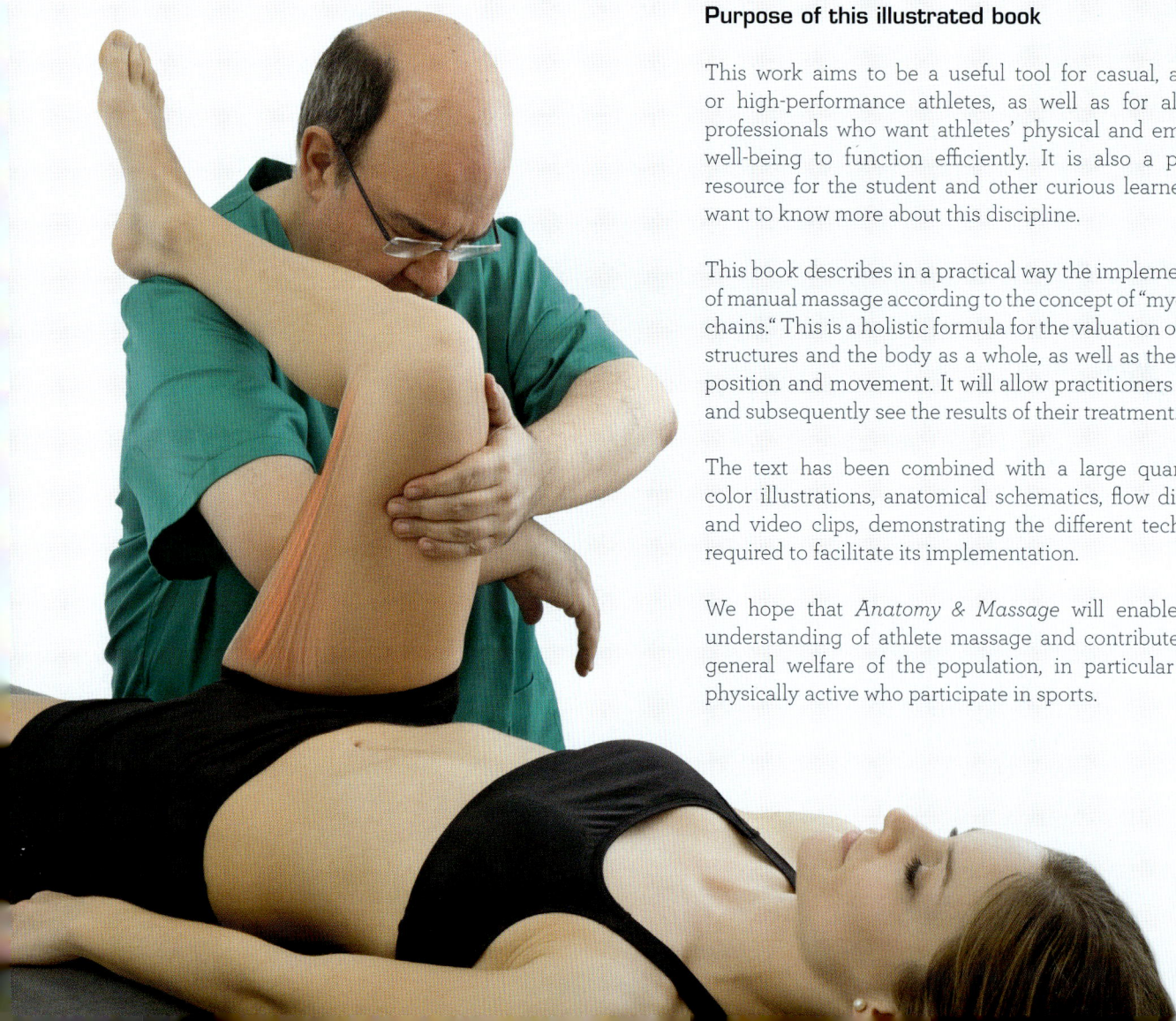

Additional content

In addition to the content presented in this book, *Anatomy & Massage* includes 21 tutorial videos, resulting in the most complete work in its field.

VIA THE WEBPAGE

Register for free on the webpage using the following credentials: **books2ar.com/pme**, then enter the following code:

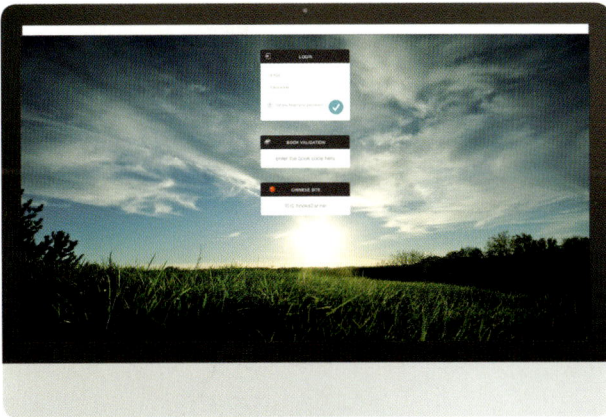

THREE EASY STEPS

1 Scratch the rectangle.

2 Access and enter the code.

3 Sign up!

VIA AUGMENTED REALITY

1. Download the free AR app from:
- **books2ar.com/pme**
- Scan these two codes:

QR for IOS

QR for Android

Available on the **App Store**

ANDROID APP ON **Google play**

■ Or search for Anatomy & Sports Massage AR in your device's official app store, Android or IOS

2. Use the app to scan the pages where this icon appears:

3. Explore the additional content.

THREE EASY STEPS

1 Download the free app.

2 Scan the image where the icon appears.

3 Explore the additional content!

Tutorial videos

Available on all pages where the icon appears.

The app requires an internet connection to access multimedia content.

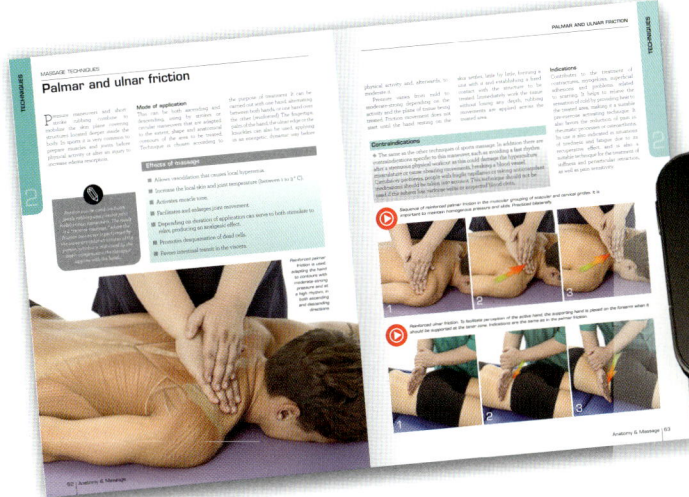

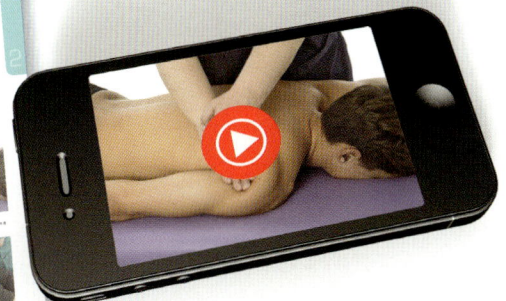

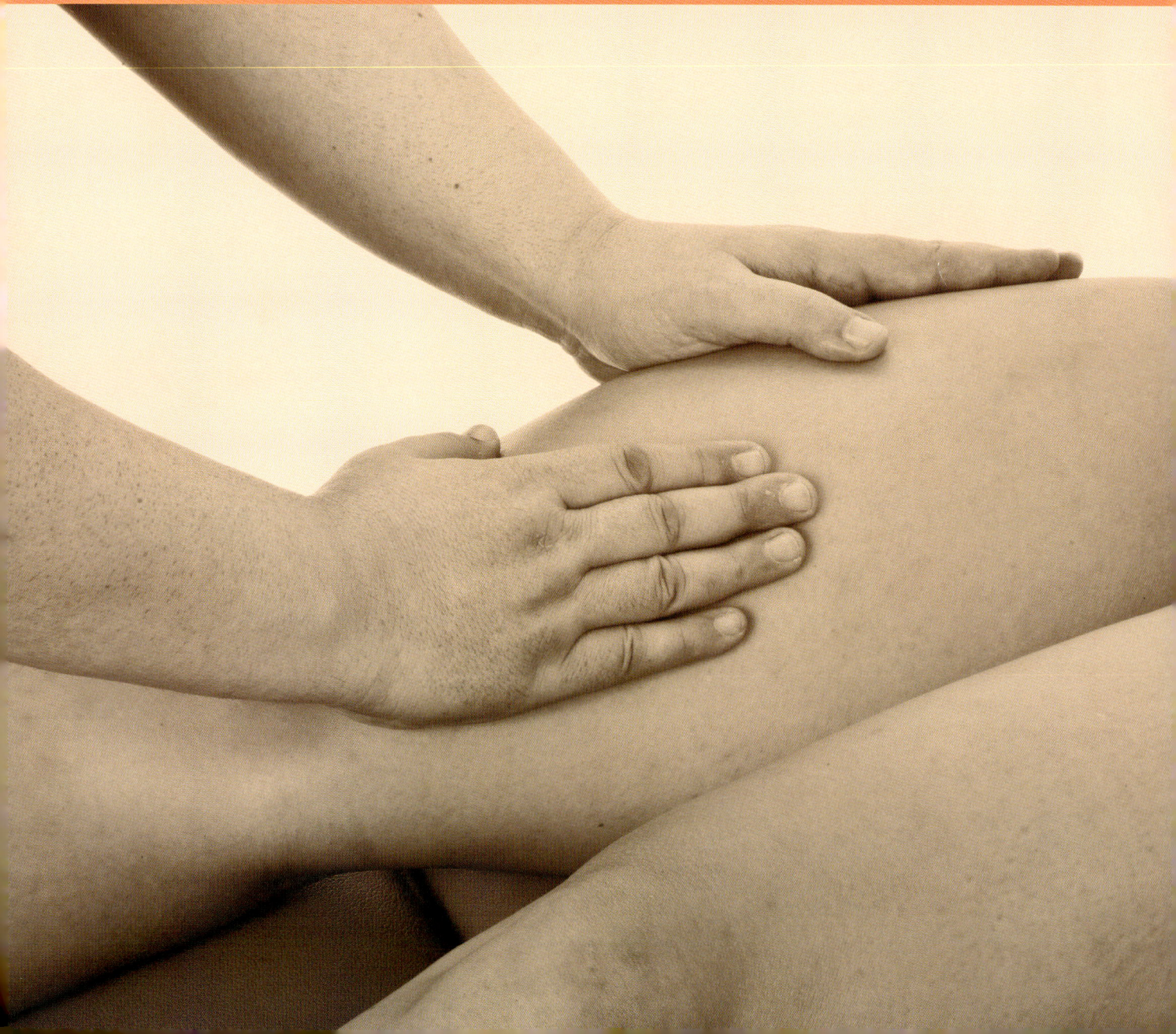

MASSAGE

1

In this first chapter the skill of manual massage is described in its most contemporary form. It underlines that both the definition of massage and the function of the masseur is not only to act on an isolated myofascial group or a certain joint, but also to work completely with a person. This is described in a comprehensive way, including diagrams and note boxes, the mechanisms of performance and the effects of the massage.

Sports massage techniques have been re-classified, according to the stage of preparation or competition in which the athlete is engaged, and presented here in three sections. The first identifies pre-competition massage. The second details the large amount of massage work during the competition. The third is post-competition massage, which includes recovery support in case of injury.

This section also uses the most recent studies and graphical representations to detail the varied fields of body massage at a psychoemotional level and its important role in body integration.

1

Different types of massage

How does manual massage work?

Massage provides a mechanical stimulus to the tissues by applying diverse forces on the skin using the hands, which in turn influences the nervous system and the myofascial framework, internal organs and circulation. But it is not limited to tissues or joints, as it is orientated to the person with full organic, intellectual and affective capacities.

Definition of massage

Massage is a form of non-verbal communication, a dialog addressing the manual treatment of the body, which is applied to the sum of the individual. It boosts physical-emotional performance and health, in general. Precisely for this reason its use should be subject to the time, activity and circumstances of the individual receiving it.

The difference between the various types of massages lies essentially in their approach to the symptoms. Another variable that distinguishes them is the anatomical plane where they apply maneuvers, rather than any increase of exerted pressure. That is why the practice of palpatory anatomy is necessary; a masseuse must identify by touch the structures to be dealt with to be sure that the massage affects them and obtains the desired results.

Massage performance areas

Manual massage allows the ongoing exploration of tissues, consciously at the start of treatment and then automatically as it progresses. The trained therapist can identify the sensitive areas with his hands and subsequently, if there are no contraindications, mobilize the tissues and promote modifications with direct reflexive or systemic results.

The combined actions of massage promote the dynamic of the bodily fluids, producing positive changes in the stability of organism and mood, reaching out to affect all aspects of a person.

The effect of massage

The benefits of manual massage are increased when used in a gradual and specific manner, moving from simple to complex. These benefits are amplified when it is part of a regular interdisciplinary program which involves reeducation of movement, nutrition plans and physical and psychological training as happens in professional sport.

Stone sculpture representing a young girl (kneeling) receiving a back massage. Khmer Empire (Cambodia), 6th century.

Hands which cure hands. A technique in which the digits are kneaded to reduce excess tension in the fascia of the palm of the hand.

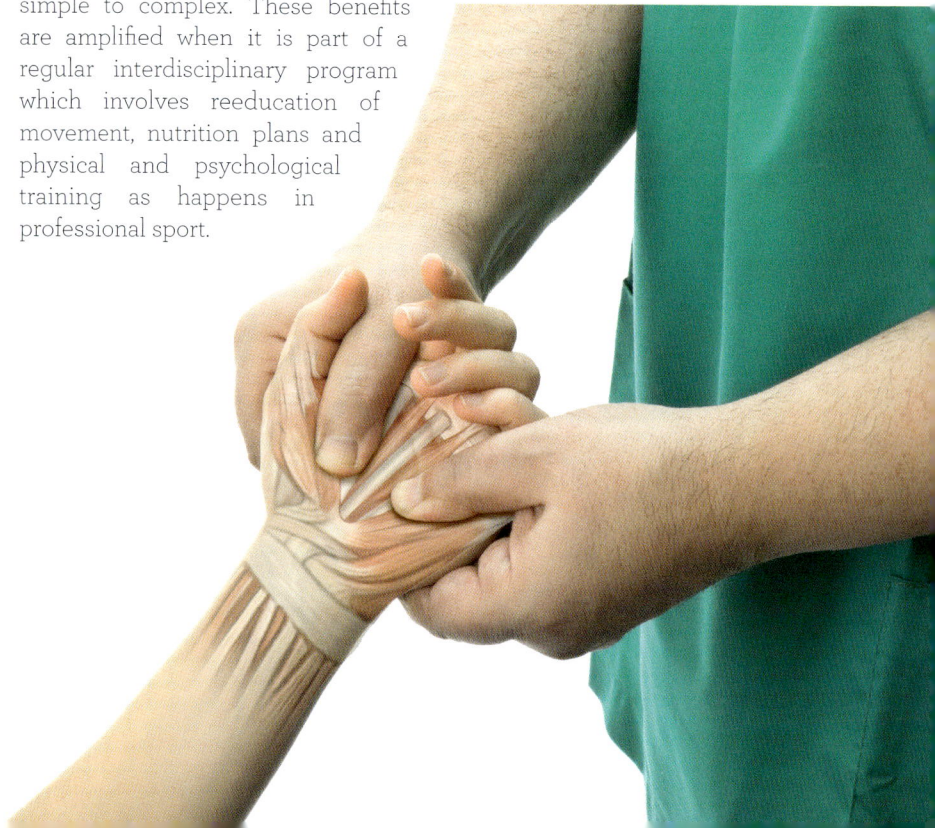

Massage has beneficial effects on mood, tiredness, fatigue and pain, although "...we must never forget that the success of massage depends on multiple effects inextricably linked to one another." Stork and Hoffa (1985).

Technique to release restrictions on the fascia and postural musculature that connect to the skull. Thumb pressure is very soft at the temples, while the other fingers hold the occipital bone, supporting the skull. Slight traction is maintained in the cranial direction at the time of expiration and is slightly loosened on inhalation, which results in a pumping movement to the whole area. This technique provides great comfort and can be used at the beginning or end of treatment.

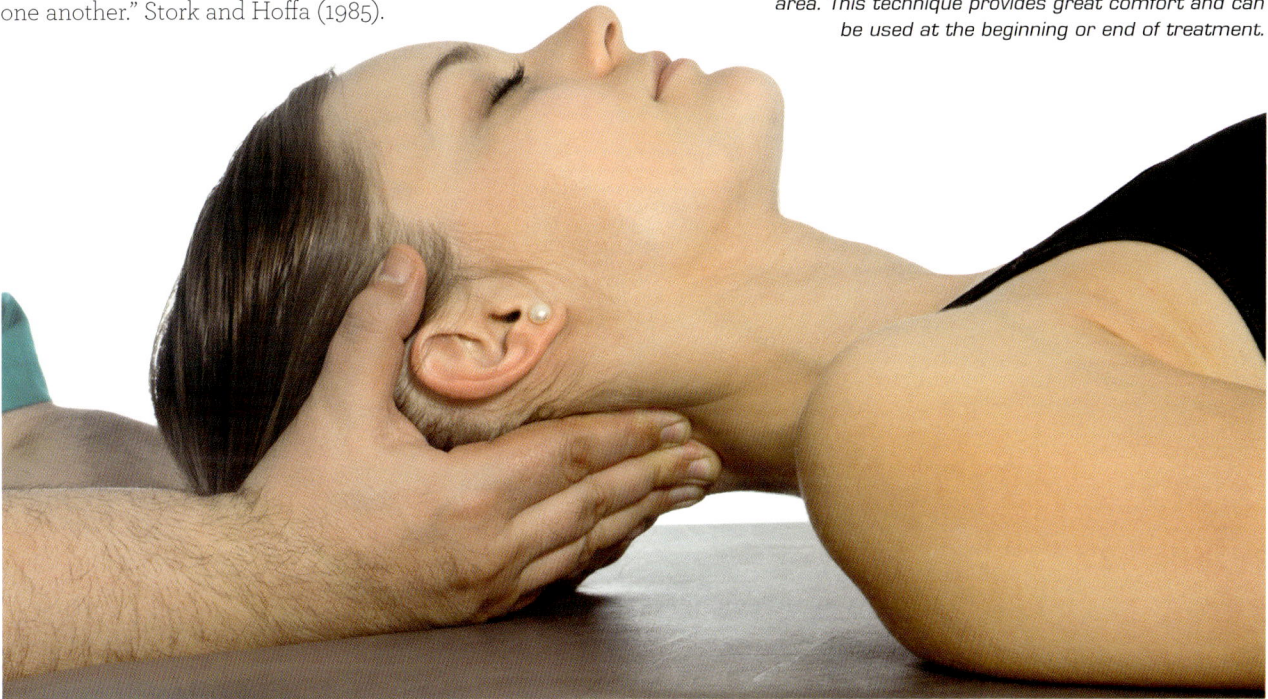

POTENTIAL EFFECTS OF MASSAGE

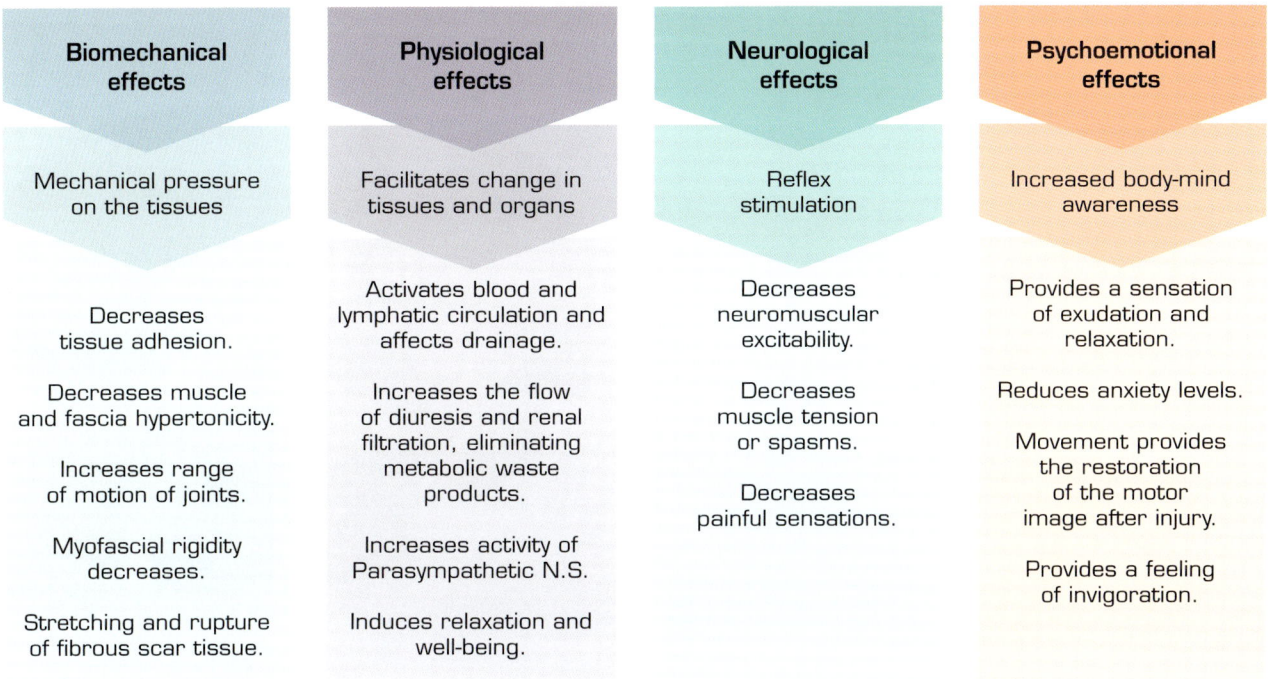

Biomechanical effects	Physiological effects	Neurological effects	Psychoemotional effects
Mechanical pressure on the tissues	Facilitates change in tissues and organs	Reflex stimulation	Increased body-mind awareness
Decreases tissue adhesion.	Activates blood and lymphatic circulation and affects drainage.	Decreases neuromuscular excitability.	Provides a sensation of exudation and relaxation.
Decreases muscle and fascia hypertonicity.	Increases the flow of diuresis and renal filtration, eliminating metabolic waste products.	Decreases muscle tension or spasms.	Reduces anxiety levels.
Increases range of motion of joints.		Decreases painful sensations.	Movement provides the restoration of the motor image after injury.
Myofascial rigidity decreases.	Increases activity of Parasympathetic N.S.		Provides a feeling of invigoration.
Stretching and rupture of fibrous scar tissue.	Induces relaxation and well-being.		

1

Massage for the athlete

During his training the athlete seeks to increase workload while maintaining a good state of health, and also requires improved results in competition. To do this the athlete must reduce the recovery time between workouts in order to accumulate the most training-recovery-overcompensation cycles in a given period, thus increasing sports performance to an optimum level in the shortest time possible.

Sports massage is one of the physical recovery treatments most in demand by athletes and the most common among their support teams. Its character is closely linked to the practice of sports; different techniques are used to promote and improve the athlete's physical-emotional performance and are oriented to prevent injury.

Its use and targets vary according to the athlete's characteristics, the training phase or competition, and the type of sport practiced.

The rhythm of maneuvers in sports massage varies depending on whether the aim is to stimulate or relax, although working deep-plane tissue is usually dominated by a combination of slow to very slow speed, especially when applied after strenuous workouts. This requires work that, by its nature, waits for a response before it can move forward, a transition that allows the masseuse to move on to another zone as the previous zone is already more relaxed or adjusted. Continuously applied massage helps to increase performance and promotes nerve activation — and therefore musculoskeletal function before a workout.

Regular application of massage to the athlete:

Allows for the exploration and identification of sensitive points and areas

Alleviates tension patterns in the body

Increases blood and lymphatic irrigation, thereby generating better cellular nutrition

Facilitates more fluid movement with less effort

Favors the phenomenon of overcompensation and adaptation

Reduces fatigue and recovery times between workouts

Aids physical and emotional recovery, preventing overtraining and risk of injury

Provides general relaxation

Collaborates in postural improvement

Carries out more intense and prolonged training, enhancing sports performance and health

---

Sports massage: compendium of techniques

In addition to massage techniques, sports massage employs a combination of techniques linked to objectives and the therapeutic indicators that they are aiming to achieve. The techniques that simultaneously influence the mobility of joints and tissues and stimulate or relax the athlete are: a) joint mobilization; b) stretching techniques; c) relaxation-breathing techniques; d) manual therapy of myofascial trigger points; e) relaxation through proprioceptive neuromuscular facilitation; and f) transverse friction techniques.

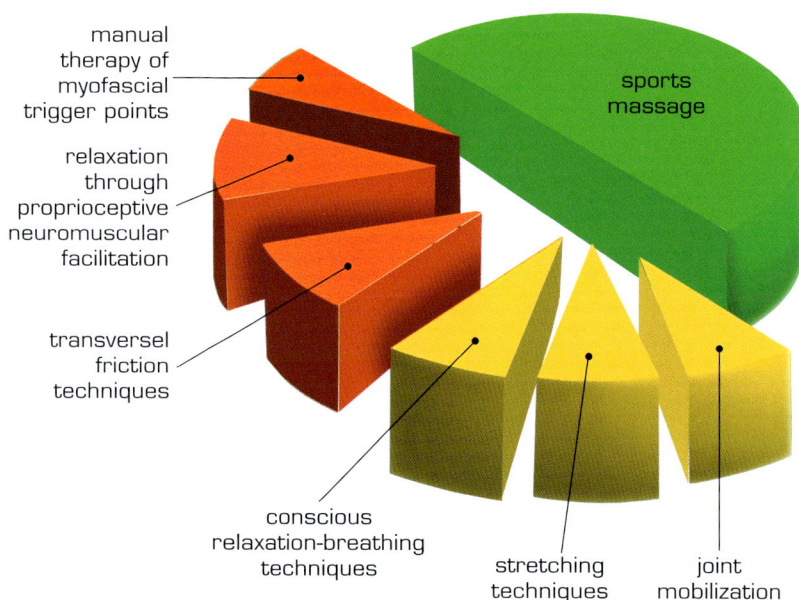

manual therapy of myofascial trigger points

relaxation through proprioceptive neuromuscular facilitation

transversel friction techniques

sports massage

conscious relaxation-breathing techniques

stretching techniques

joint mobilization

Sports massage: classification

Massage for the athlete is planned according the preparation phase or competition and in accordance with the objectives it is intended to achieve. The first classification divides sports massage into three interrelated groups: massage before competition, massage during competition and massage after competition

1. Massage before the competition or pre-season

Intended to promote fitness in the athlete before entering the competition stage.

2. Massage during competition

When entering the competition phase, the athlete requires differentiated care aimed at preventing injury. Depending on the training aims involved a distinction is made between:

Massage before or immediate to the competition, which is applied locally and quickly and seeks to activate.

Massage during the competition and during **rest periods**, a medium or short discharge massage that does not reduce tone and keeps the attention focused on the event.

Inter-competition massage, which is performed when the athlete has completed an event and must re-compete. It is a specific discharge massage directed to the areas with greatest overload.

3. Massage after the competition

Employed differently depending on the situation:

Post-competition massage. For the athlete who has finished the competition or after heavy workout.

Recovery massage. Aimed at the athlete with an injury that prevents him from making a technical movement.

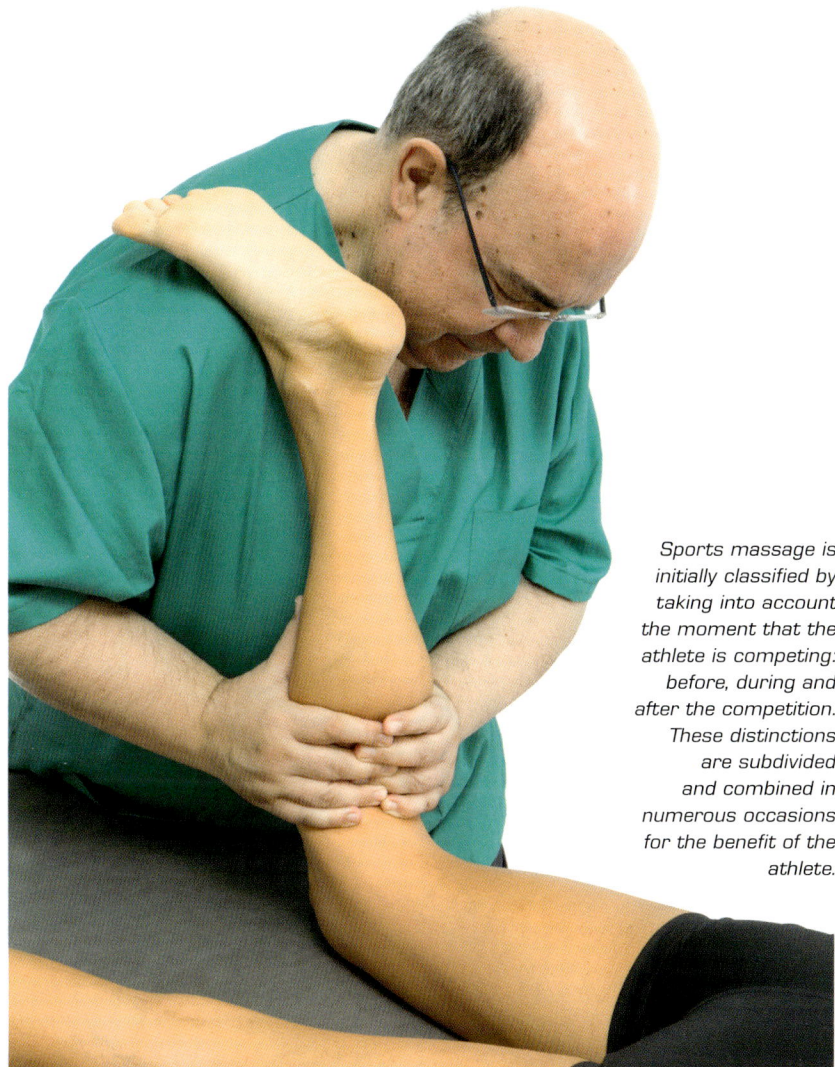

Sports massage is initially classified by taking into account the moment that the athlete is competing: before, during and after the competition. These distinctions are subdivided and combined in numerous occasions for the benefit of the athlete.

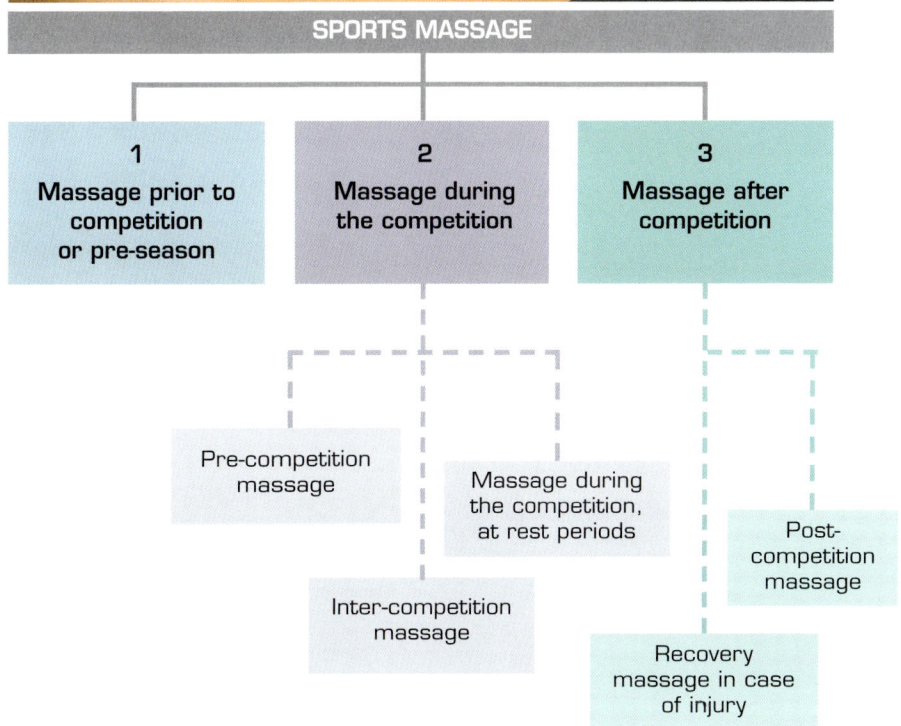

SPORTS MASSAGE

1 Massage prior to competition or pre-season	2 Massage during the competition	3 Massage after competition

Pre-competition massage

Inter-competition massage

Massage during the competition, at rest periods

Post-competition massage

Recovery massage in case of injury

Massage and competition

The main functions of sports massage are to enhance the athlete's performance and prevent injuries. It is applied on a regular basis over the course of the season, depending on the workload and training intensity or the event in which the athlete is participating. It discharges pressure and physical-emotional tension after heavy workout sessions.

Massage adapted to training and rest periods

In addition to following basic parameters to improve performance, such as planning the training loads and controlling rest periods and adequate nutrition in order to achieve overcompensation, the athlete uses sports massage as an aid to eliminate fatigue and reduce recovery times. This allows the athlete to sustain an increased number of training cycles.

During the recovery phase of the training cycle (see graph), the massage should be deep discharge and structure modeling. In contrast, massage in the overcompensation phase, when approaching a new training target, should be short and intense (4-6 minutes), stimulating the tissues and preparing other structures so that they can perform maximum workouts without relaxing their tone. Massage after competition is a restoration treatment, where maneuvers are applied slowly while tissues are being explored.

General adaptation syndrome

Training is a planned process for improving performance. Rest periods after subjecting the body to stressful workout sessions facilitate recovery and improvement of physical characteristics. The way that the body reacts to workouts and is reconditioned after them is called the "General Adaptation Syndrome," known by the acronym GAS.

Benefits of sports massage

In the course of a competition, after each stage or track race, massage is an indispensable means to avoid vascular congestion and the accumulation of fluid between tissues. Its use reduces or prevents overloads that can lead to injury and is very useful to treat its aftermath. It is an important aid when regulating tone and continued myofascial stability.

Regularly applied it improves dexterity and facilitates ease in the execution of complex movements, allows the athlete to maintain rhythm and concentration (see the previous example) and reduces anxiety, very common before a time trial stage or team competitions. Through reflex the massage promotes a deep relaxation that facilitates rest and sleep, while also favoring the optimal regulation of stimulation levels ahead of the next event.

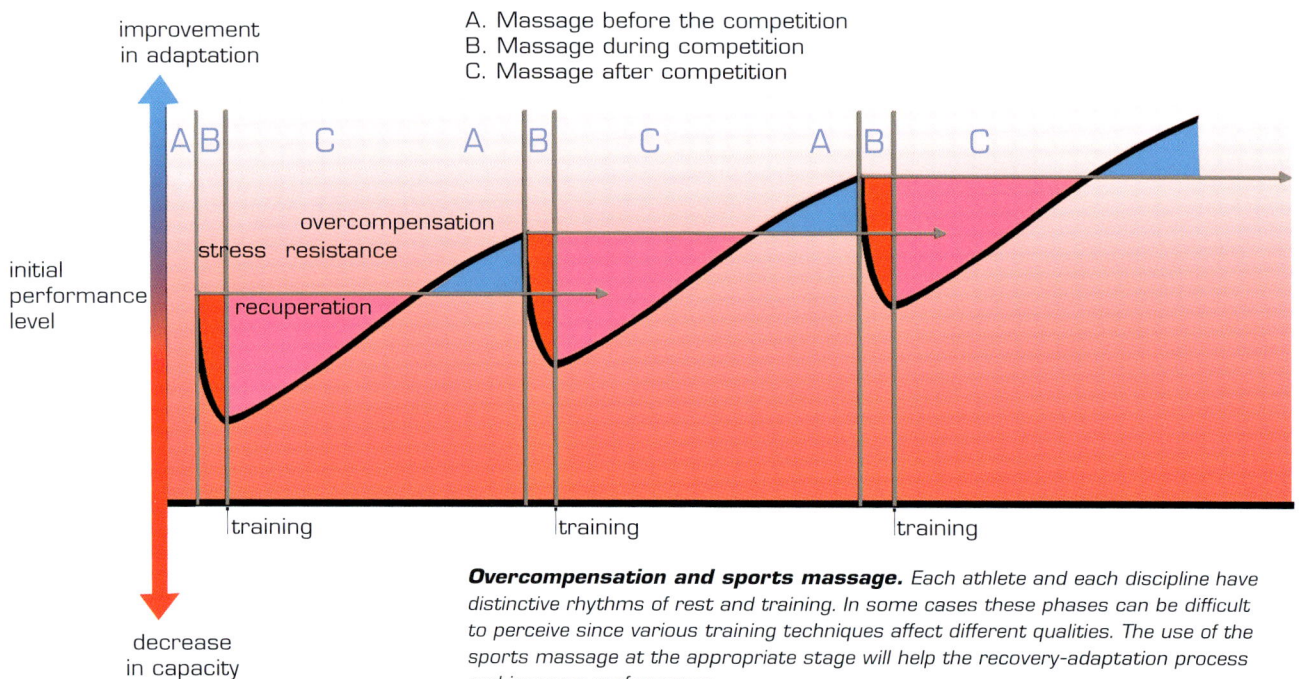

A. Massage before the competition
B. Massage during competition
C. Massage after competition

Overcompensation and sports massage. *Each athlete and each discipline have distinctive rhythms of rest and training. In some cases these phases can be difficult to perceive since various training techniques affect different qualities. The use of the sports massage at the appropriate stage will help the recovery-adaptation process and increase performance.*

Before an important competition, high-performance swimmers can accumulate daily workloads of up to 16,000 meters in the pool, plus 3,000 meters in the gym, simulating the strokes using pulley exercises with weights. Maintaining this volume of training for six or seven days and several weeks makes the daily loosening massage after the last workout essential so that the athlete can face the next exercise cycle.

The body reacts to workouts and recovers after them. Fatigue after exercise is due to:

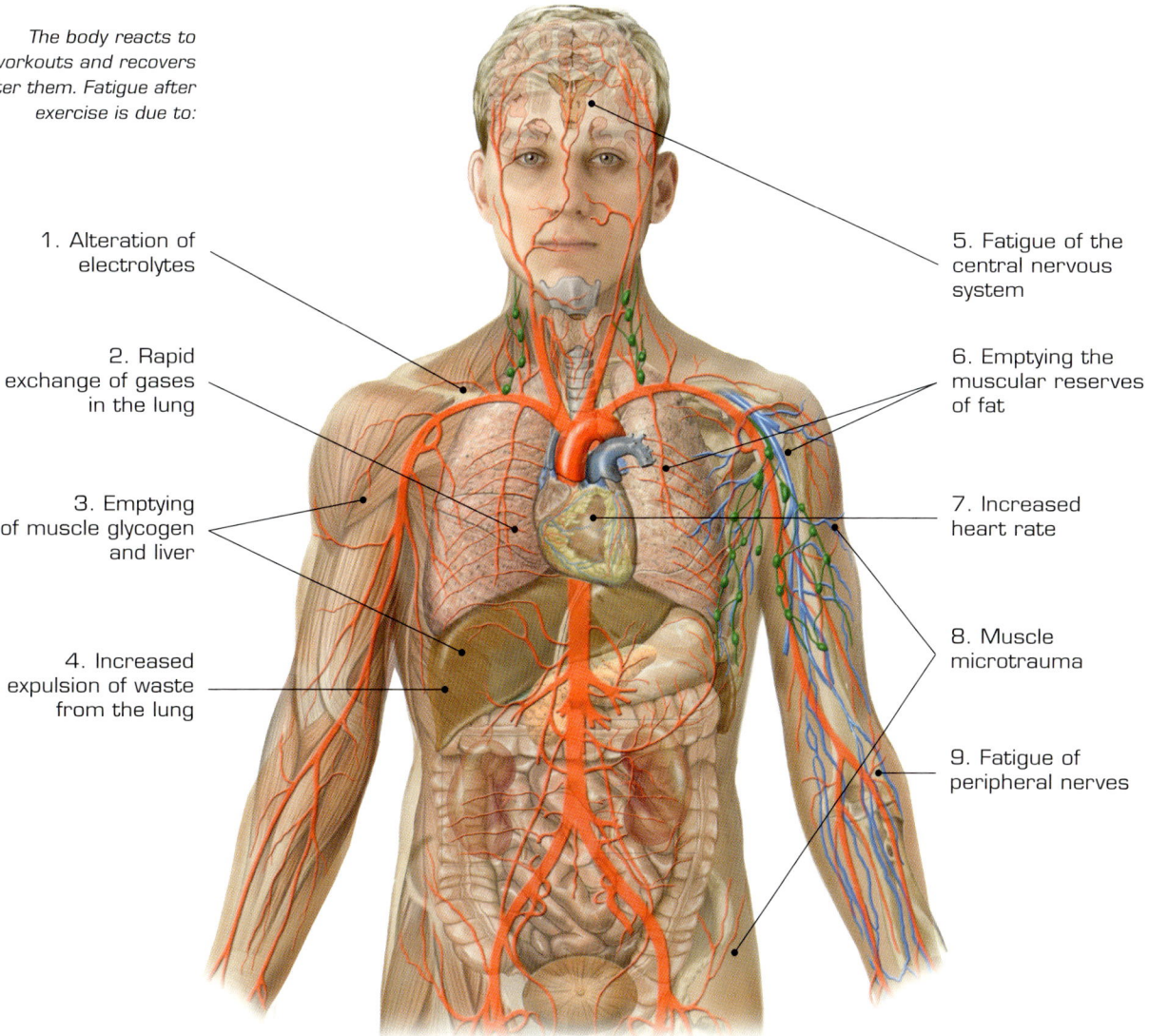

1. Alteration of electrolytes

2. Rapid exchange of gases in the lung

3. Emptying of muscle glycogen and liver

4. Increased expulsion of waste from the lung

5. Fatigue of the central nervous system

6. Emptying the muscular reserves of fat

7. Increased heart rate

8. Muscle microtrauma

9. Fatigue of peripheral nerves

Massage before competition

This is the pre-season massage and is it applied after a rest period (end of season, holidays, etc.). In this period, the athlete's priority is to improve the quality of their workouts, a need to condition their body for the effort. Its function is to provoke active drainage and renewal that vitalizes the metabolism, facilitating performance and inhibitory pain pathways.

Conditioning and adaption

Prolonged massage is necessary; maneuvers must reach the deep tissues with moderate-low dynamism. It will influence the patterns of movement the sport requires to improve proprioception. This pre-season phase is the best time to apply massage and induce changes in myofascial chains, allowing the athlete time to adapt.

Continued evaluation

During this period the massage allows the identification of alterations in tissue states and the location of areas that are potentially susceptible to overload long before they present symptoms. This is an advantage as prevention methods improve movement patterns and prevent injuries.

Listen, act and prevent

An example of the above is an aphorism cited by sports masseurs: "Fatigue, more fatigue, equal to contracture; contracture on top of contracture, leads to rupture." Or in other words, fatigue is a warning sign that indicates the likelihood of injury or irreversible damage to muscle fibers, which can be prevented by manual massage.

The sports masseur must adapt to environmental conditions (clothing, weather, playground, sports center facilities, etc.) and be able to improvise and act: a bench can become a couch, a jersey a pillow, a plastic sheet a raincoat or overcoat. Ask for help from another athlete to hold a body part, etc.

Final phase of discharge massage after pre-season training. The most overloaded area is affected, in this case the femoral quadriceps.

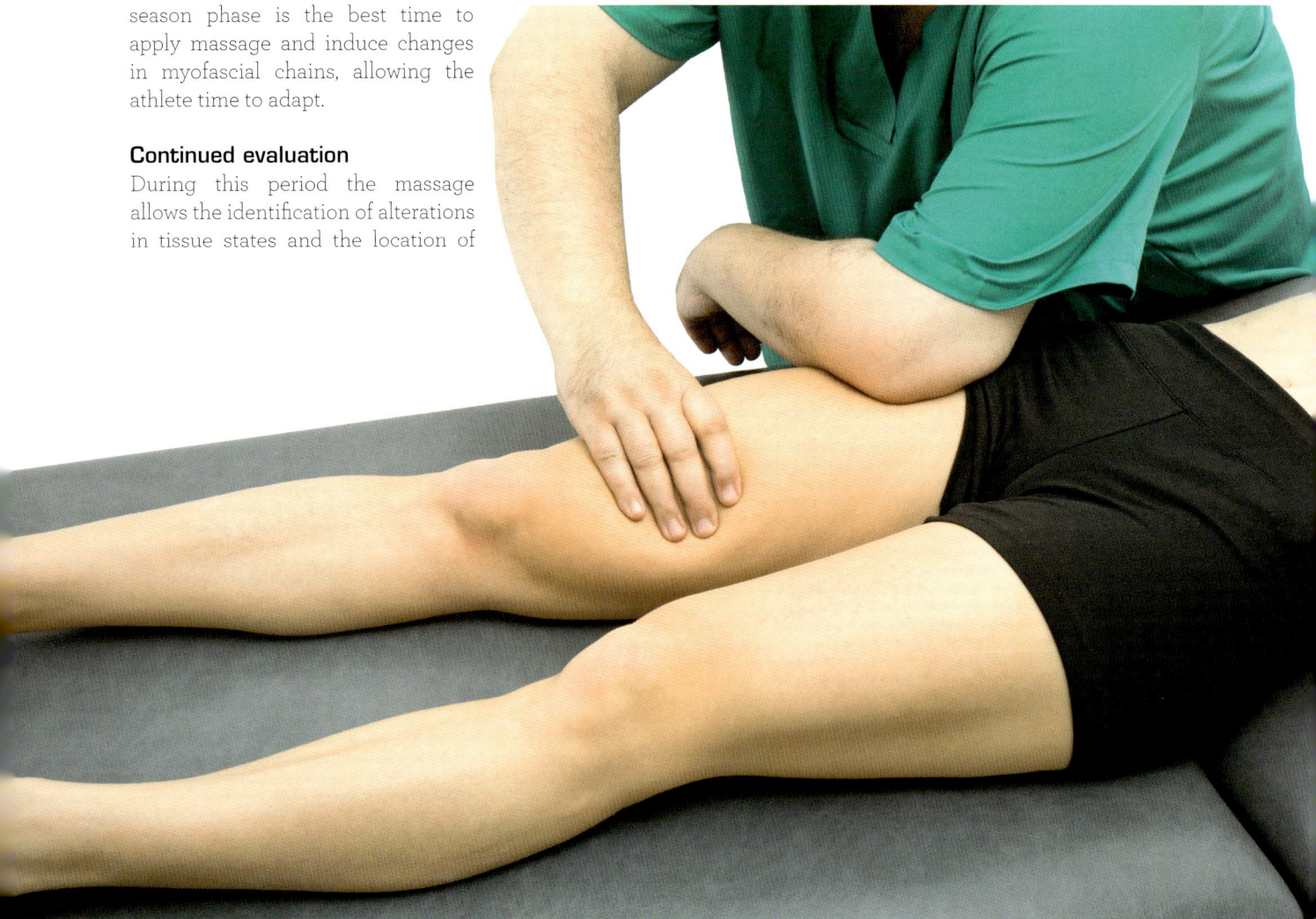

During pre-season massage problems that the athlete has suffered from in previous seasons are frequently detected. It is the ideal time to evaluate an athlete's condition. The athlete, in his eagerness to resume sporting activities, can despise those small problems that could influence the current season or future ones.

Guidelines to follow in pre-season massage

One of the functions of the physiotherapist, the masseur and the physical trainer is to observe the athletes during practice in order to detect possible problems or anomalies in their performance. During subsequent training or treatment they will be able to comment and to apply the measures that they consider appropriate.

Time trial techniques are part of conditioning after the summer break. It's the right time to adjust pace and create team cohesion. The therapist and the physical trainer are present trackside, observing the behavior and physical response of each athlete.

Guidelines to follow in pre-season massage

PRINCIPAL OBJECTIVES
Detect problems or anomalies. Explore and locate possible discomfort or tension to prevent injury. Help reduce recovery times between workouts and increase tolerable working times. Reduce motor and coordination constraints.

POSITIONING
Different decubitus positions are combined to achieve the successful application of a technique with the relaxed muscles.

PRINCIPAL MANEUVERS
Venous renewal, kneading, pressure, pumping, depth application, prolonged application across wide areas.

OTHER TECHNIQUES
Passive stretching and passive joint mobilizations.

DYNAMISM
Moderate-low.

TIMING AND DURATION
15 minutes minimum for a partial massage and 40-45 minutes maximum for a longer one.

PERIODICITY/PLANNING
Weekly, after workouts. Do not apply if there are less than two days before competition or strenuous workouts scheduled.

MOOD BOOSTER
Coexistence during pre-season promotes the cohesion of a group of athletes, and massage can also be used to further encourage this.

Massage during competition

This massage is directed at the athlete who is active and follows the exercise plan put forward for the competition. Depending on the course of training in which they participate, this is subdivided into pre-competition massage, massage during competition or rest periods and inter-competition massage.

Pre-competition massage

A massage that prepares the athlete immediately before participation, optimizes bodily functions and provides a greater consciousness of the body before a workout.

Preparing for the competition

Correct preparation for competition includes adequate physical and technical preparation, managing sleep habits, nutrition and hydration, maintaining attention during the event and the application of physical types of manual treatment, such as sports massage.

Massage can never replace but should boost a warm-up, especially those muscle groups which have not been exercised because of climate conditions or injury.

Precautions

◆ One way or another, massage decreases muscle tone. Therefore before receiving massage the athlete must actively warm up and afterwards retrain in order to recover lost tone and retain attention on the event.

Bimanual kneading technique in the thigh's quadriceps femoral area, after active warm-up before competition.

Guidelines to follow in massage before the competition

PRINCIPAL OBJECTIVES

Favor a positive attitude toward performance and competition, activate the proprioceptive and nervous systems, increase blood circulation, warm muscles, favor hyperemia, oxygenate and make flexible musculature stimulate muscle tone.

POSITIONING

Standing or sitting are preferable. The decubitus is avoided so that the athlete remains focused on the competition. The limb or region to be treated is placed below the heart to improve its irrigation.

PRINCIPAL MANEUVERS

The most activating maneuvers are prioritized. This is a short-duration, energetic massage without depth (stimulation). Friction. Pressure maintained. Kneading. Stimulating and prolonged beats. To conclude, shake the entire limb or area being treated.

DYNAMISM

High/vigorous (carried out quickly).

TIME

Short duration, about 4-6 minutes.

MODE

Superficial/shallow; stimulating and painless.

PERIODICITY/PLANNING

Apply 45 minutes before competition as some muscle tone is lost. Don't forget to allow time for technical and tactical training ahead of the event.

MOOD BOOSTER

This massage is predisposed to the workout.

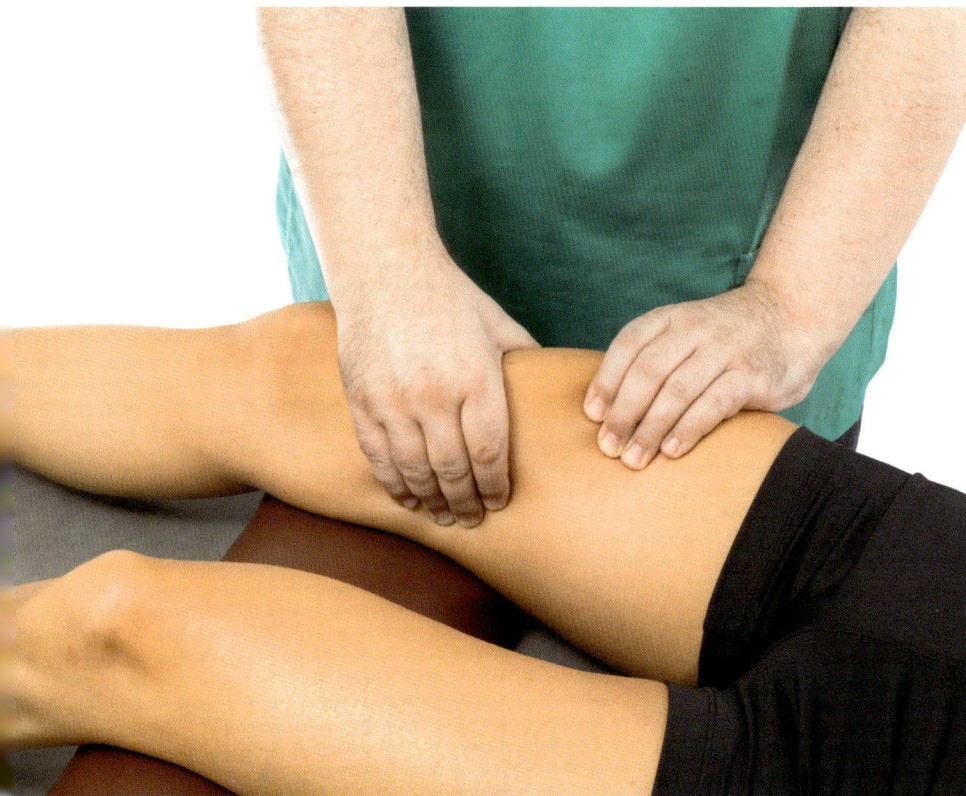

Massage during rest periods within the competition

The main objective here is to activate and download "media," maintaining attention on the event. It is essential that the athlete does not disconnect from the competition. It is a massage that adapts according to how the competition develops, and is applied (if possible) at the halfway point of the event, during rest periods or between series.

Inter-competition massage

This type of massage is applied when competition lasts more than one day, with one or two days of active rest. It is applied in the locker room after afternoon training, or if you are traveling, before dinner in the hotel.

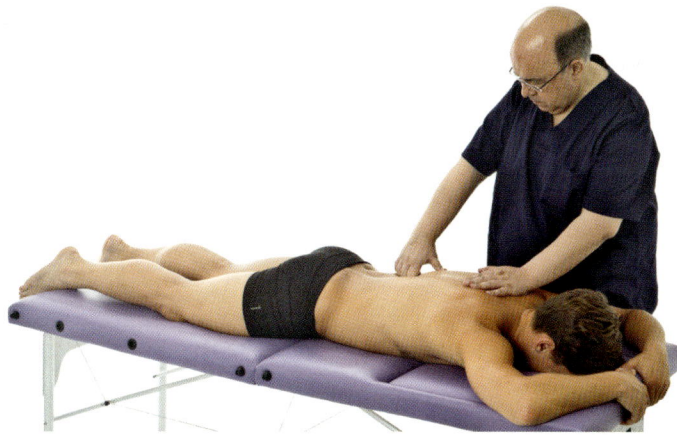

Precautions

◆ The therapist will wear disposable gloves to avoid leaving traces of active substance creams on the hands that may bother other athletes or players, or even himself when he has to perform other techniques.

Avoiding the use of ointments can alter the thermal response of the skin and irritate the athlete, making them feel sticky during the event. Small amounts should be used.
If it is cold use oils with substances that give heat, and in very hot climates, creams that cool.

Guidelines to follow in rest-period massage

PRINCIPAL OBJECTIVES
Maintain attention during competition, and favor lymphatic drainage and blood renewal. Soothe, by activating pain inhibition. The areas most prone to muscular overload are treated.

POSITIONING
The member or region to be treated will be placed above the heart to favor drainage.

PRINCIPAL MANEUVERS
Pressure, kneading, very soft rolling.

OTHER TECHNIQUES
Stretching.

PERIODICITY/PLANNING
Performed in the locker room, surrounded by other colleagues. Must be very fast-acting as there is little time and in a few moments the coach must speak or is already speaking.

DURATION
From 5-8 minutes.

MODE
From surface to shallow.

SPEED
Slow to fast.

INTENSITY
Moderate/medium and varied.

MOOD BOOST
Download and prepare for efforts.

Guidelines to follow in inter-competition massage

PRINCIPAL OBJECTIVES
Explore and treat areas damaged by bruises, strains, overloads, etc. Assess the general condition of the athlete. This is a drainage massage to facilitate venous return and eliminate metabolic by-products. Myofascial discharge and muscle groups with contractures, cramps and fatigue. Attempts to stall fatigue.

POSITIONING
Preferably decubitus.

PRINCIPAL MANEUVERS
Pressures, kneading, bimanual and tenar, friction, soft percussion.

OTHER TECHNIQUES
Myofascial trigger points stretching techniques.

DURATION
Extensive, minimum 35 minutes.

INTENSITY
Intensive, from superficial to deep as the session progresses and according to tolerance.

MOOD BOOST
Discharge and decrease pain, allowing the athlete to execute movements with greater security.

Massage after competition

Post-competition massage

This massage serves the athlete who has finished competition or a strenuous workout. It provides a more intense discharge than previous massages to facilitate post-event relaxation and help the athlete to disconnect. It works slowly, profoundly reaching different anatomical planes. It influences restorative and sedative-analgesic functions.

Its main objective is the restoration of fatigued structures. In addition, it favors the elimination of waste substances produced by muscular metabolism and helps to control hyper-toning of the muscles.

It is to be applied 2 or 3 hours after the event, after the athlete has warmed-down; that is, after "active washing," when heart and respiratory frequencies are restored to normal.

An antispasmodic massage reduces the sensation in the thighs or DOMS and troublesome cramps. It prioritizes the drainage effect and promotes muscle oxygenation. It lasts no longer than 20 to 35 minutes, and maneuvers are of medium to low intensity.

Occasionally, after competing three days in a row, this massage is performed 24-36 hours after exertion. In this case, it can be applied for a duration of 35-45 minutes.

Precautions

◆ Massage practiced 2 or 3 hours after the competition is used to reduce muscular hypertonia, but this should not be eliminated in a single day and in a single massage; the intention instead is to control it. In addition, the massage should not cause pain.

After the competition the athlete is sweaty and restless, and muscles are swollen, hypertonic and hardened. It is advisable the athlete stops sweating before taking a shower, avoids getting cold or experiences sudden changes in temperature. It is advisable to do stretching exercises beforehand.

Sequence of treatment of muscle hyper-toning in the thigh through palmar pumping maneuvers.

1. Palmar pumping maneuvers allow the adjustment of excess tone (hyper-toning).

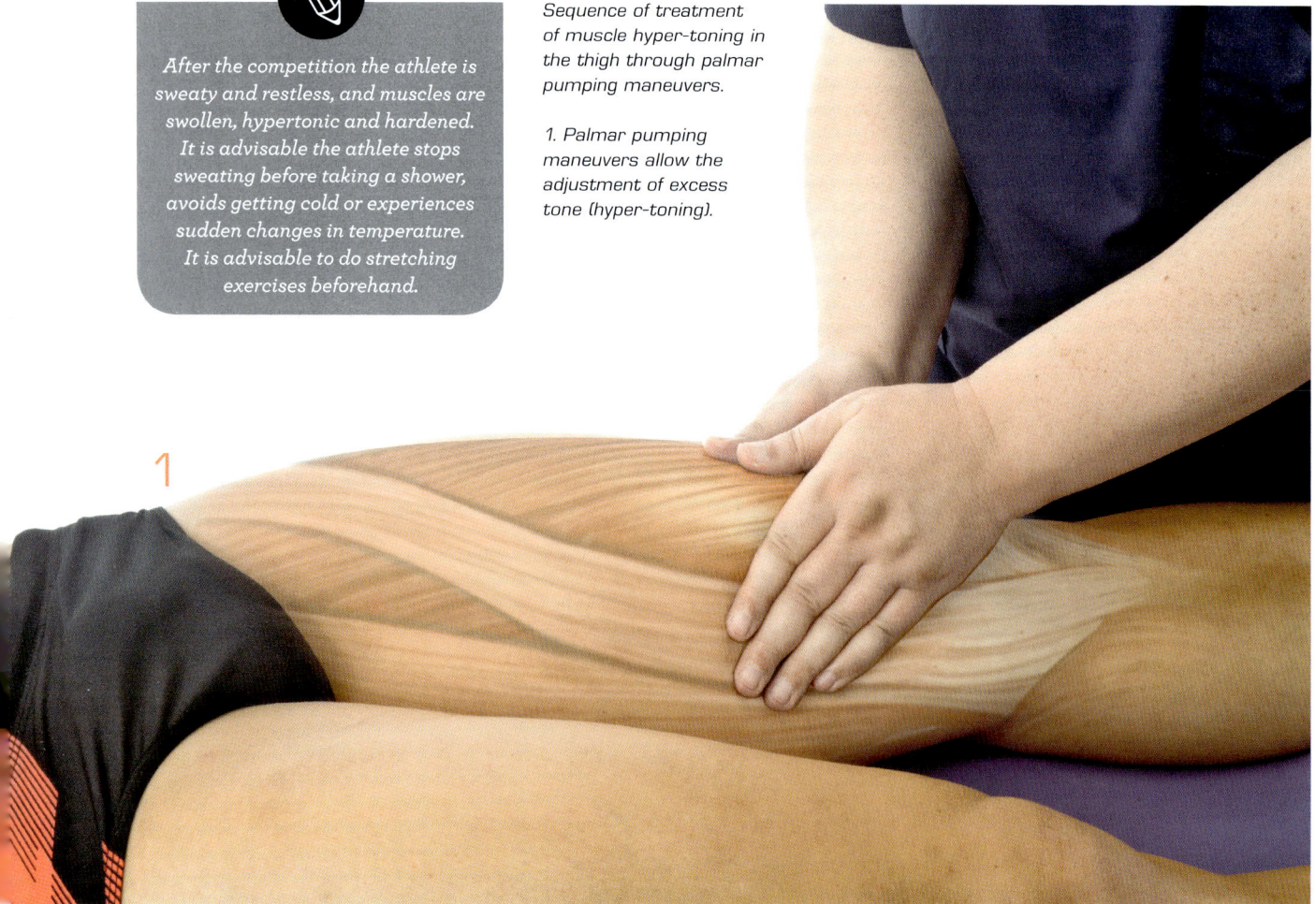

1

Guidelines to be followed in post-competition massage

PRINCIPAL OBJECTIVES

Explore possible overloads as soon as possible. This is the right time to treat and control excess local tension, contractures, and suppress possible spasms. This massage seeks to promote venous-lymphatic drainage and local blood replacement, eliminating metabolic waste (lactate, etc.). It also serves to stimulate pain-inhibiting pathways.

POSITIONING

You can and should use supine positions that favor drainage and athlete comfort. Breathing and relaxation techniques are also employed.

PRINCIPAL MANEUVERS

Venous renewal, palmar pumping, medium-intensity friction, soft to moderate kneading, palmar pressure.

OTHER TECHNIQUES

Very soft passive stretches (according to tolerance) and gentle, unforced joint mobilizations.

DURATION

This is a slow, adaptive and progressive massage that takes a minimum of 35 minutes. However, depending on tolerance, it can be anywhere from 20 to 45 minutes.

MODE

Because the athlete is sore, a large number of maneuvers are applied from superficial to shallow and repeatedly in the area being treated.

SPEED

Slow and smooth.

INTENSITY

Moderate/low and varied.

PERIODICITY

To be used after 24 or 36 hours after the last post-competition massage.

MOOD BOOST

It is preferable that the athlete is alone with the physiotherapist or the masseur. This is the "confessional massage," typical in cycling; both parties should listen and speak to each other at all times.

2. Palmar pumping is applied by slow movements, maintaining a uniform pressure and avoiding friction.

Precautions

◆ In team sports it is advisable for athletes to walk or jog gently for a while after competition, thus taking advantage of slow- and deep-breathing cycles.

Take into account changes in the thermal response of the skin; cold or de-fatting creams or natural gels should be used sparingly.

2

Recovery massage in case of injury

One of the least desirable aspects of sports is injury. This massage is aimed at the athlete who presents with an injury that prevents him from making a technical movement. It is used to favor and help the athlete during the rehabilitation period.

Manual massage is a very useful therapeutic tool to reduce pain and support therapeutic exercise during this phase. Since at this stage the injured anatomical structures are coerced until they perform their normal function, the massage must be adapted to target the injured area, loosening tissues and achieving greater mobility. It also reduces swelling and residual pain that may appear after recovery exercises.

Massage techniques in rehabilitation

A good example of this massage is when it is employed after the removal of immobilizing bandages, since it allows the tissues to maintain tropism, improves their mobility and eliminates rigidity, simultaneously helping the athlete to maintain a positive mental state, another factor which accelerates recovery. The massage has to adapt to the athlete's rehabilitation phase, respecting the phases of immobilization, partial supports, the occurrence of pain, etc.

It is preferable to combine this massage with joint mobilizations, stretching and other therapeutic techniques. The goal is to avoid stiffness, atrophies, adaptations by shortening tissues and improve functionality of the injured areas.

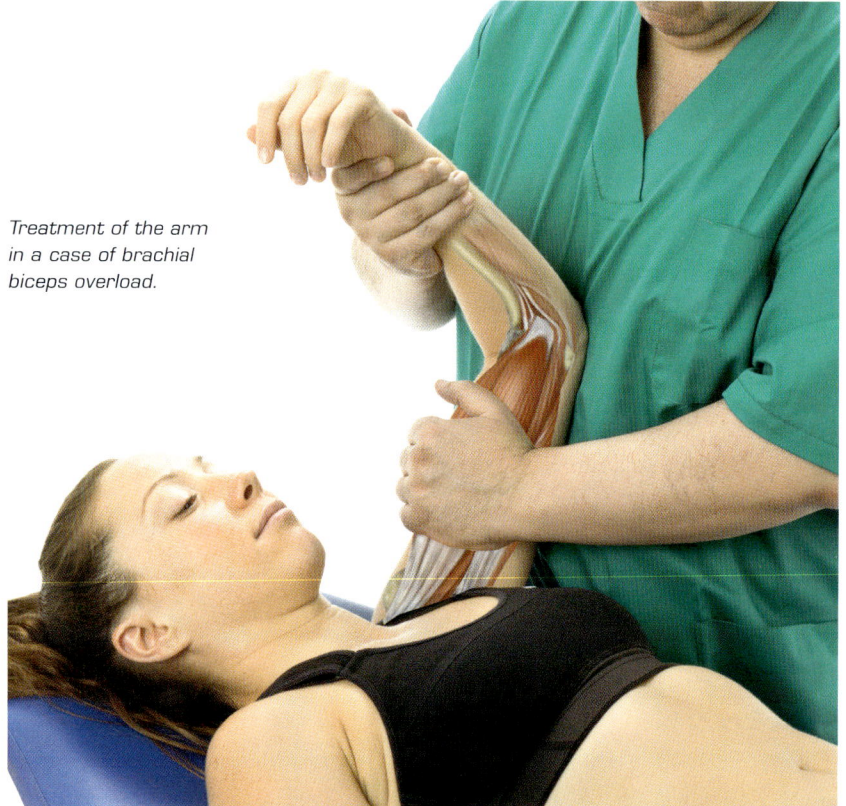

Treatment of the arm in a case of brachial biceps overload.

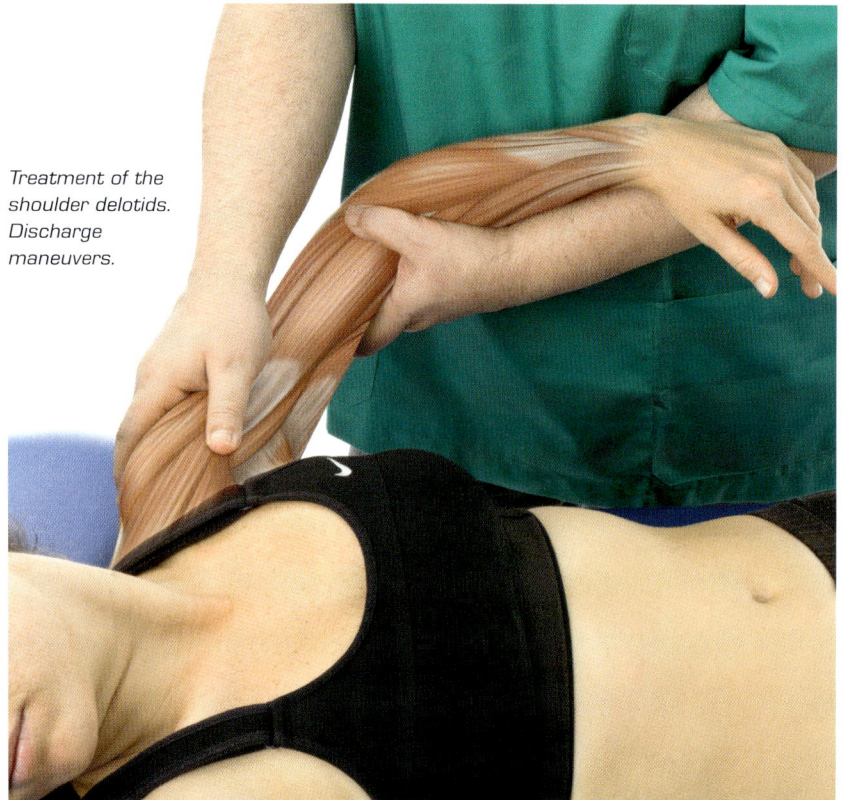

Treatment of the shoulder delotids. Discharge maneuvers.

Rules to be applied in case of injury

When an injury occurs the application of massage is subject to a number of rules. If there is edema or a loss of tissue integrity, massage is not applied at the initial stage of the injury, but when repair and healing allow. This occurs between the first 48 hours and 10 days in the case of injury, as it could increase the rupture. The existence of calcifications must be ruled out. In this case, intense friction massage should not be used. As a rule, in the inflammatory phase no massage is applied.

Exceptions to the rule

If calcifications occur it would be advisable to use very subtle manual lymphatic drainage and apply a little cream or oil. Pain and a contracture response should be avoided, which would delay or hinder the rehabilitation process.

Direct and sustained pressure in a contracture with trigger points.

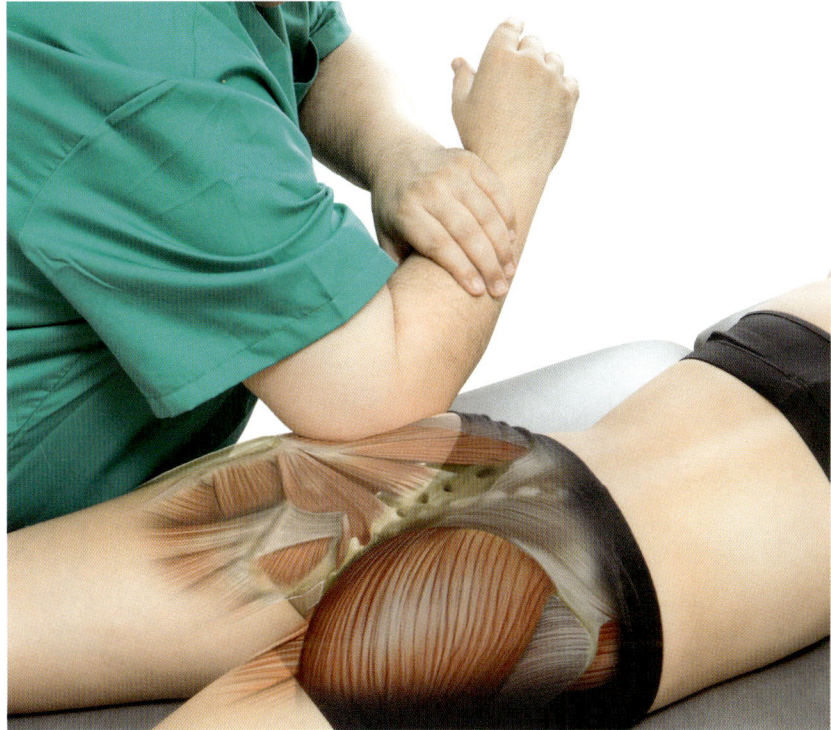

Transverse friction technique in the area of the piriformis muscle made with the elbow. The friction is performed by positioning the olecranon over the affected area point and using small and controlled movements of the body, swinging between anterior and posterior but never rubbing with the elbow.

Combining techniques

The characteristic cramps and spasms during and after exercise require massage, preferably sustained pressure maneuvers combined with stretching techniques. When the cramp decreases, a soft massage kneading will help the area to improve. Hydration is also recommended.

In cases of contracture, especially painful contracture when manual trigger point techniques are applied, it is preferable to build up pressure, gradually increasing it according to the athlete's tolerance. Avoid increasing pain.

Tendinitis is another pathology that may appear during the continuous practice of sport, manifesting itself suddenly with acute pain and functional affectation. Once diagnosed (taking into account calcifications), Cyriax massage, combined with stretching techniques, is very beneficial.

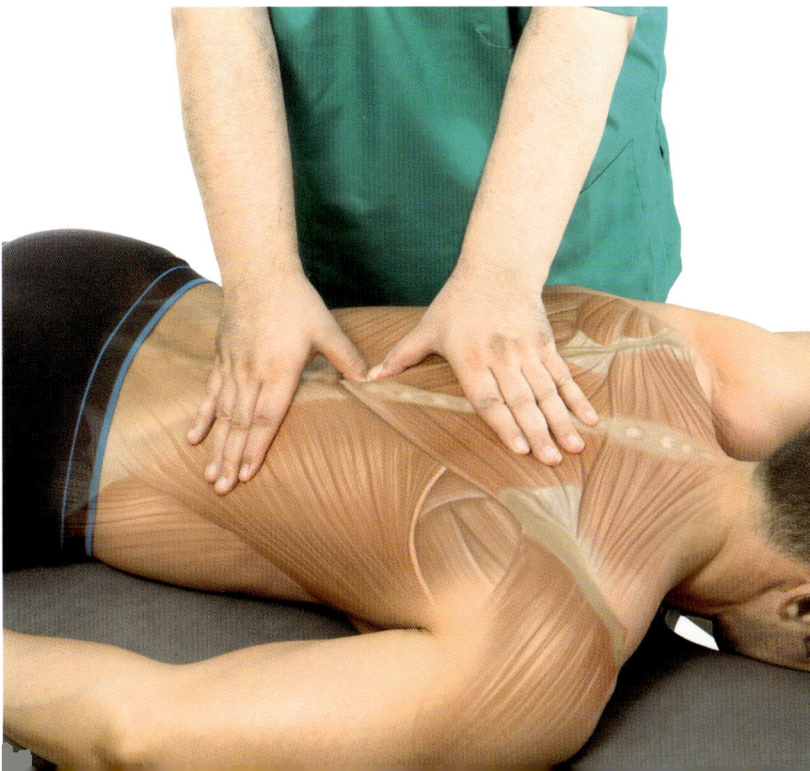

The skin

The Skin is an organ that protects us and communicates with the outside world. It is a gigantic communications network that is responsible for receiving and transmitting the environment, perceptions and external stimuli to the internal organs. Through sensory nerves it gives us sensations of touch, vibration, pressure, temperature and pain, which allows the skin to activate mechanisms of defense and adaptation to the environment.

The skin and superficial fascia act as a functional unit; skin intervenes in movement and, at the same time, the fascia transmits to all underlying and peripheral elements.

The skin not only reports the external environment, but it is also a reflection of the internal state of the body mediated by the nervous system, which provides information that can be seen and felt. In each medullary segment, the nervous system receives, interprets, coordinates and transmits information about certain areas of the skin, muscles and organs. This division by segments can help us to detect alterations in different tissues with which the skin contacts the same medullary segment (dermatome).

There is also an exchange of substances: The skin breathes, transpires, absorbs, produces and communicates. It is a dynamic system that is in constant renewal.

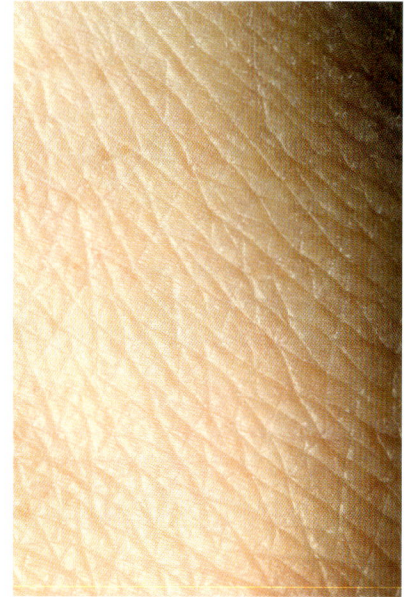

A cross-section showing the multilayer structure of the skin. Epidermis and dermis sit over a layer of connective tissue that links the skin to deeper-lying structures such as the fascia, muscle, periosteum and other tissues and vessels.

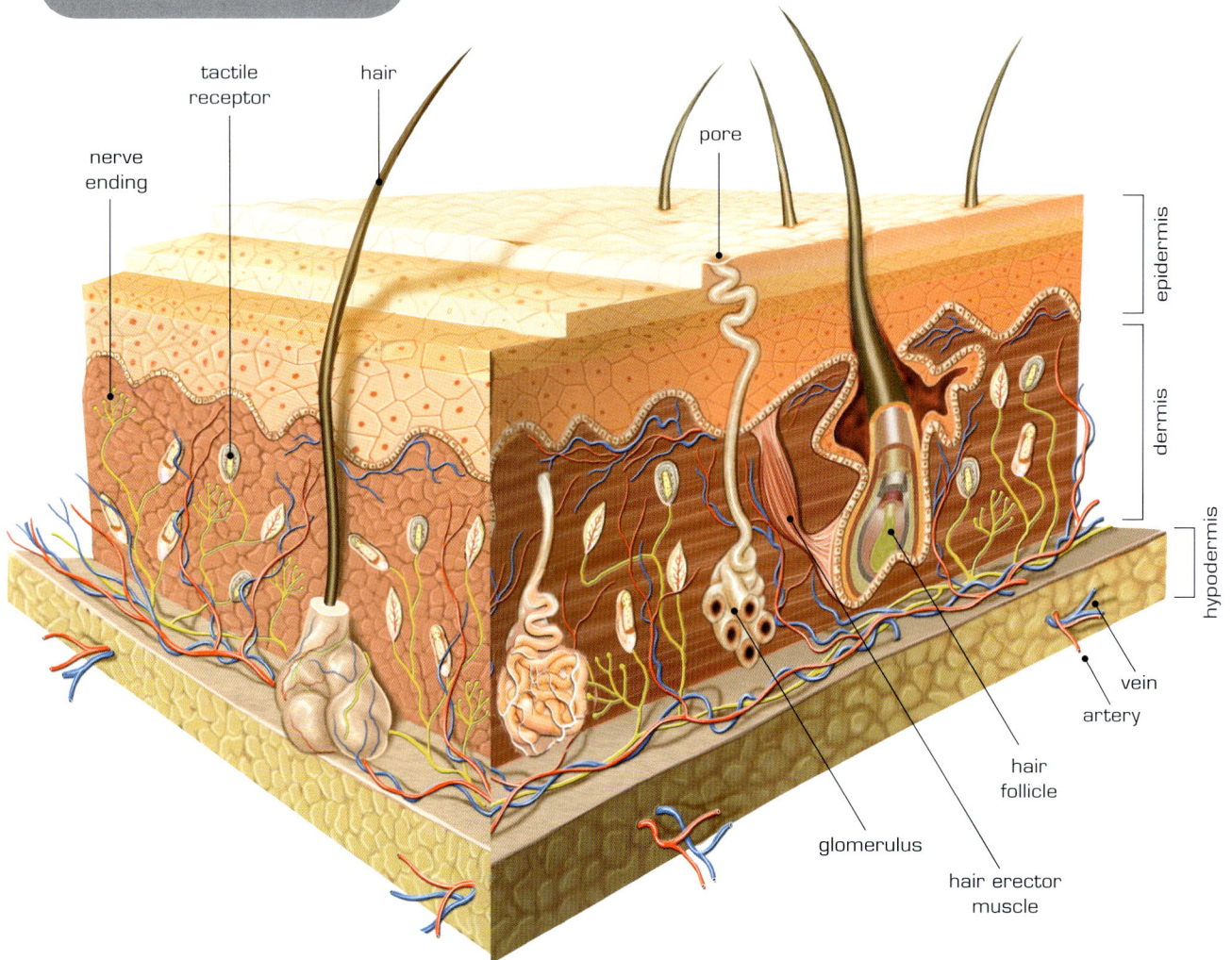

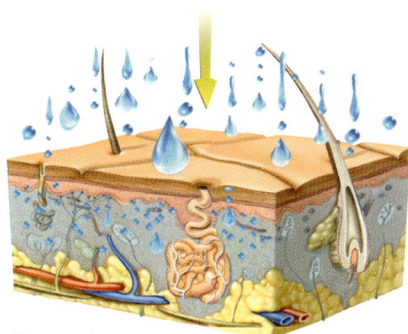
Absorption

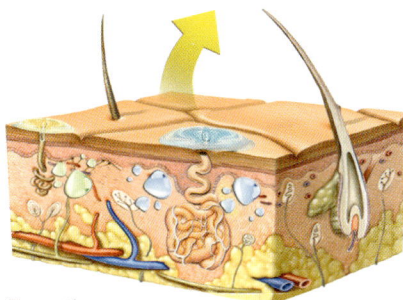
Excretion

Metabolism

Functions of the skin

Absorption of water and some substances, such as fatty acids from oils and their properties.

Excretion of water and waste in the form of sweat.

Secretion of sebum, which protects against the loss of moisture in the skin and forms part of the protective layer against possible pathogenic organisms.

Protection against mechanical, chemical, biological or physical aggression (such as the sun's rays).

Communication with the environment. The skin has receptors sensitive to touch, pressure, pain and temperature.

Thermal regulation. With the coordination of the circulatory system, blood can circulate more superficially in the dermis to cool by convection, or circulate through the inner layers, where the adipose tissue will retain heat.

Respiration. Small scale of carbon dioxide (CO_2) and oxygen (O_2).

Metabolization and production of vitamins.

Effects of massage on the skin

- Increased vasodilation and hyperemia, with increased local temperature.
- Improvement of cellular nutrition.
- Improved disposal of waste and toxins.
- Increased protective glandular function.
- Increased sebaceous and sweat secretion.
- Favors desquamation.
- Increased flexibility.
- Absorption of fatty substances, with greater penetration of oils and their properties.
- Decreased pain due to saturation/occupation of the nerve pathways by skin receptors (gate-control theory).
- Sedative effect of the central nervous system through the psychoemotional mediation established with the physical contact between the therapist and patient.

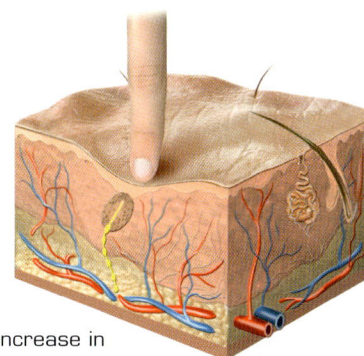
Increase in flexibility

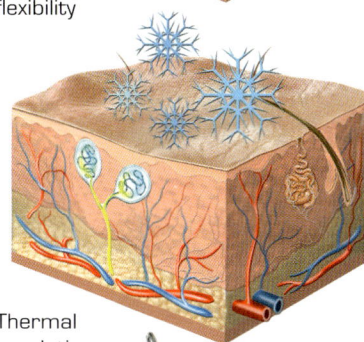
Thermal regulation

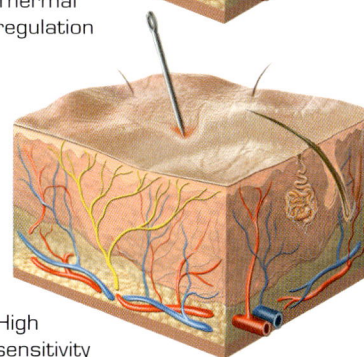
High sensitivity

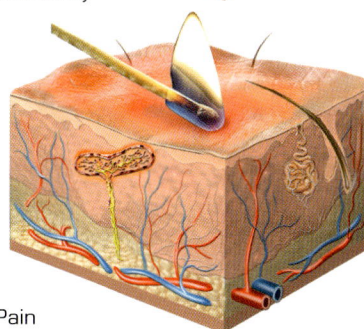
Pain

The muscular system

The tissues formed by cells with a contractile capacity constitute the muscular system. The contractions can be mediated by voluntary stimuli (the skeletal muscle, responsible for movement and posture) or involuntary (the heart muscle, the muscles of organs and viscera, those of the circulatory system, some sphincters, etc.).

The skeletal muscle

A muscle is divided into fascicles, and these into muscle fibers formed by tissue cells. Each of these divisions is wrapped in a layer of connective tissue. This connective tissue receives a different denomination according to the structure that surrounds, unites or communicates with it. Skeletal muscles attach to the bone through the periosteum and tendons fuse to the epimysium. This extends to the interior of the muscle and surrounds all the fascicles (forming the

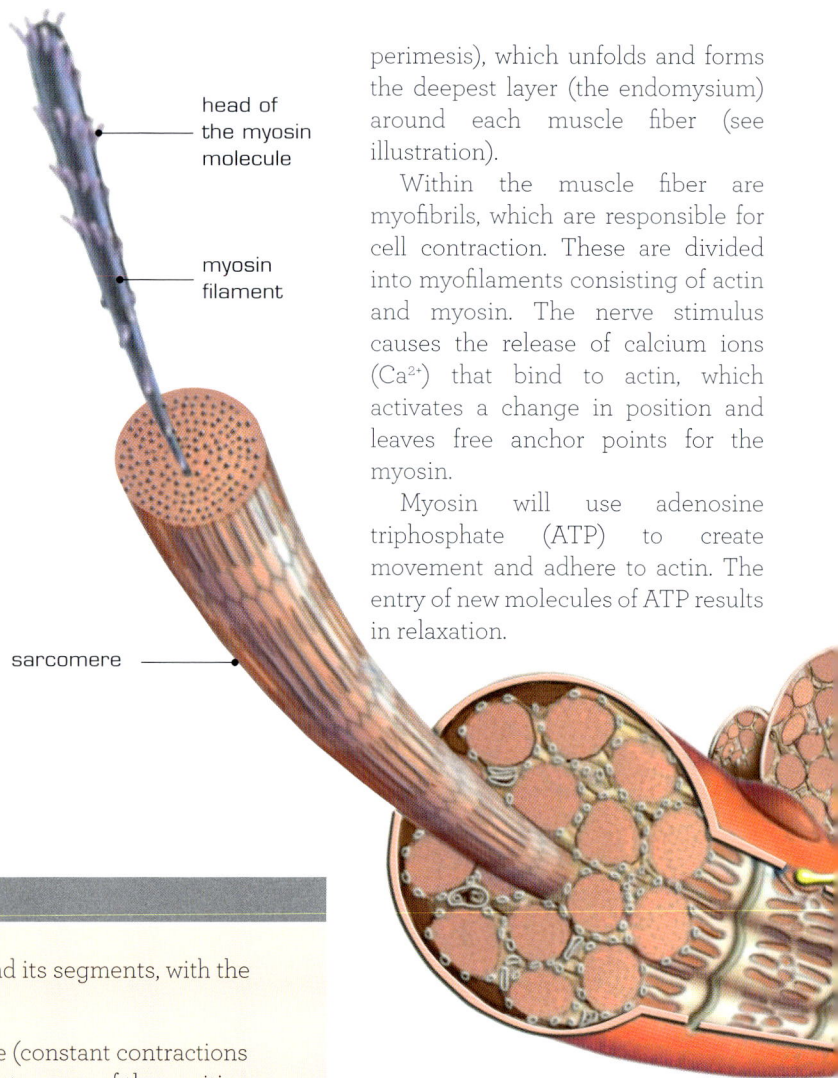

perimesis), which unfolds and forms the deepest layer (the endomysium) around each muscle fiber (see illustration).

Within the muscle fiber are myofibrils, which are responsible for cell contraction. These are divided into myofilaments consisting of actin and myosin. The nerve stimulus causes the release of calcium ions (Ca^{2+}) that bind to actin, which activates a change in position and leaves free anchor points for the myosin.

Myosin will use adenosine triphosphate (ATP) to create movement and adhere to actin. The entry of new molecules of ATP results in relaxation.

head of the myosin molecule

myosin filament

sarcomere

Functions of the muscular system

Movement and displacement of the body and its segments, with the support of the bones.

Stability and posture. The basal muscle tone (constant contractions in a small part of the muscle) permits the maintenance of the position of structures and posture. Forced postures will require a greater number of fibers to maintain, which will lead to overload, increased energy and circulatory expenditure. The system maintains body shape in certain areas.

Thermogenesis. The generation of muscular contractions is the main producer of body heat.

Circulation. The heart muscle drives blood in the arteries that maintain blood pressure thanks to their muscular walls, which are able to change diameter to regulate the flow according to requirements. During contraction, skeletal muscles increase internal pressure, thereby improving venous return.

Protection. At a signal of external aggression, the muscles will contract to create a movement of estrangement (nociceptive reflex). In case of an internal injury, the muscles of the affected area will shrink to prevent movement of the area in order to avoid further damage.

Visceral **containment and protection.**

What is a contracture?

A contracture supposes an alteration in the relaxation of muscle contraction: a failure in the Ca^{2+} pump inhibits its transport to the reticulum and causes a cessation of muscle contraction. May occur due to depletion of ATP.

This spasm causes contraction of blood vessels, resulting in decreased nutrient supply and waste disposal, making recovery difficult. If it lasts some time, connective tissue will thicken around the fiber.

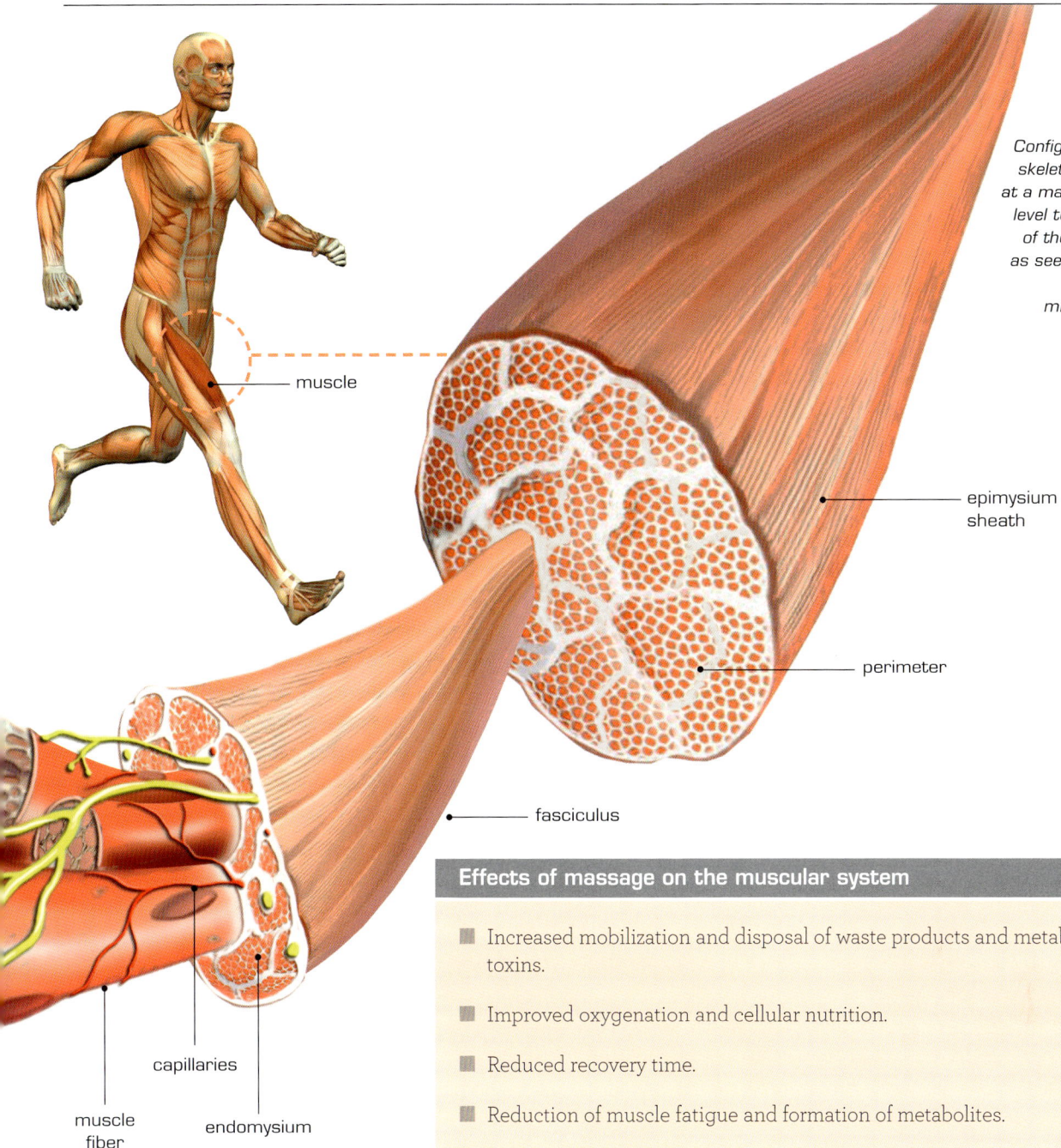

Configuration of skeletal muscle at a macroscopic level to the view of the filament, as seen through an optical microscope,

muscle

epimysium sheath

perimeter

fasciculus

capillaries

muscle fiber

endomysium

Late-onset muscle pain occurs between 12 and 48 hours after intense or very prolonged exercise and may be accompanied by stiffness and a feeling of weakness. This pain can last for several days, although its intensity diminishes until it disappears, unless intense activity is carried out again.

Effects of massage on the muscular system

- Increased mobilization and disposal of waste products and metabolic toxins.
- Improved oxygenation and cellular nutrition.
- Reduced recovery time.
- Reduction of muscle fatigue and formation of metabolites.
- Reduction of the formation of connective tissue accumulations.
- Decrease in muscle tone due to action on muscle spindles and other tension receptors.
- Decreased contractures due to pressure exerted on them (tight bands or trigger points).
- Improved muscle function, range of mobility and strength.
- Increased ability to stretch.
- Reduction of the onset of spasms and cramps.
- Pain relief.

Connective tissue

The fascia is a form of connective tissue that specializes in joining, connecting (connective) and distributing forces throughout the body, and is made up of cells, a matrix and fibers. The cells lend their properties in each area, making "cables," adhesives, lubricants or elastic elements. The extracellular matrix is a solution of water and glycoprotein molecules that possesses a consistency of gel that nourishes and organizes. Most fibers are composed of collagen, are very resistant to traction and give the fascia its principal mechanical capabilities. Reticulum and elastic fibers also serve to repair and maintain tissues.

Connective tissue specialization

The fascia is different depending on the amount, order and packaging of collagen. For example, the tendons integrate into the muscles and transmit force, while the ligaments join the bones: They are "cables" of collagen. Bones are densified, rigid parts, enclosed within the fascial mesh. In engineering this is called a "tensegrity" system: a set of structures with resistant parts that connect (integrate) and, at the same time, distribute the forces of tension and compression. The fascia has a viscoelastic behavior, that is to say, it is a tissue with the capacity to modify its shape, length and consistency as they affect forces.

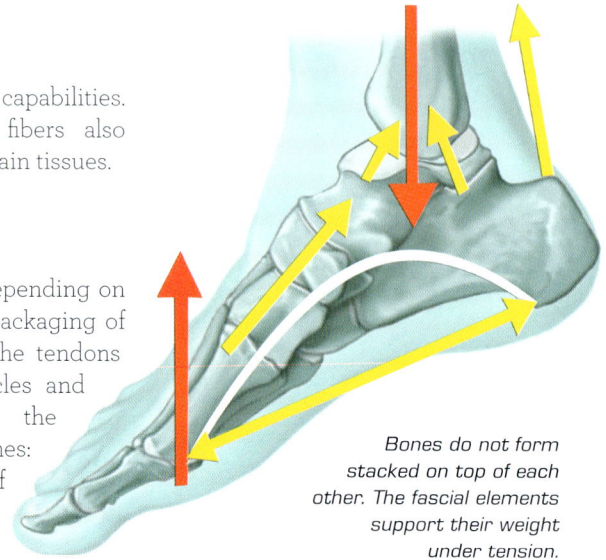

Bones do not form stacked on top of each other. The fascial elements support their weight under tension.

Connections for sports movement

Athletes endure significant impacts during sports movement, so they require, in addition to strong bones, a system of "tensioners" ("cables") that serve to keep the bones in place. The fascial system is a way to transmit, distribute, create continuity and modify the tensional forces of the tissues, as well as external impacts. Forces run along this pathway of collagen using the bones as points of support or transmission. Muscles associated with fascial tissue act together as a motor and regulator of the tension of the whole system.

The myofascial network is a continuous path through which the body's movements traverse. For example, when a blow is struck with the fist (action), the driving force comes from the ground (reaction according to Newton's third law). Tension rises from the foot, plantar dome, leg and thigh, the tendons of the pelvis and thoracolumbar aponeurosis until reaching the arm and hand.

Functions of fascia

The fascia is a tissue that provides support and intercellular nutrition. Its functions include: support, protection, separation, cellular respiration, elimination of toxins, metabolism and flow of fluids and lymph. It has a profound influence on the immune system and cellular health, but, above all, it is responsible for creating continuity and providing the body with its shape. The collagen of the fascia has piezoelectric properties, meaning it transmits electricity. Any tension generated in a given area produces electrical and mechanical information that is transmitted to the body as a whole. This information serves to organize structural changes, for example, where there must be more bone density (Wolff's law).

Effects of massage on the fascia

Sports massage should be oriented to induce changes in the overall system, not only in local structures. Lesions of the fascia alter the environment of surrounding cells and the way tension is transmitted. If there are restrictions, the forces of an impact will not disperse properly and cause overloads and injuries. In addition, there will be excess pressure in a given area with abnormal activity, reduced overall performance and recurrent injuries. When a movement fails, the substitute movement creates an adaptation, with its clinical signs and symptoms.

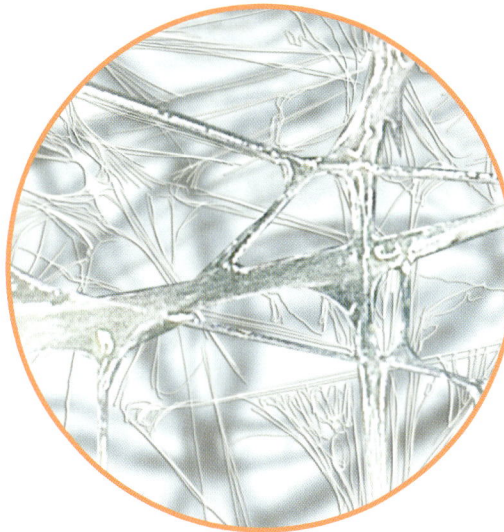

Structure of the fascia seen through a microscope.

The term "fascia" includes the family of aponeurosis, membranes, cartilage and joint capsules, ligaments, tendons and sheaths. It forms the meninges and cerebral support cells, the lens of the eye, the periosteum, the lining of the liver, etc., as well as the nearly microscopic coatings of vessels, nerves, muscle fibers, etc. The fascia is truly a global presence throughout the body.

Traction technique on thoracolumbar fascia. When working on the back, be mindful of the relationship between the tissues of the back and those of the upper extremities, as well as between the bones and the joints of the pelvic and scapular girdles. The relationships between the thoracolumbar fascia and the contralateral gluteus must also be taken into account.

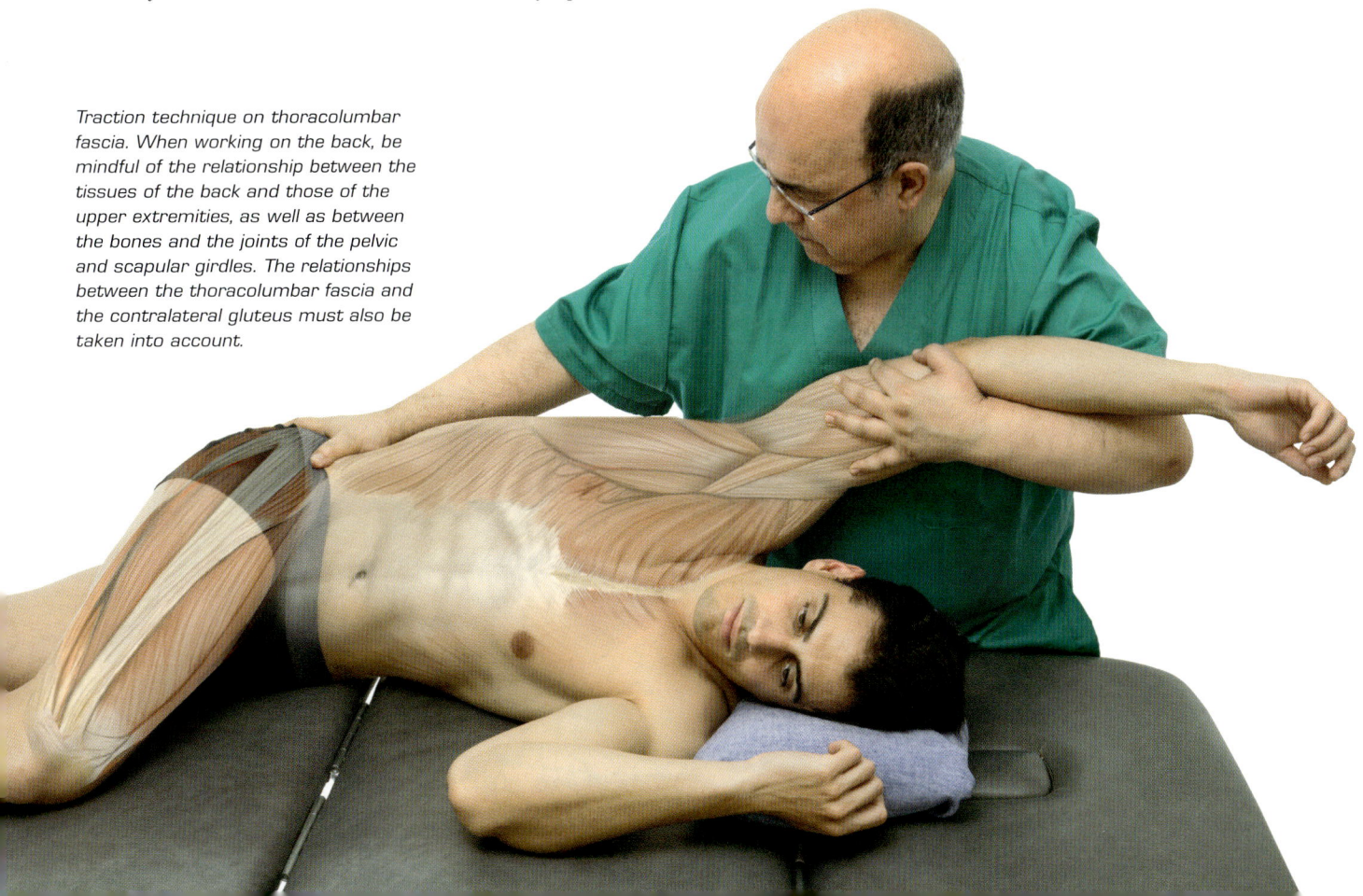

The nervous system

The nervous system collects and transmits electrical information via neurons and their infinite network of fibers within the nerves. The sense organs (sight, hearing, etc.) collect external "data" from the environment, and the kinesthetic or proprioceptive system, which is the body's internal "sense," reports temperature, pain, tendons and joints, movements, etc.

A global communication system

Information from the sensory receptors arrives via nerves to the spinal cord and on to the brain (inputs). This information is processed in nuclei when the neurons come into contact with each other. Based on the sensations and our memory, a motor command is emitted to move (outputs). This order is complex because it not only sets in motion the muscles needed for movement, but also includes activity to stabilize the body, adjust the movement accurately and react to unforeseen changes. Major movements in sports are voluntary, although much of their preparation and fine adjustment occur involuntarily thanks to the nervous reflexes.

Training skills and abilities

The athlete is constantly perfecting movements while training. The repetition of these movements trains the physical capacities, the circuits and the nerve connections, making the movement more and more fluid and efficient. Learning is the process by which we acquire and improve new knowledge or skills. The memory serves to retain this learning over time and it is associated with a generation of new reflexes.

Motor skills are learned by repetition and by trial and error. Perception is fundamental for the correct execution of motor actions and for storing the best-learned neural patterns. Improved perception occurs when working with different combinations of stimuli and is somewhat similar to what would be "training" of the sensory system.

Training produces changes (even hypertrophy) in the cortex, where the senses are "represented." The reorganization of the cerebral cortex is constant as it reinforces the part that is most stimulated

Nervous pathways:
1) External senses
2) Proprioception
3) Sensory medullary pathway
4) Conscious feeling
5) Motor area (voluntary)
6) Cerebellum (motor patterns)
7) Motor medullary pathway
8) Motor neuron
9) Movement adjustment (reflexes)

Nerve centers:
A) Brain
B) Core

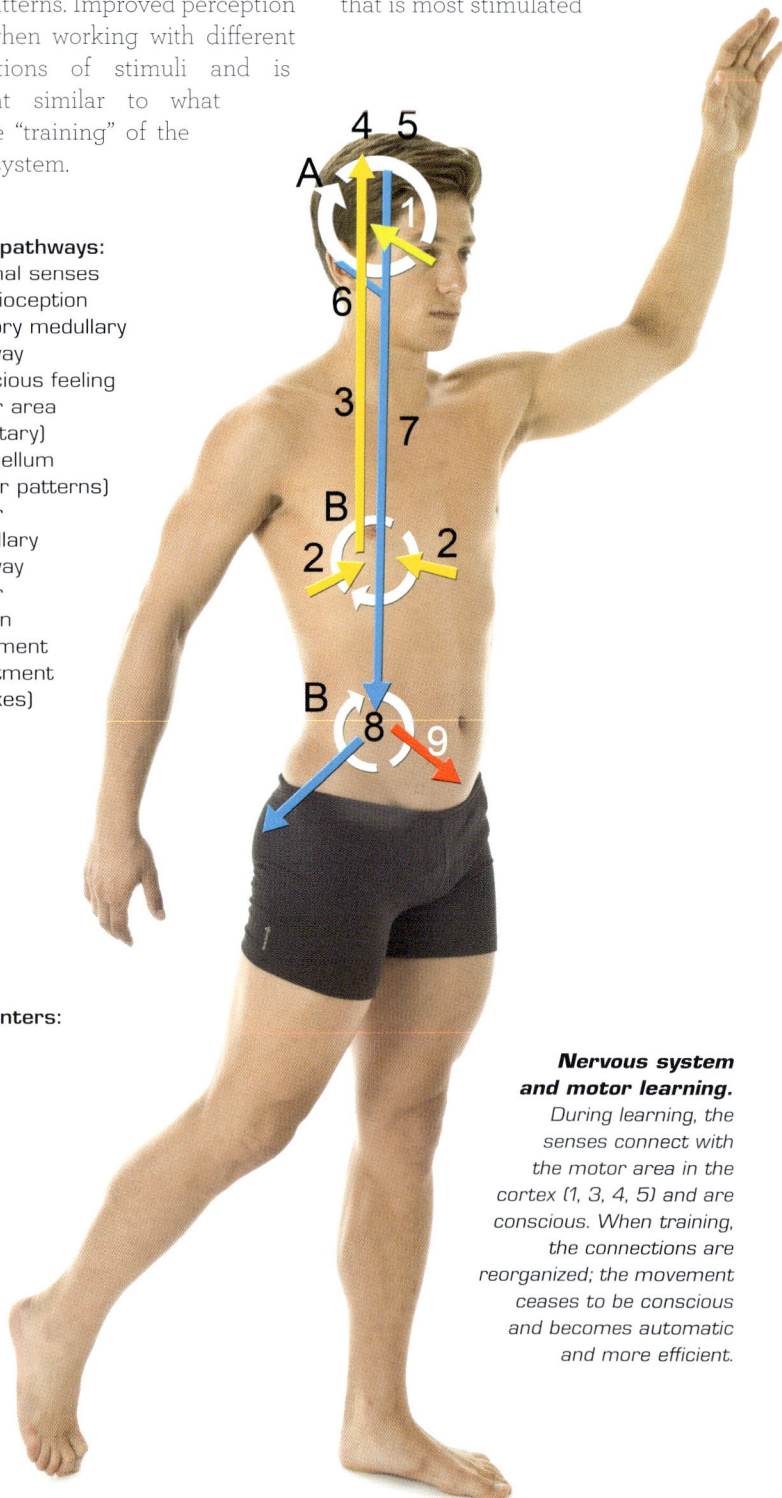

Nervous system and motor learning.
During learning, the senses connect with the motor area in the cortex (1, 3, 4, 5) and are conscious. When training, the connections are reorganized; the movement ceases to be conscious and becomes automatic and more efficient.

nerve cell: neuron

1

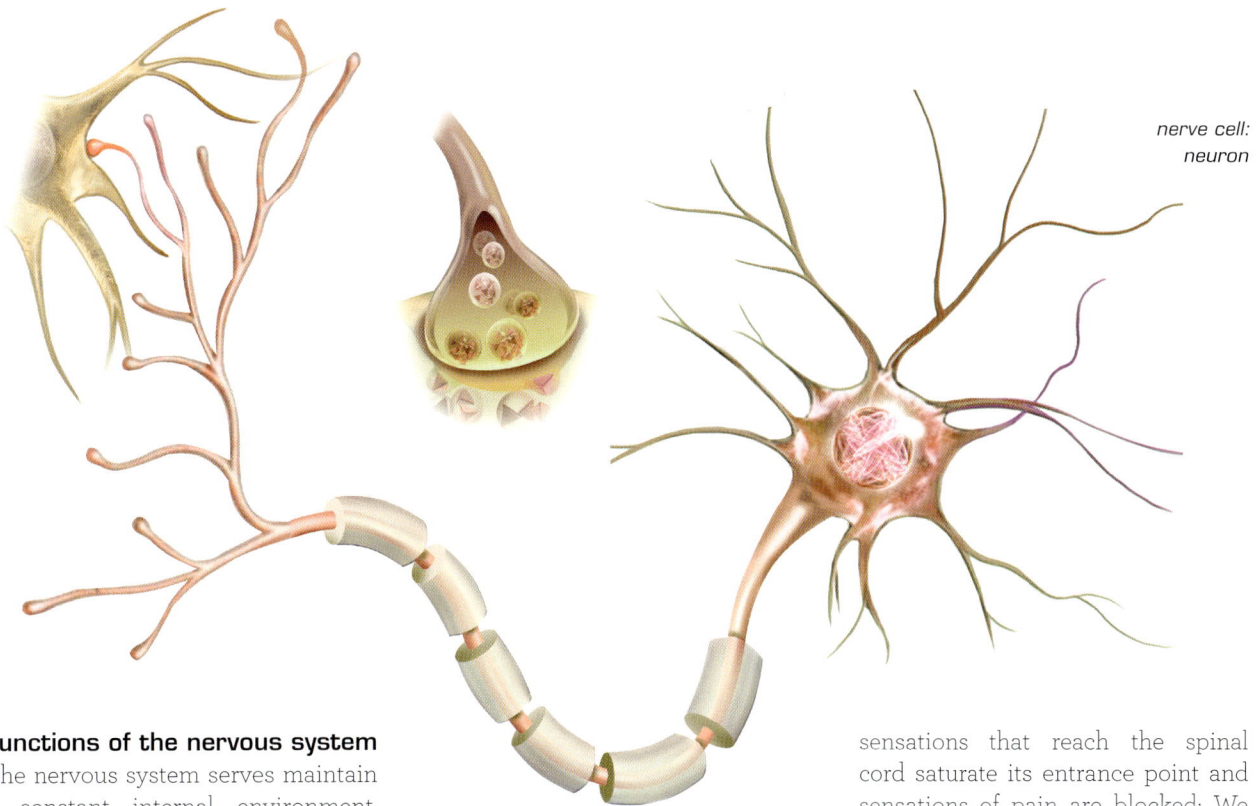

Functions of the nervous system

The nervous system serves maintain a constant internal environment. Strengthening or resistance forms of exercise must be alternated with rest and recovery; these cycles are regulated automatically and involuntarily by the sympathetic or parasympathetic vegetative nervous system.

Activity is related to responses such "fight or flight," ancestral reflexes that rely on the sympathetic nervous system. Warm-up routines in sports

Diagram of a neuromuscular junction, or motor plate, between a motor neuron and a skeletal muscle fiber. This serves to activate contractile machinery.

consist of preparatory movements that progressively activate the "sympathetic" state of the body to train or to compete.

Recovery, controlled by the parasympathetic nervous system, is associated with rest and digestion, and consists of a reduction in overall tone, nutrient absorption and tissue reconstruction, the same function as recovery massage.

Massage of the nervous system

Sports massage can activate or relax the nervous system.

Vigorous, intense and rapid maneuvers increase the excitability of neurons and send many nerve impulses to the spinal cord. This activation has several consequences: a) The body as a whole enters a state of alert, predominantly sympathetic, and releases adrenaline. The organism is now prepared to work. b) The excitability of the nerves makes contractions easier and faster, making sporting movements more explosive. c) The large number of

sensations that reach the spinal cord saturate its entrance point and sensations of pain are blocked: We are "anesthetized" by a gate control mechanism.

In recovery massage, slow and wide maneuvers relax and cause the body to enter a state of parasympathetic predominance. The nervous information produced by the massage reaches the limbic system (called the reptilian brain), where mechanisms of reward and satiety are found. The limbic system allows us to distinguish between well-being and anxiety, and through massage we can act on it according to the athlete's needs.

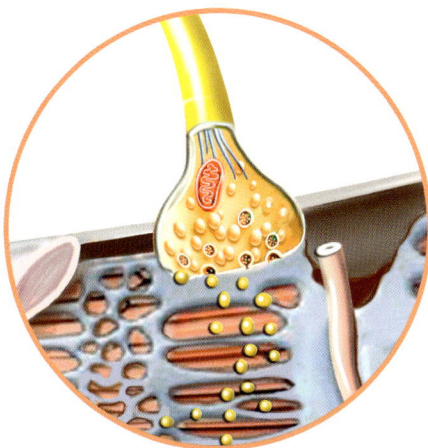

The nervous system connects with muscle fibers in the motor plates, which function as switches that transmit electricity from the neurons to the fibers. Calcium enters them, releasing energy that sets muscle contractile proteins in motion. Intense workouts tear the motor plates, which recover with great difficulty.

The circulatory system

The circulatory system is one of the global organism systems and is formed by a set of tubes that direct the venous and arterial blood and the lymph. Its function is to transport nutritional substances or waste and gases from one tissue to another throughout the body. It is also responsible for transporting "information" molecules (hormones, neuropeptides, immunoglobulins), markers which integrate bodily systems.

The working heart

The amount of blood that the heart pumps in one minute is called the cardiac output. The blood circulates through vessels, reaching all tissues depending on heart rate and volume of blood driven by the ventricles in each heartbeat. At rest, the heart pumps 5 to 6 liters of blood per minute, rising to 25 liters per minute during intense effort.

Physical exercise and blood volume

Physical activity moves the precise myofascial tissue to make the sport movement and, for that reason, it is necessary to circulate more blood through active muscles, providing them with oxygen and nutrients. The circulatory system adapts to the rising demand of blood by accelerating heart frequency, propelling a greater volume of blood in each beat toward the musculature, especially to heavily-working muscles. The flow through small arteries (arterioles) that bring the blood to the cells increases in speed because some of those that are inactive or at rest reopen.

The muscle fibers receive gases and nutrients from the blood through the capillaries that surround them. Note the greater use of oxygen by the muscular fibers, which increase from 30% to 70%. Blood that leaves an active muscle is much poorer in oxygen; the oxygen gas has been extracted. This combination during exercise means that the local oxygen and energy consumption of active muscles is multiplied by one hundred!

Redistribution of blood flow during exercise.

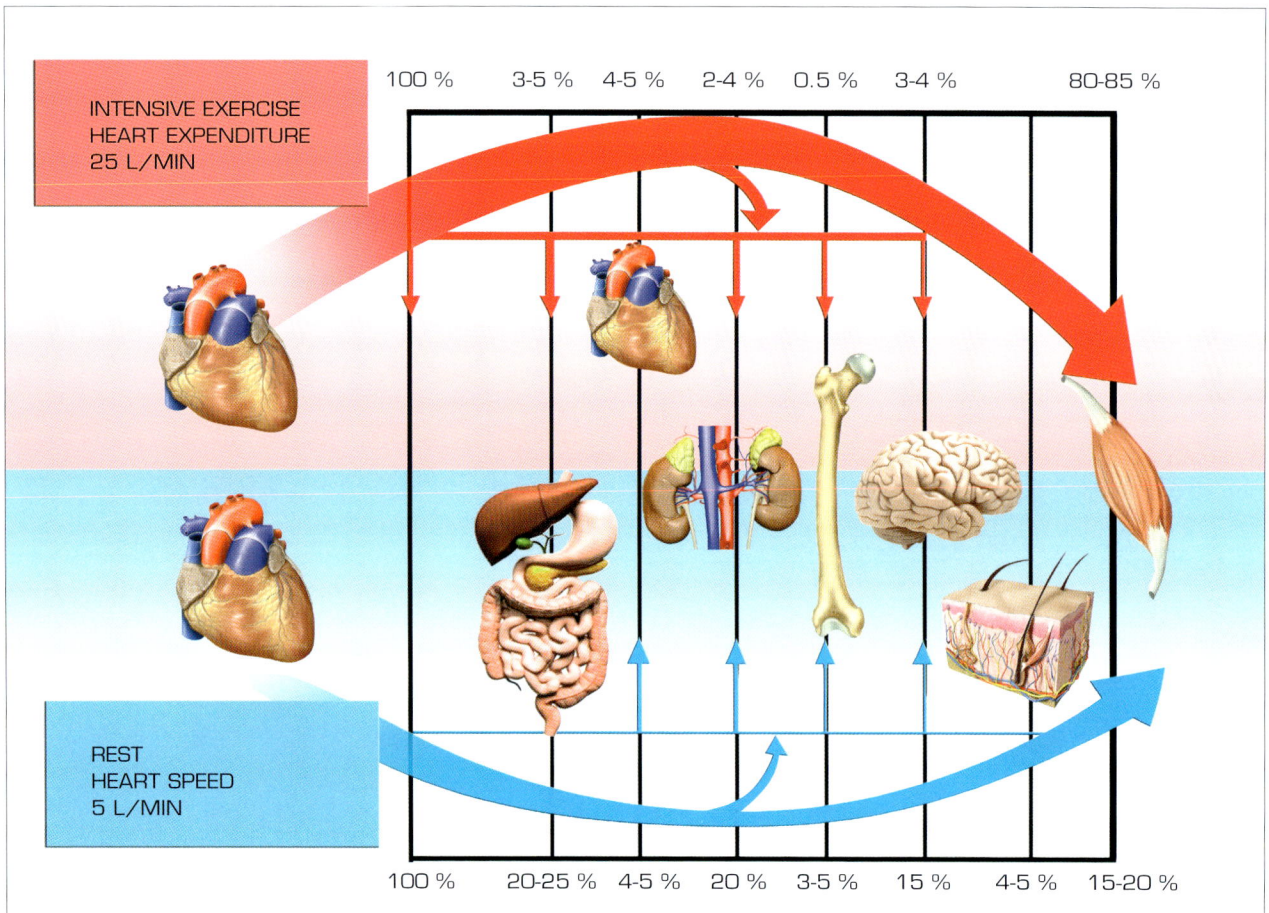

Influence of massage on blood and lymph

The application of massage produces compression and rhythmic decompression of muscle fibers, which causes the exchange of liquids with the extracellular matrix. The mechanical effect on the tissues generates the renewal of fluids, with a transfer of intercellular fluid to the blood and lymphatic capillaries, thus facilitating the movement of blood and lymph, which regenerates tissue.

Effects of massage on circulation

Massage has important effects on circulation before and after exercise. Before exercise, mechanical action heats the area and causes vasoconstriction, the action of the nerves that sets in motion the sympathetic system, which is a vasodilator. These effects pump the blood into the muscles which contract, preparing them for physical activity.

After exercise the "inflamed" muscle needs to recover, and, as a result, blood continues to flow through it. The blood supplies energy and nutrients to fill reserves and provide materials for muscle reconstruction. Venous blood and lymph remove waste and fragments of broken muscle fibers.

Training and the circulatory system

Endurance training is the ability to sustain prolonged exertion and depends on the amount of oxygen the athlete can retain (aerobic system). The greater the oxygen uptake capacity, the greater the endurance.

The oxygen reaches the muscle fibers by a process that develops in the lungs; the oxygen crosses the alveolo-capillary barrier and passes into the blood, where it is transported by hemoglobin. The heart drives oxygenated blood to the active muscles, passing through the arteries, arterioles, and capillaries. Gas exchange between blood and fibers occurs at the microscopic level.

The aerobic power that a person can develop depends on the efficiency (economy) of the sports movement, its cellular metabolic capacity and the adequate transport of oxygen.

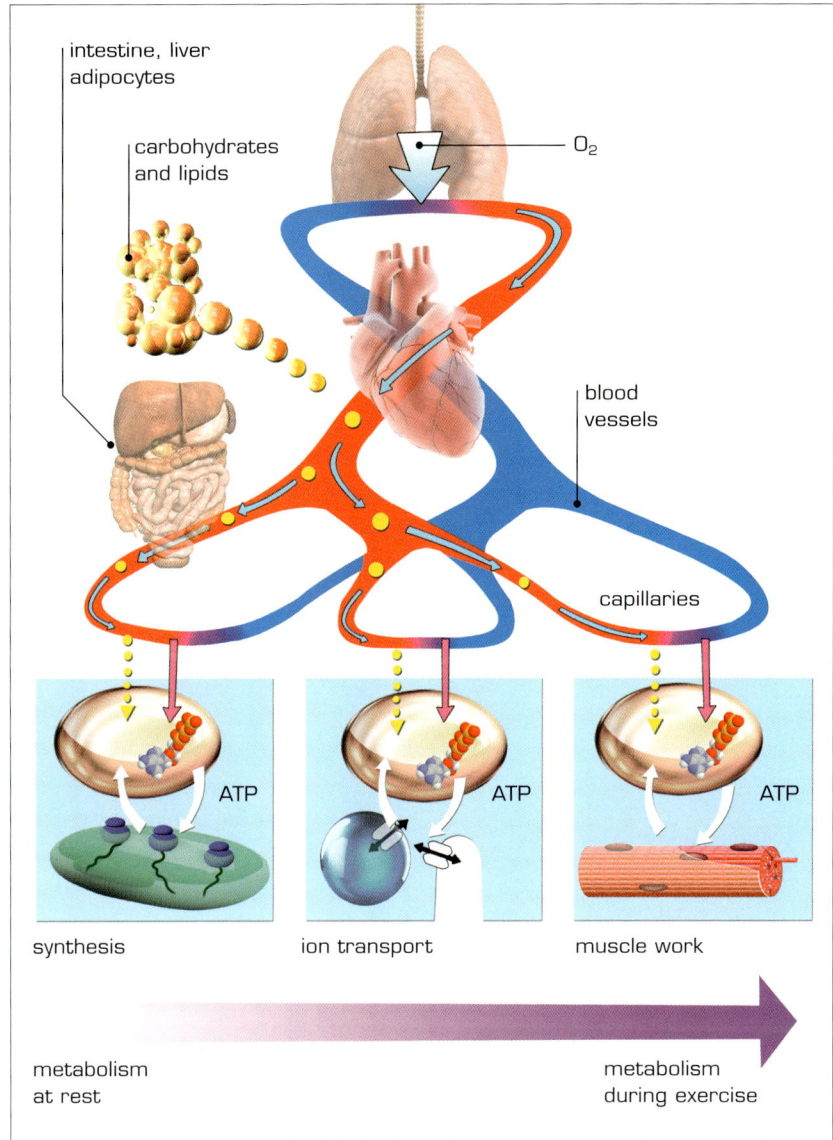

Circulation between the metabolism at rest and during exercise.

The irrigation of muscular fibers is carried out by a system of capillaries (between 3 and 8) running parallel to or located around them and immersed in the muscular connective tissue. Aerobic training increases the number of capillaries surrounding the fibers, increasing their capacity for oxygen uptake.

The organs

Systemic effects of massage

Post-exercise fatigue manifests itself in the organ systems and the musculature itself. Muscle inflammation is associated with an increase in blood acidity and concentration changes in electrolytes. The lungs accelerate the exchange of gases, expelling more CO_2 and forming more bicarbonate. This compensates for post-workout acidity.

On the other hand, the remains of muscle fibers and their coatings must be metabolized and expelled. This cleaning work is carried out by the liver and kidneys, so circulation is accelerated.

Work of the organs and exercise

The heart pumps extra amounts of blood into the muscles, lungs, liver and kidneys. Heart rate is accelerated after exercise. During rest the blood carries glucose to fill the glycogen stores of the liver, and does the same to fill triglycerides of the worked musculature. Eventually physical activity causes central fatigue (psychological effect) and peripheral fatigue (conduction, coordination, motor plaques, etc.) in the nervous system.

The massage in practice

The massage acts on the skin, dermis and superficial fascia, and through them transmits to the muscles, bones and joints, also reaching the vessels and nerves between these tissues. In their action of compression and decompression, they nourish and clean damaged muscle fibers. Venous renewal activates blood from the large veins, thereby increasing cardiac output, lung flow (gas exchange), and kidney filtration. Action on the nerve endings helps the recovery of the central and peripheral nervous system.

From large to small

These actions can be understood at the macroscopic level. At the microscopic level, deep tissue massage modifies genetic replication, the state of the internal environment of the cells, the transport of materials by fiber membranes and the state of the fundamental matrix (Stecco, 2016; Pilat, 2007). Therefore at the macroscopic level, massage has systemic effects on tissues and, at the biochemical level, on cells.

Electrical and biochemical effects of massage: mechanotransduction

The mechanical action of massage causes microscopic physiological responses in the body. The mechanism by which these effects occur is called "mechanotransduction," the conversion of mechanical signals into electrical potentials in the cell membranes. It converts a mechanical stimulus (massage movements) into an electrochemical response (tissue regeneration).

At the molecular level these stimuli cause fascial proteins to connect to the basal membrane and cell walls. Their structure consists of "microtubules" that hold the organelles and regulate the intracellular environment. By tensing this protein framework, the membrane is electrically polarized, the flow of material is changed, the cytoskeleton is adapted and the genetic material starts to rebalance the internal environment.

Effects of mechanotransduction.
Massage pulls at the tissues in three dimensions. Mechanotransduction via massage is the way to interrelate from a "fractal" conception of the organism.

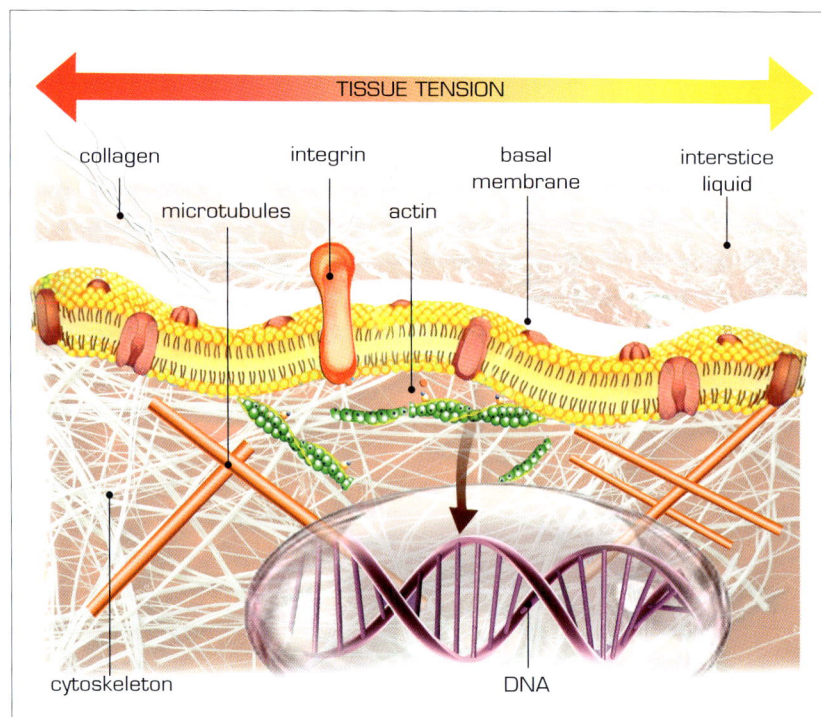

Myokines: the muscle as an endocrine gland

Muscle/organism integration is performed, in part, by the secretion of hormones called myokines. Skeletal muscle, in addition to its known functions as part of the locomotor system, is also classified as "a secretory gland" (Pedersen and Febbraio, 2008).

Myokines are responsible for stimulating metabolic processes locally and globally. Locally, the muscular hormones act on the fibers that capture more glucose, oxidize internal fat and, in general, provoke muscle hypertrophy, mature satellite cells or myoblasts and increase and repair their vascular network. Myokines

prepare the muscle for physical activity. Globally, they increase the lipolysis of adipose tissue, release glucose from the liver, activate cortisol in the adrenal glands, act on the digestive tract to activate insulin secretion (pancreas), stimulate growth and repair of blood vessels and create new bone.

In short, myokines tell the body that the muscles are active and put performance-improving mechanisms in place. They mobilize energy products, improve blood flow and activate states of alertness. In addition, they regulate the growth of their own muscle as an adaptation to the training.

The integral systems of the organism correspond to the circulatory system (transport of matter and energy), to the fascial system (movement and support of the internal environment) and to the neuroendocrine system (systems integration).

Red circles
Circulatory system
1) Arteries, capillaries (micro)
2) Veins
3) Lymphatic vessels
4) Heart

Blue circles
Fascial system
1) Subcutaneous tissue
2) Tendons, ligaments, aponeurosis
3) Meninges, glia, lining of the nerves
4) Pericardium, peritoneum, mediastinum
5) Cartilage, periosteum

Yellow circles
Neuroendocrine system
1) Neuronal nervous system
2) Neuropeptides
3) Endocrine System

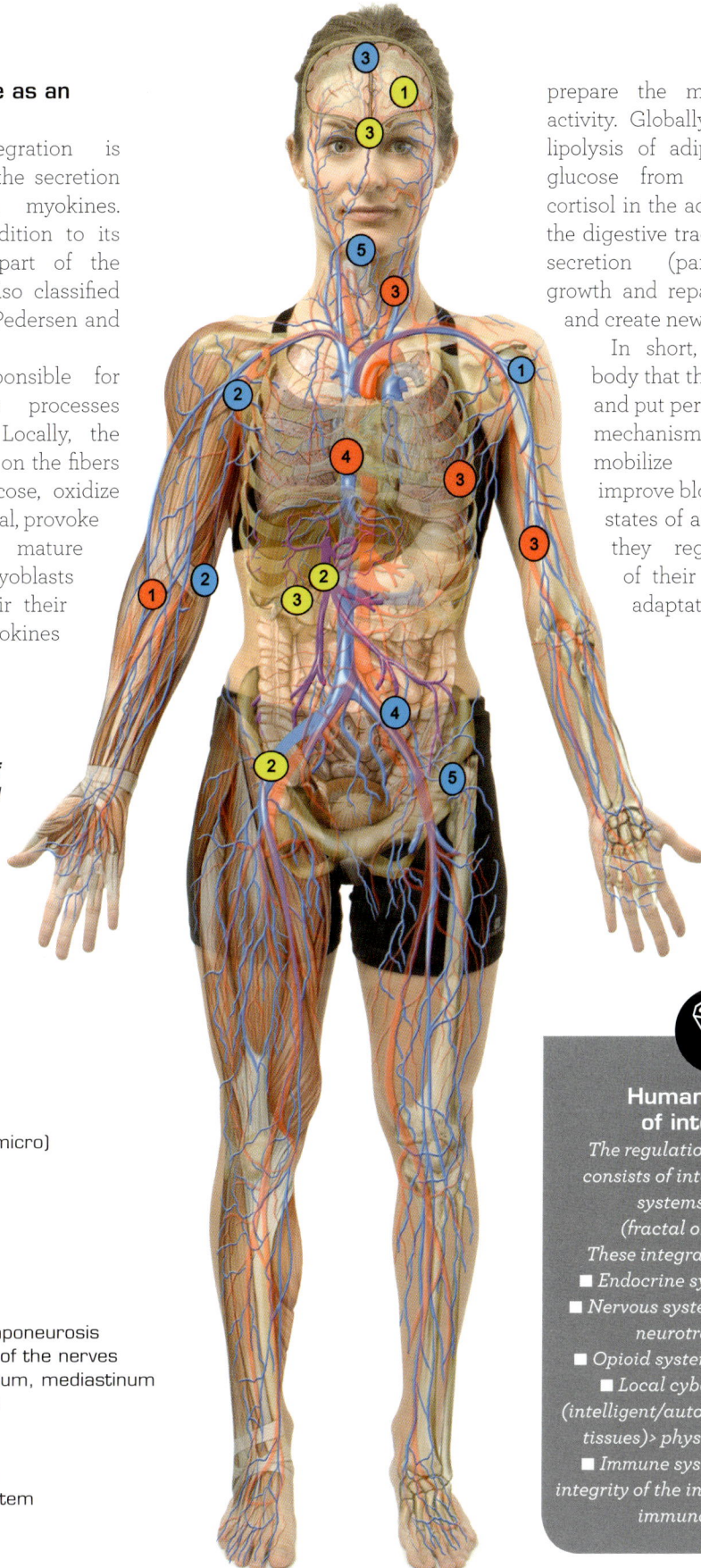

Human systems of integration
The regulation of an organism consists of integrating different systems and levels (fractal organization). These integrating systems are:
■ *Endocrine system > hormones.*
■ *Nervous system > electricity and neurotransmitters.*
■ *Opioid system > neuropeptides.*
■ *Local cybernetic systems (intelligent/autonomous response of tissues)> physical mechanisms.*
■ *Immune system (identity and integrity of the individual) > cells and immunoglobulins.*

The osteoarticular system

The osteoarticular apparatus provides stability and mobility to the body. In sports movement, body segments stabilize to give a solid base to other moving elements. To achieve this relationship between mobility and stability we need the "mechanical" action of the bones, joints, fascia and muscles.

The locomotor apparatus can be regarded as a lever mechanism with the point of support in the joints. Arms are the bones and apply forces (momentum), corresponding to the weights and muscular interventions. This is the mechanistic model of biomechanics. However, it is an insufficient model to truly understand how human movement functions.

Fluid biomechanics

A "biological lever" does not have a fixed fulcrum, but varies constantly with movement and is located at different points in the articular cartilage. The axis of rotation of the joint changes location and orientation while the displacement of one bone occurs with respect to the other. In fact, the "shape" of the articular surfaces changes continuously; they are fluid, rather than solid. The bones are not loaded directly on top of each other but rather float on the articular cartilage, vibrating, slippery and adaptive.

Variability and adjustment of sports movements

An example of this versatility is flex-extension of the elbow when it follows different "angular paths" as it moves and then returns, taking a different route every time we repeat the action; we never repeat exactly same gesture twice. This variability depends on changes in the environment and serves us in adapting to those changes, making the movement more accurate and effective.

Brain motor impulse is dictated by sensory information. As we execute a gesture, the environment changes, and the fluidity of the joints with their "micro-movements" allows us to fine-tune the position of the bones. Thus the movement is precise, economic and efficient. The force lines "vibrate," multiplying in a constant adjustment to the needs of the joint loads at each given moment.

"Macro" movement depends on the freedom of "micro" movements. A very small blockage at the joint can block the large and perceptible arc of movement, reducing the range of motion (ROM) in that joint.

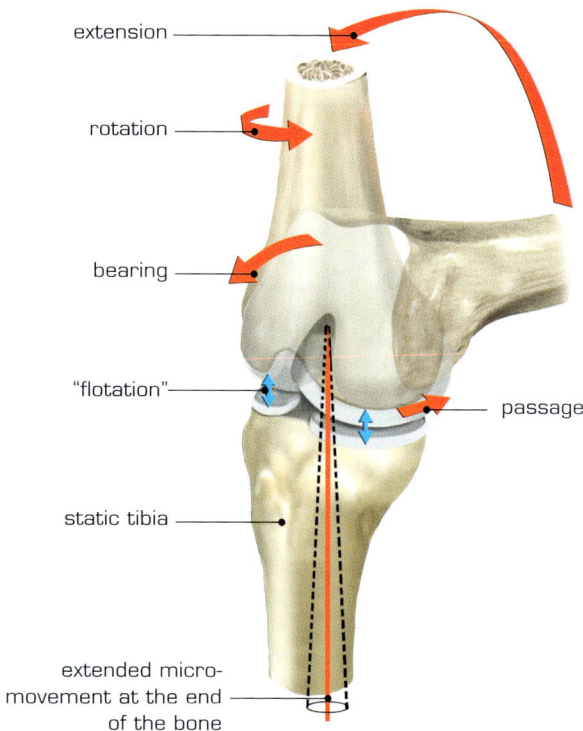

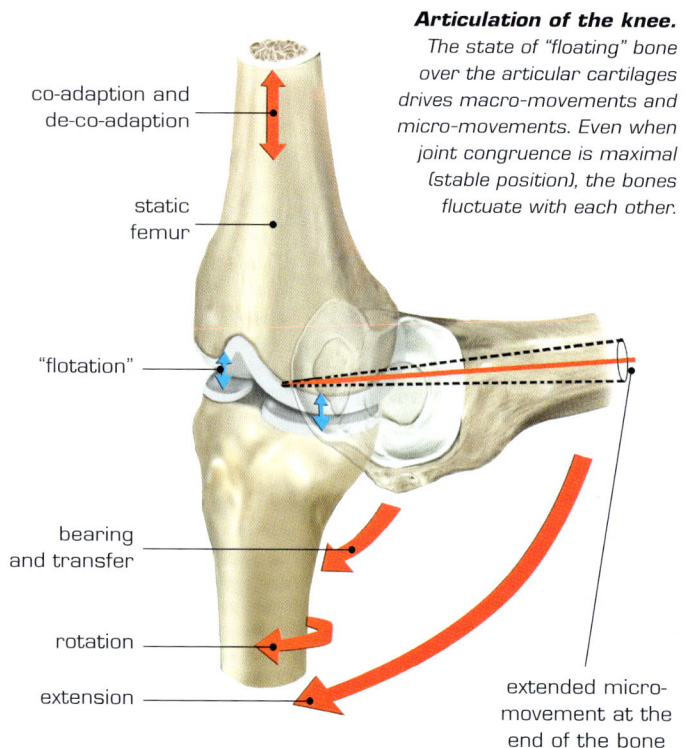

extension
rotation
bearing
"flotation"
static tibia
extended micro-movement at the end of the bone
passage

Articulation of the knee.
The state of "floating" bone over the articular cartilages drives macro-movements and micro-movements. Even when joint congruence is maximal (stable position), the bones fluctuate with each other.

co-adaption and de-co-adaption
static femur
"flotation"
bearing and transfer
rotation
extension
extended micro-movement at the end of the bone

lower extremity osteoarticular structures

body weight

hamstrings

tibio-peroneus ligament

interosseous membrane and compensatory tension

peroneal deformation

tibial deformation

femoral quadriceps

sural triceps

tibia anterior

body weight

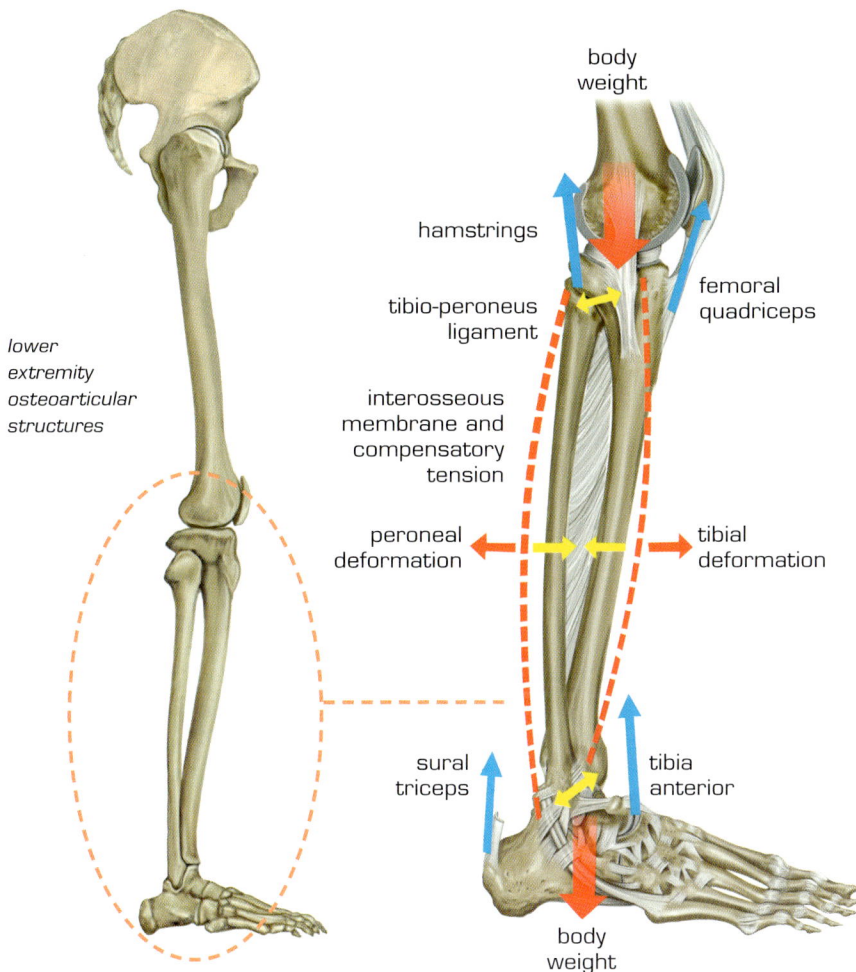

Bones are flexible.
Load capacity is characteristic of bone tissue. But thanks to their internal design they are also very slightly flexible; they will deform with heavy loads but are quick to respond and regain their shape. Myofascial tension counteracts these deformations during impacts, damping them when running, falling, etc.

Biotensegrity mechanism.
Bone elasticity and tension of the interosseous membrane, together with the myofascial system, act as a shock absorber as the foot impacts on the floor.

Micro- and macro-movements

The joints can move in many different ways. The enarthrosis achieves this in six ways: flexion-extension, lateral inclination (right-left) and rotation (right-left). These are "macro-movements," or visible movements.

However, articular faces move subtly and almost imperceptibly in "micro-movements," in which the surfaces glide together, compressed or decompressed (MacConail and Basmajian, 1977). The joints may have anteroposterior and lateral (right-left) slippage and decompression-compression. There are twelve movements in total: six macro-movements and six micro-movements.

Articular cartilage

Although the articular cartilage seems rigid, it is in fact endowed with a great plastic/elastic capacity. It contains water retained in hydrophilic molecules, a kind of bubble or sponge whose external surface supports huge tension. The structure is assumed thanks to the internal collagen that supports the pressure of the "swollen" tissue. The "micro" and "macro" movements of cartilages occur when adapting an elastic surface to its complementary surface.

Joint mobilizations

The term "mobilization" refers here to movements in which one or more forces are applied, with rotational, transferring or inclined movements, achieving greater range of movement in a joint. Joint mobilizations act on capsules and ligaments, which are slightly stretched by "displacing" the bony ends of the joint.

If a joint suffers from a functional disorder (is restricted and reduced in its range of motion), it has a pathological stop point that is perceived as non-elastic.

Mobilizations with torsions and bearings on the bones pressing against each other produce slips, twists and displacements that unlock the joint, release the adjusting micro-movements and increase the amplitude of the movements of that joint.

Muscle pain and inflammation

1

Pain is "information," a symptom, an internal and subjective sensation. It is caused by tissue injuries, but is not the injury itself. For human, the sensation of pain is a protective mechanism, a signal to react and to suppress the origin of the pain.

Sense of pain

Indicators of pain depend on emotional state, experience and sensory information processed by the nervous system. The "consciousness" of pain is not proportional to the injury and depends on stored experiences (memory), attention paid to the damaged area, mood, etc.

The effect of this information on the "neurological self" elicits responses such as: how pain is perceived; associated emotions (anxiety, fear, etc.), voluntary and involuntary movements, stress (cortisol, adrenaline, etc.), and the immune and neuropeptide response. The overall response to pain is aimed at keeping our identity intact.

The origin of pain

The athlete perceives pain for a number reasons aside from purely trauma. Intense physical activity accelerates muscle metabolism, lactic acid accumulates and muscle irrigation decreases: This is the painful sensation of fatigue.

Muscle spasms also cause pain because of the direct effect of contracture (stimulus of pressure receptors) and indirectly because they constrict, strangulating blood vessels. A spasm impairs muscle metabolism due to relative ischemia.

Muscular inflammation

Intense muscle contractions, especially eccentric muscle contractions, tear muscle fibers at the microscopic level, and the contractile apparatus ruptures. Fragments of actin, myosin, internal

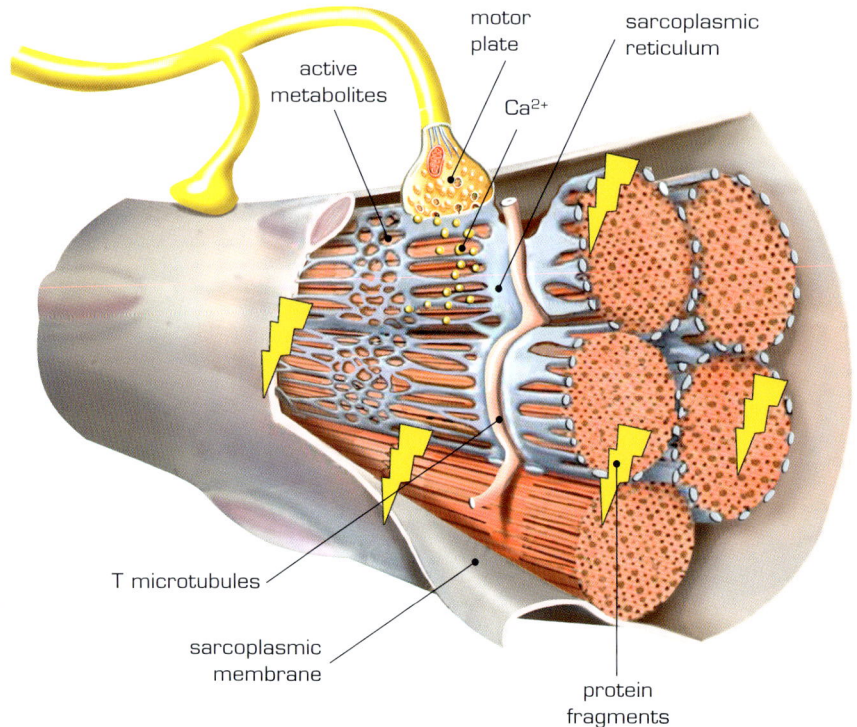

motor plate
sarcoplasmic reticulum
active metabolites
Ca^{2+}
T microtubules
sarcoplasmic membrane
protein fragments

cell membranes, etc. appear, and inflammatory muscular mechanisms are set in motion. Chemicals that trigger pain appear. Inflammation is a process of defense, with biochemical, vascular and systemic reactions whose purpose is to put into action the mechanisms that try to reconstruct damaged tissue.

The four signs of inflammation

Inflammation manifests with clear signs of heat and redness from vasodilation, swelling from edema due to increased permeability, functional impairment (or impotence) and pain. There is evidently a close relationship between exercise, inflammation and pain.

Muscle inflammation and cramps

Inflammation due to intense workout sets repair processes in motion. The immune system is activated and its white cells eliminate dead tissues and waste. At the same time, blood

Injury and muscle inflammation.
Muscle tensions break the contractile structure, sarcoplasmic reticulum and fascial muscle envelopes.
This triggers an inflammatory process, sometimes in the form of "stiffness", beginning the recovery process.

vessels dilate and fluid flows out of the capillaries into the affected areas (edema). Muscles swell, becoming stiff and painful. Tensions tear and rupture elements of muscle tissue and connective tissue (Schoenfeld, Contreras et al., 2013); microtraumas occur that trigger local inflammatory processes in order to initiate repair. This results in familiar "cramps," which manifest (variation depends greatly on the individual) within 24-48 hours. At present, this phenomenon is known by its acronym DOMS (delayed onset muscle soreness).

Transmission of pain signals

Pain signals (nociceptive) arrive at the spinal cord and are either processed and transmitted to the brain or discarded. Reflex medullary connections can generate flight movements before signals arrive: We can remove a hand from the fire before we are "conscious" of being burned. This mechanism allows us to react quickly to danger.

The brain receives and analyzes the information it receives from the nociceptors in the thalamus, hypothalamus and limbic system, producing (or not producing) the "sensation" of pain and activating defense mechanisms. In turn, the brain sends nerve and molecular signals (endorphins and enkephalins) to modulate the sensation of pain.

Massage effects

Massage stimulates the skin and myofascial receptors, sending massive amounts of information to the spine marrow. Different sensations (cold, pressure, vibration) arrive at the same point and must be ordered and avoid a medullary collapse. This gate-control mechanism causes anesthesia and relieves symptoms. The athlete has a window of rest; this is the time to apply mobilizations or other restorative techniques.

Ascending and descending pathways of pain. The sensation of pain is transmitted upwards, from the damaged tissue to the brain, where it becomes conscious and is modulated downwards by modulating mechanisms.

The stress of training

Immediate effects
- Fracturing of fibrils, sarcomeres and motor plaques.
- Tearing of the fascia and muscular membranes.
- Fragmentation of the tensegrity cytoskeleton.
- Accumulation of calcium and acidity: lactic acid, ammonium, etc.
- Intracellular molecules (kinases, actin, mycochines, etc.).
- Inflammatory molecules (interleukins, prostaglandins, cytokines …).

A posteriori effects
- Attraction of macrophages, lymphocytes, etc.
- Hormonal stress.
- Motor Decomposition. Muscle spasms.

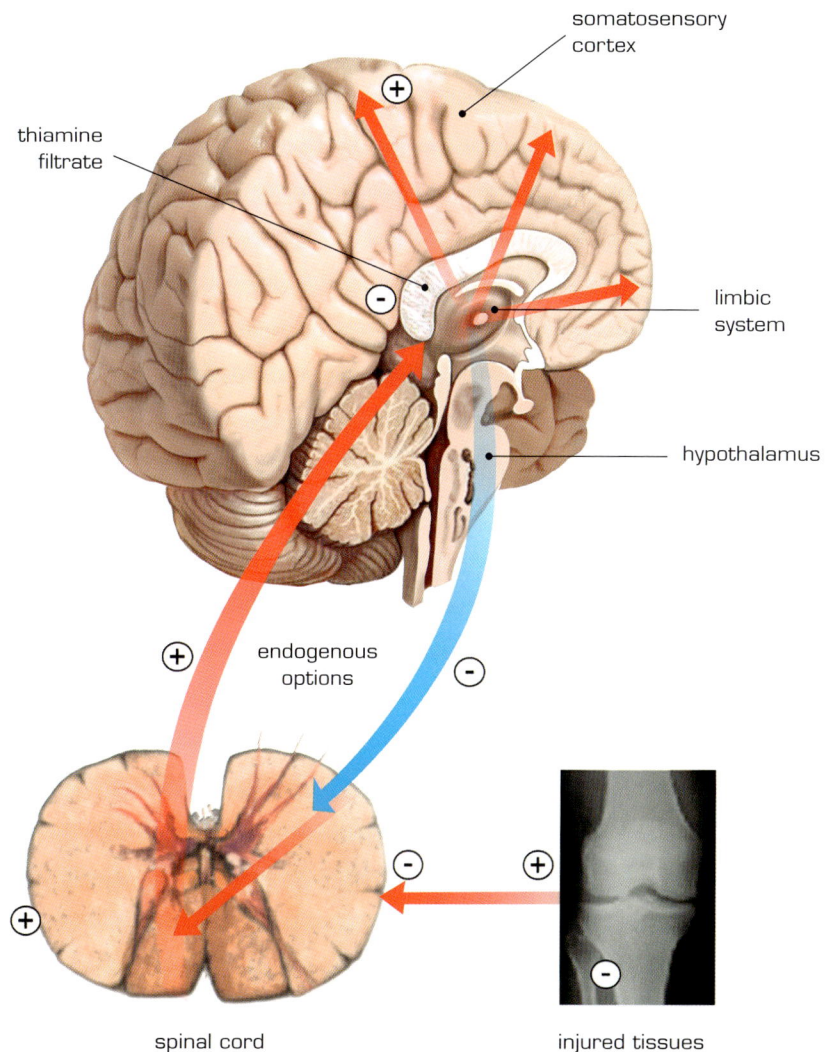

somatosensory cortex · thiamine filtrate · limbic system · hypothalamus · endogenous options · spinal cord · injured tissues

Physiological and psychoemotional effects

The effects of massage on the body are described at a general level, since there are multiple modalities of massage and the results are as varied as their techniques are different. In addition, depending on the technique used, an anatomical or other plane will be accessed, and this will also generate differing results.

Effects of massage: direct, reflex, late and integration

An organism is influenced by two types of physiological effects, direct and reflex. In addition both result in the late effect. It is also necessary to take into account the effects of body-mind integration (described on page 41), so central to the athlete's internal stability.

These direct or pure biomechanical effects arise at the site of application and are the result of the use of compression forces, friction or direct mechanical torsion on the tissues. For example: the increase of the heat in the skin after rubbing maneuvers or friction and the return of circulation after a pumping and kneading cycle. The technique of deep transverse friction helps to break down adhesions of scar tissue, etc.

Indirect or biomechanical reflexes, or segmental reflexes, arise when forces are applied in a given area of the body and produce results at a distance from where they were used. They correspond to changes linked to the stimulation or inhibition of various nerve and endocrine pathways. Some of them are: general relaxation, increased microcirculation, relief of pain, stability of the nervous system, etc.

Permanence of results

The retrospective effects of massage refer to the ability to retain benefits sometime after treatment, for example to achieve pain reduction or analgesia, be antispasmodic, provide vascular trophic adjustments, etc. Massage treatment in itself brings positive results for the organism thanks to multiple, closely related beneficial effects.

Massage searching for direct effects. *Kneading work on the forearm. The massage is directed to the first and second plane of the forearm flexors' anterior group.*

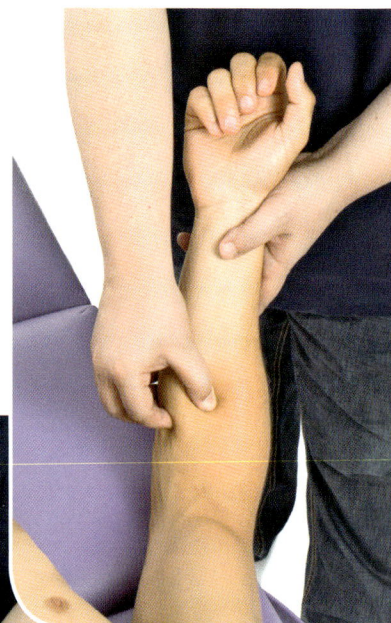

Massage searching for a reflex effect. *The athlete undergoes a massage session to treat leg pain, receiving kneading and pumping maneuvers in the lumbar region. The masseuse looks to create a reflex effect by reducing excess lumbar and sacral strain that will reduce sciatica in the area.*

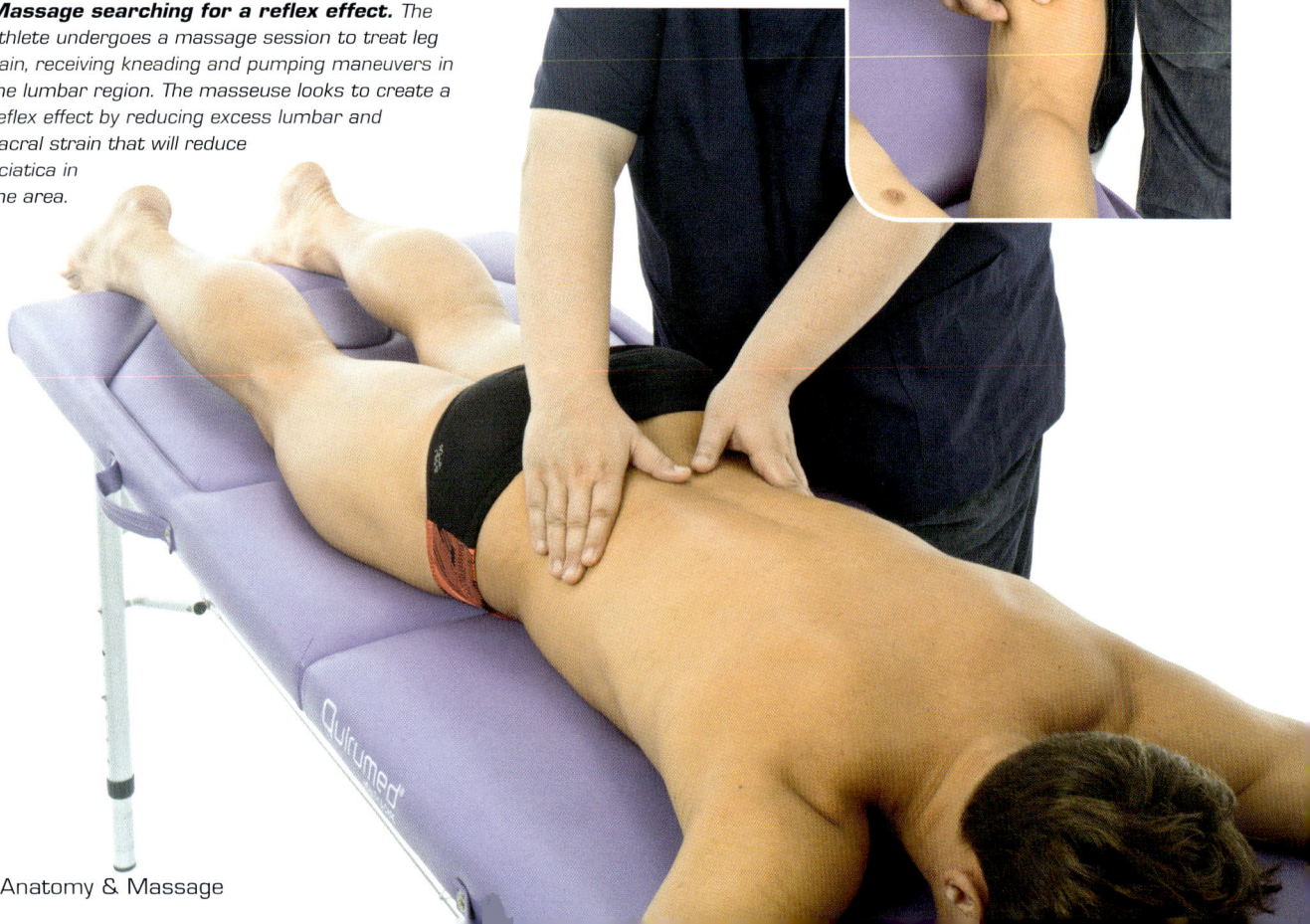

Psycho emotional and integration effects

The competing athlete is subject to many physical demands that can cause an excess of tension, and this manifests in the exhaustion, depression or anxiety that will hinder concentration and effective use of skills that influence sports performance. In addition, there are recent studies that describe anxiety as a factor that considerably raises the frequency of injury.

Sports massage focuses on regulation and prevention and in recent years, by including myofascial tissue massage techniques, is linked to body-mind integration work.

Habitual, slow-paced and progressive application of massage at the "tempo" that the athlete's body needs to react stabilizes and integrates the changes it brings. It causes a reflexive response of relaxation and general well-being that modifies the athlete's emotional state, replacing anxiety and restlessness with calm and stability.

Non-biomechanical effects

Massage is a good way to regulate both the degree of general and localized tension of myofascial tissue — considering among other factors that have already been described — thanks to the confidence that a firm hand contact inspires, coupled with an active listening attitude and a few supportive words during treatment.

Massage techniques provide a sensation of relief and are employed in a way that keeps the athlete relaxed throughout the session. It is recommended that the athlete harmonize breathing in order to integrate and accept tissue change during treatment, decrease heart rate and blood pressure, stimulate digestion and regulate sleep disorders derived from anxiety and stress.

Passive cooperation

The athlete who trusts the masseuse without resistance, who is predisposed to the benefits of massage and adopts an attitude of "passive cooperation" will achieve results that exceed expectations.

Young gymnast displaying clear signs of stress due to the demands that result from competition.

Respiratory rate. *The way we breathe and our capacity for relaxation are closely related. It is important to include work on conscious breathing techniques in sports massage sessions, as they favor body-mind integration and relaxation.*

Position, space and tools

The basic conditions for efficient massage are: to apply the correct techniques with the minimum effort, giving each maneuver and the whole treatment adequate "tempo" and rhythm. This control facilitates the athlete's relaxation, relieving anxiety and predisposition to improve.

The athlete's position

A comfortably positioned athlete favors the effective application of maneuvers. The position will vary according to the area to be treated. **Prone decubitus** (lying face-down) is used to treat the back of the body. If lumbar discomfort occurs, a pillow is placed at the level of the iliac crest to relieve it. The head is rotated to one side or positioned above the face opening of a massage table. The arms are positioned according to the athlete's preference or the need to access the scapula girdle or shoulder. A round cushion can be placed under the feet to increase comfort.

Supine decubitus (lying face-up) allows treatment of the anterior face. A towel is placed under the knees to avoid excessive curvature of the lower back, and another under the head.

If for any reason the athlete cannot tolerate the aforementioned positions, or for easy access to the side of the body, the **lateral decubitus** (lying on one side) is used, with the hips and knees flexed. A towel can be placed between the knees to increase the athlete's comfort. Another position that is used is **seated**. During competition this position keeps the athlete focused on the event while the massage is taking place.

Most commonly used positions during manual massage treatment:
1. prone decubitus; 2. supine decubitus; 3. lateral decubitus, and 4. seated. The variety offered by these different positions avoids discomfort for the therapist as much as the athlete, while improving access to tissue areas.

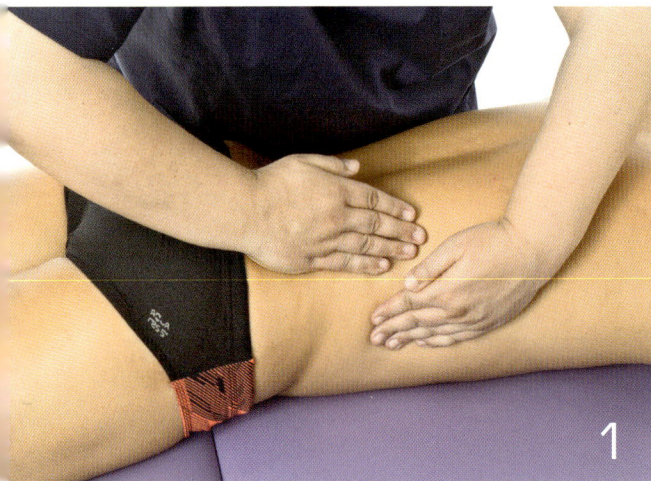

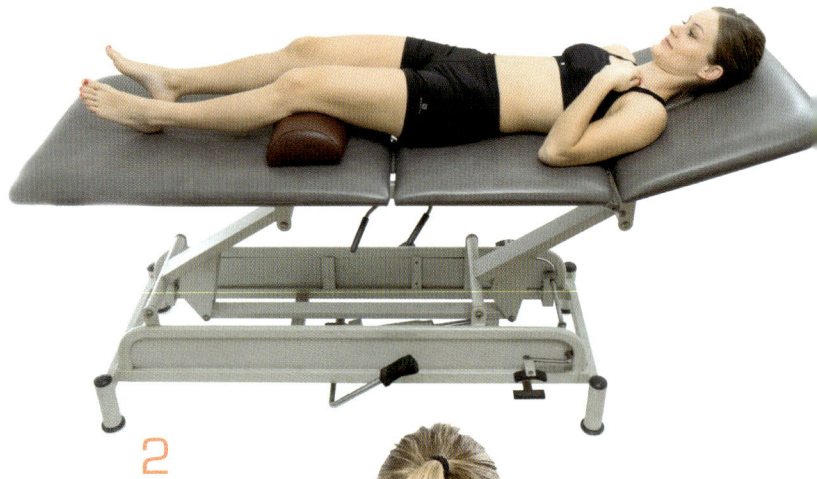

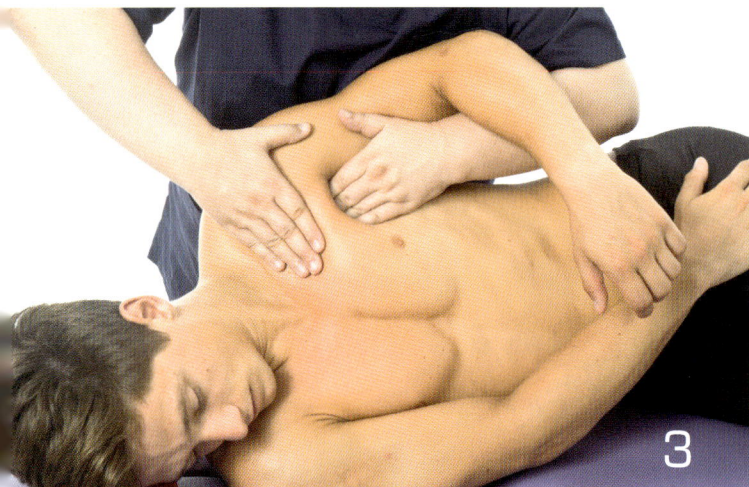

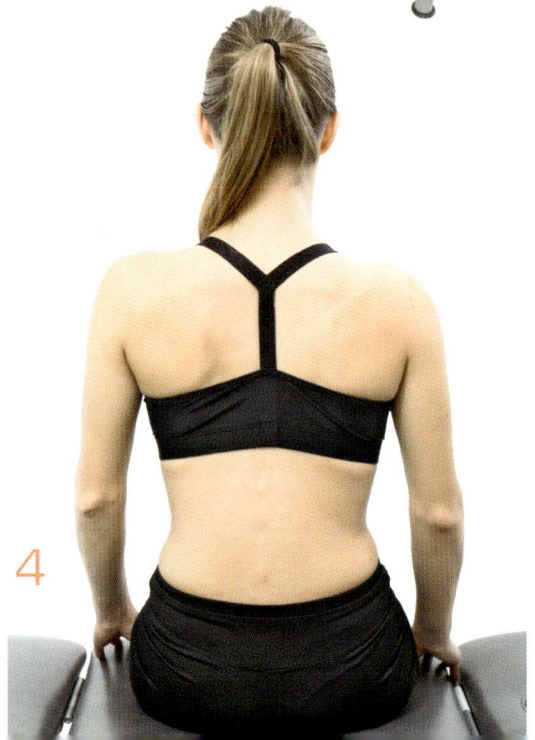

1

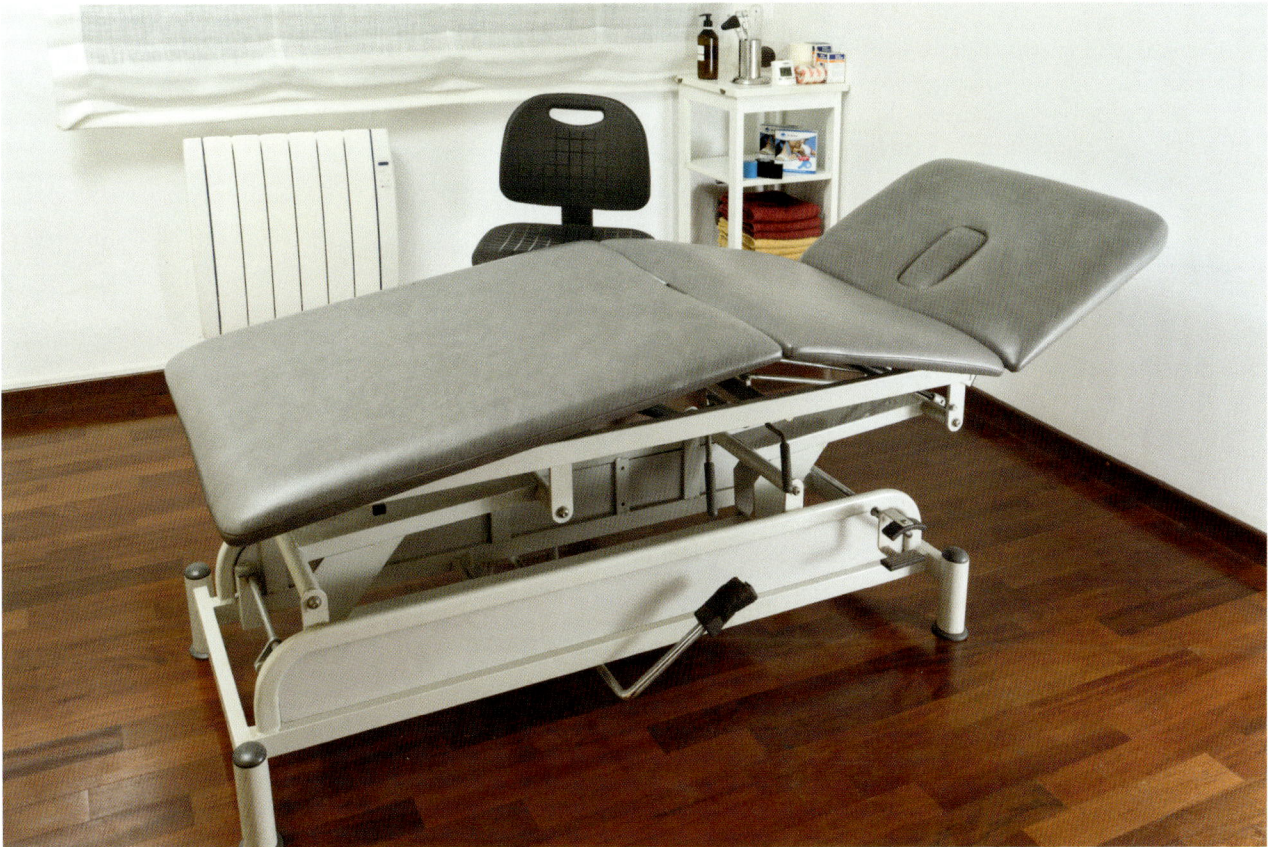

Massage space

It is favorable to have a space dedicated exclusively to the massage and the care of the athlete. This will vary depending on where the sporting event takes place; it could be the visitors' dressing room, a hotel room or outdoors. An outdoor area should offer some privacy while the treatment is carried out. A mattress can be placed on the floor or a portable couch used.

The "home" treatment area should be equipped with a couch and the usual utensils (rolls, towels or blankets to cover the athlete, oils and massage creams, alcohol and soap for hygiene), as well as a comprehensive and updated medicine cabinet. It should be well differentiated from other areas, easy to ventilate and to control light and temperature. There should be a toilet and a handheld massage tool for hand and arm hygiene.

Massage tools

It is important that the couch is comfortable and consistent, as the athlete will be lying on it for a minimum of 15 to 20 minutes. This prevents forces applied during the treatment from dissipating. We advise that it is height-adjustable, adjustable at different points along the body and has an opening for the face. It can be portable or fixed, electric, hydraulic or manual, but its minimum width should be 60-70 cm.

These characteristics will ensure greater comfort for the athlete and allow the therapist to be located near the work area, avoiding discomfort, overload or injury.

Massage space inside a sports facility. The couch is hydraulic with adjustable segments at different points along the body, a stool with wheels for the therapist, towels, exploration instruments and massage material, medication, and hygiene liquids and bandages.

Space/ambience

The working space can be adjusted with stimuli that favor massage and invoke the desired state in the athlete (ambient relaxing or stimulating music, aromatic oils that induce one or another state, etc.). The masseur must be similarly receptive and adopt a tone and frequency of voice to match the ambience of the working space.

Position of the therapist

Sports massage is a technique that demands great physical effort due to the daily repetition of maneuvers. The therapist experiences a considerable accumulation of mechanical stress.

In order for the massage to be applied effectively, body alignment must be correct. If the massage is carried out carelessly, it can increase stress (in both parties) and can even injure the therapist.

Body mechanics and massage

The masseuse uses the entire body in every technical movement. For this reason, in addition to trained hands, they will need to learn a body-working dynamic that guarantees maximum efficiency with minimal effort, to avoid fatigue, overload or injury. The masseuse must adopt a dynamic that ensures optimum application and results.

Relaxed, flexible, "listening" hands

The way the hands are used is as important as body posture. The hands must be loose and relaxed and sensitive to perceive any tissue disorder. This sensitivity is acquired through the daily practice of manual massage. An initial form of training is exercising mobility, relaxing the hand while working, and sensitivity. The improvement of tactile sensitivity allows the therapist to "listen" to changes in tension between the different anatomical planes, from the superficial (the skin) to the deepest (the skeleton), and without pressuring one area too much if a perceived problem can be solved at another level.

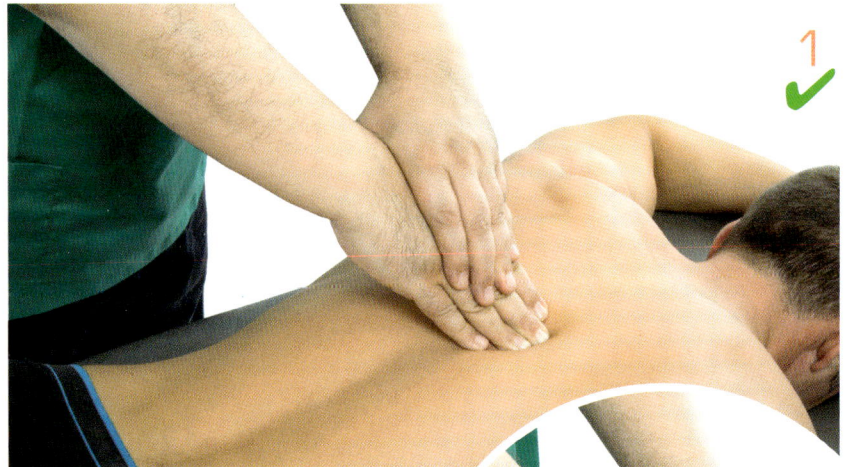

1. Correct and effective alignment to transmit the force of the press-and-drag maneuver.
2. Placement of the hands show an incorrect alignment of arm, wrist and fingers.

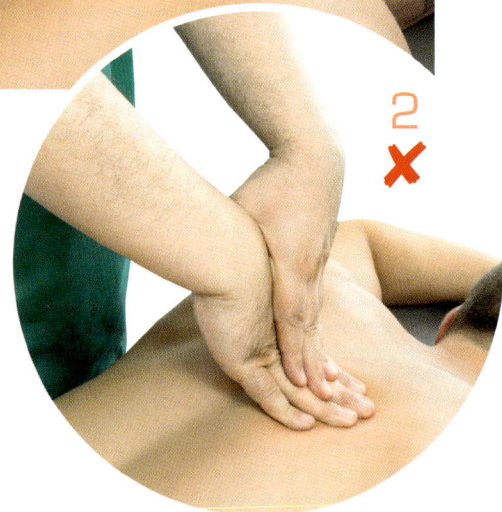

Precautions

◆ Regular breaks are necessary. Find moments to relax and stretch for a minimum of three minutes every two or three hours.

In this position the therapist is very close to the athlete, his body is squared and his arm is excessively extended, so that the shoulder is elevated and obstructs any attempts at fluid movements.

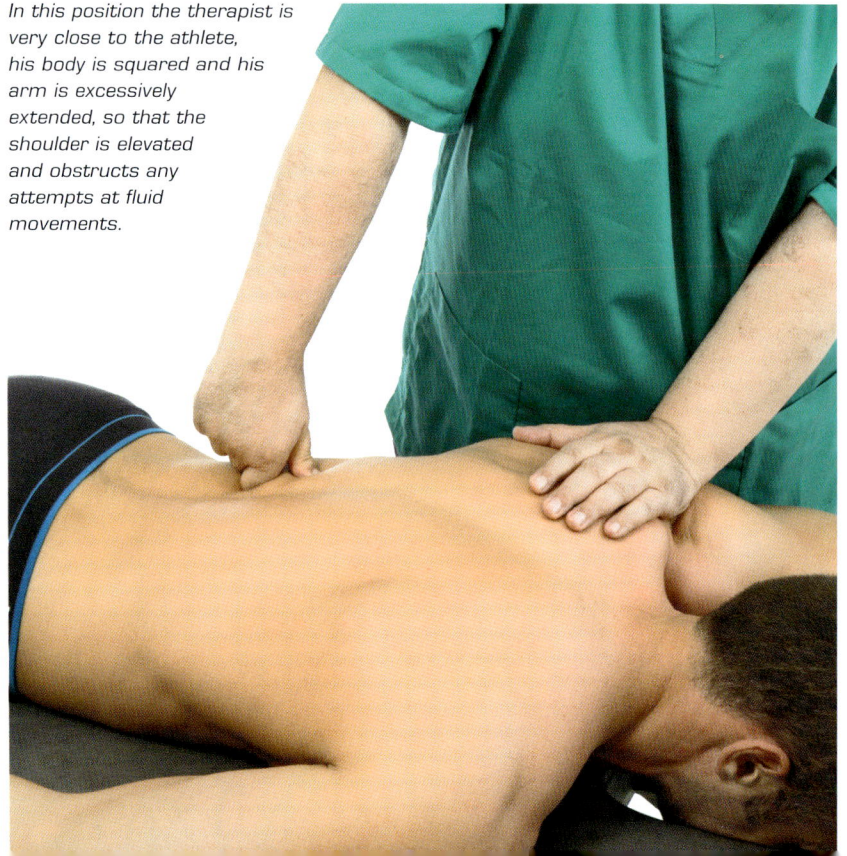

Finger flexion.
The tip of the thumb touches each fingertip, from the index finger to the little finger. This exercise can be continued by touching the base of each finger with the tip of the thumb.

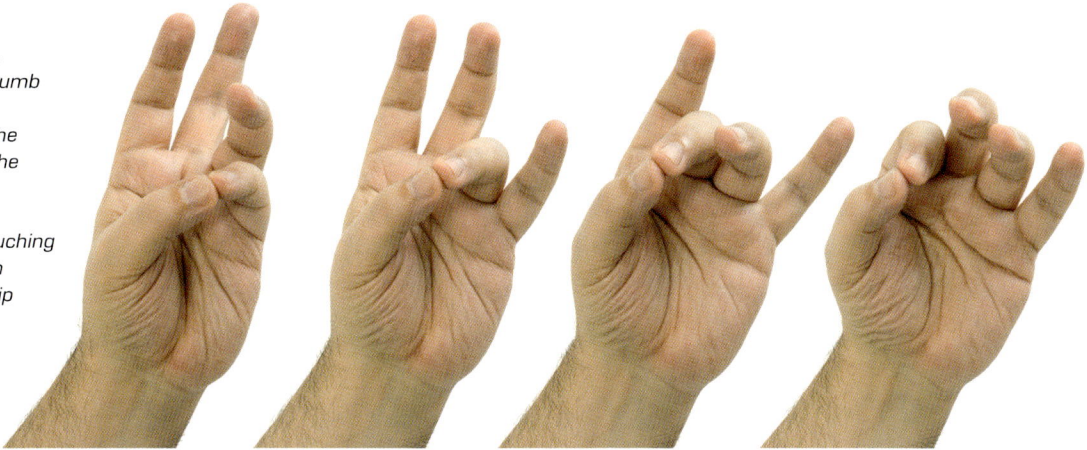

Hand exercises

It is best to incorporate a series of exercises for the hands, arms and the scapular girdle during daily practice, such as rubbing both hands to warm tissues and stimulate circulation, pushing the thumb into each of the fingers to facilitate mobility, or hand resistance using both hands to increase muscle tone.

Flex-extension exercises and lateral deviations are used in order to improve wrist mobility. Improved mobility of the scapula girdle and shoulders can be achieved with controlled shoulder twists.

These exercises release excess tension in the upper limbs and scapula girdle, preparing the therapist for massage work.

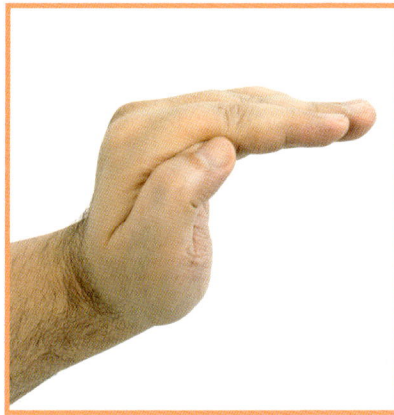

It is important to be aware of posture and how the body is used during the application of massage maneuvers. Correct posture with scapular and pelvic girdles aligned, together with correctly applied, properly aligned force determine the angle at which pressure is exerted during the massage, allowing efficacy in treatment and preventing therapist fatigue.

Intrinsic position plus. *This exercise promotes sliding of the hand's intrinsic tendons located in the face of the palm, the superficial tendon moving more than the deep tendon.*

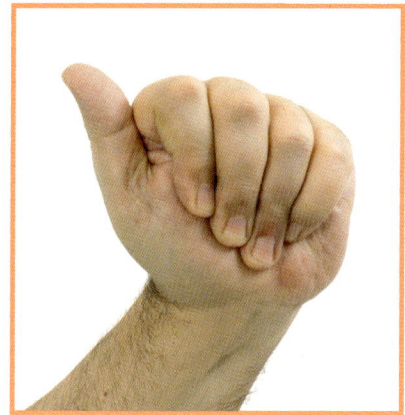

Straight fist. *Exercise to facilitate sliding. This position allows the superficial tendon more flexibility with respect to the sheaths and bones.*

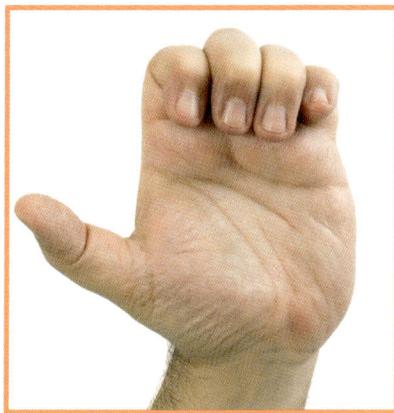

Hook position. *This position allows maximum slippage between the superficial and deep tendons, maximizing the flexor deep fingers (or FPD).*

Closed fist. *This position allows the deep tendons to move with respect to the sheaths, bones and surface. Maximizes work of the superficial finger flexor, or FSD and FPD.*

Health questionnaire and data collection

During the interview and before contact with the athlete, it is advisable to formalize a "health questionnaire." This personal data and a short clinical history will place the case in context.

Data and details for the questionnaire

The questionnaire lists the history of injuries, especially those that may have been suffered by the athlete, as well as surgical precedents and any treatment received in the past.

If an injury occurs, it is vital to know when and how it occurred, and also if it occurred at the beginning of a training session or during competition, halfway through or at the end, as these parameters relate it to the athlete's state of fatigue. Physical examination will include assessment of movement patterns and range of joint mobility (ROM), as well as muscle tone, asymmetries and possible limb dysmetria. Among other parameters it may also be of interest to note psychobiological habits and history of family pathologies, as reflected in the health questionnaire on pages 48 and 49.

In the case of an amateur athlete it is important to detail workout types: hours of workout per day and the physical characteristics that such work requires. These data allow the analysis of the daily workout's impact on the development of physical activity.

Test and adjustment data

Mistakes and program errors can be detected in the methodologies of sports training during interview and examination, especially in amateur athletes, as well as technical errors that may be responsible for overload and injury. If the therapist is knowledgeable about the subject or works with physical trainers, he can guide the athlete and adjust performance-affecting faults.

Interview prior to treatment is carried out in a methodical way. Avoid rambling and be clear and focused in your questions. The therapist must possess skills such as an ability to communicate effectively and manage time and interview duration. Lead the meeting and direct it toward the issue that concerns the athlete. A good interviewer should not confuse or mix different themes, must be able to listen and interpret and must always use intelligible words, even if they lack a certain rigor.

Interview. *Collection of data places the injury in context and is essential for exploration and treatment. In addition it allows the storage of data that will be very useful in subsequent follow-ups.*

Bending the lumbar and sacroiliac spine, standing. *The athlete is asked to lean forward until pain manifests. This informs the therapist which joints are working in excess to compensate for a lack of mobility. This test should be performed along with the pain provocation test in order to deliver reliable data.*

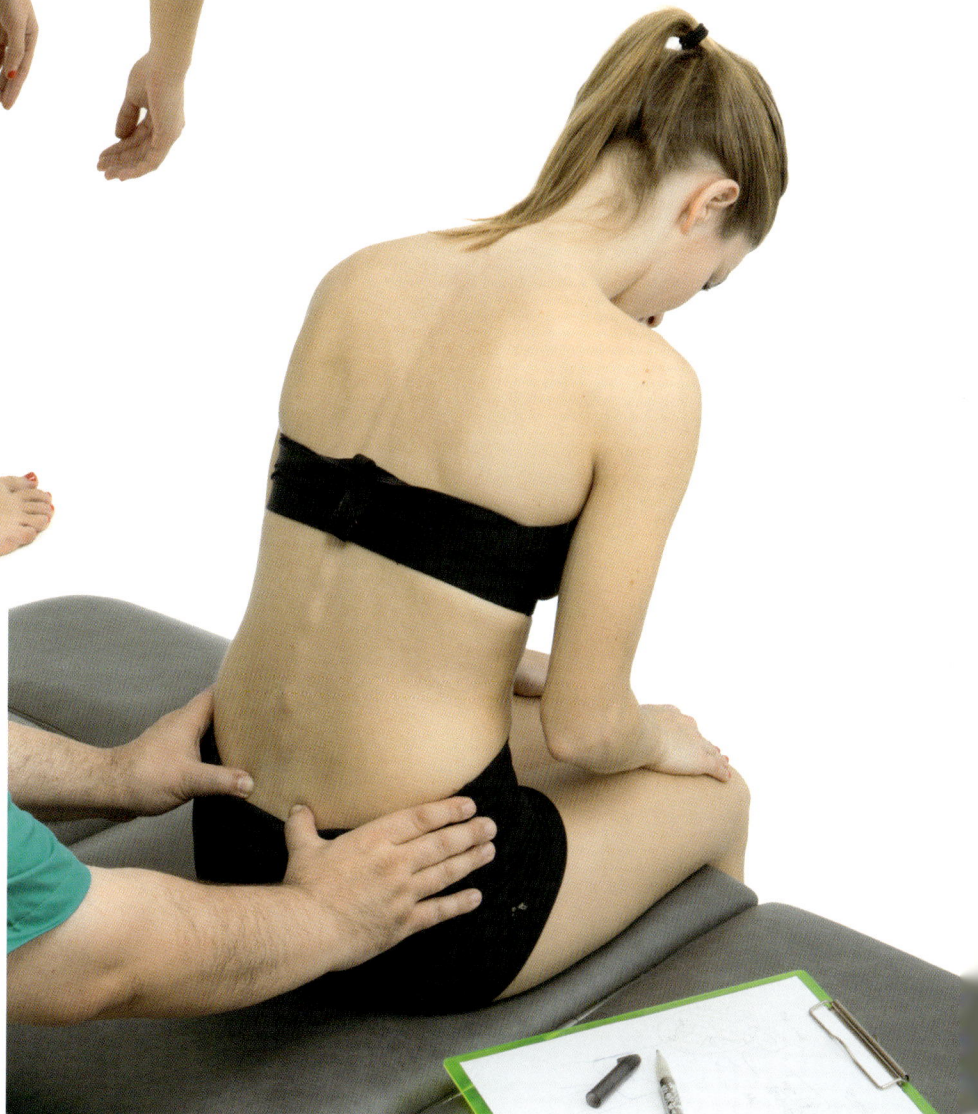

Flexural test of the lumbar and sacroiliac spine, seated.
Exploration of the sacroiliac joints reports the quality of flexion and extension movements in the trunk and lower back; if blocked, they could cause local pain and pain in nearby joints. This test will deliver reliable data if performed along with the pain provocation test.

1

The preparation of an "Athlete's Sports Questionnaire" requires the therapist to test different formats until they find the format most appropriate to the type of treatment. An example questionnaire is given on these two pages; it is a model that can be printed out or entered into a computer database. Note: This is a "minimum item" guideline.

ATHLETE'S SPORTS QUESTIONNAIRE: EXAMPLE

Case number: .... Appointment: .... Date: .... / .... / ....

First and last name: ....

Tel.: .... Contact tel.: .... E-mail: ....

Date of birth: .... / .... / .... Nationality: .... City: ....

Sport or physical activity: .... Weekly frequency: .... Daily hours: .... Length of time practiced: .... years

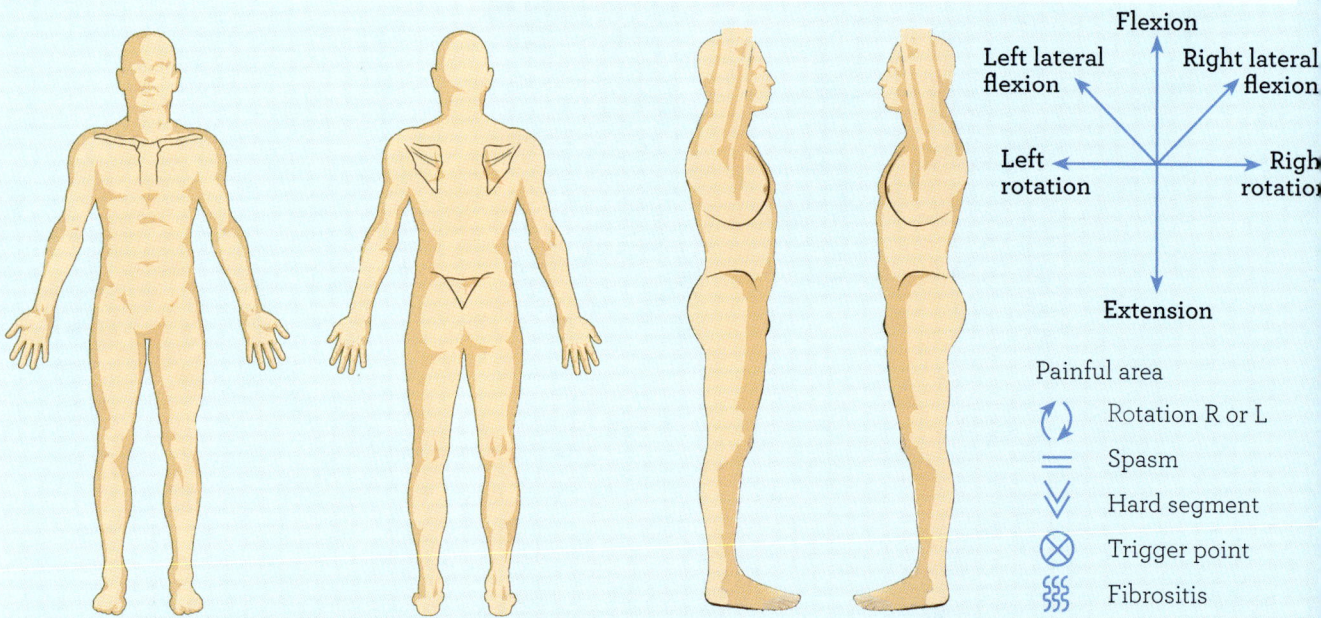

Height: .... Weight: ....

Flexion

Left lateral flexion   Right lateral flexion

Left rotation   Right rotation

Extension

Painful area

Rotation R or L

Spasm

Hard segment

Trigger point

Fibrositis

Current assessment of pain using visual analogue scale

No pain   Worse pain

Family background: ....

REASON FOR CONSULTATION

Where does it hurt? ....

How long have you had pain? Date: ....

What were you doing when you first felt the pain? .... How did it occur? ....

Could you carry on with your workout session? ....

When do you feel pain? .... What causes it to increase? .... What alleviates it? ....

Have you ever had a similar injury? .... Date: ....

What does it feel like? .... Do you feel numb? Yes/No   Where? ....

Current or previous treatment: ....

Anything else? ....

OVERALL RATING
General health status: ....

Social environment: ....

Recent weight changes? Yes/No   How much? .... kg

Weakness? Yes/No   Where? ....

Fatigue? Yes/No   When? .... Febrile syndrome? Yes/No   Temperature: .... °C

Allergies? Yes/No   What type? ....

Skin condition: ....

Dental health: .... Bruxism: ....

Nose: .... Nosebleeds? Yes/No

HABITS
Sleep: .... h   Hydration: L   Coffee: Yes/No   Tobacco: Yes/No   Alcohol: Yes/No

EXPLORATION
ROM joint movement range   Vertebral column: ....

      Extremities: ....

Tissue tension and texture: ....

Painful areas: ....

Can you perform active contraction? Yes/No

Can you perform a small resistance exercise with your hand? Yes/No

CONCLUSIONS 1st visit
Injury caused by:
Fatigue, overuse, mechanics altered, repetitive stress, entrapment, contusion, whiplash, joint syndrome, entorsis, distension, partial/total rupture, fracture, surgical intervention, etc.

Nature of injury::
Increased or decreased tone, ecchymosis, tight muscle band, trigger points, contracture, sprain, edema, alteration of muscle mass profile of, bursitis, adhesion, etc.

OBESRVATIONS 1st visit
Treatment to be performed: ....

End-of-session observations: ....

End-of-session statement: ....

Appointment made for 2nd visit day .... of .... 2-- at ....

INFORMED CONSENT
I ...., have been informed in an understandable and satisfactory manner of the benefits and also the risks involved in different applications of massage treatment, stretching techniques, joint mobilizations, bandages. Therefore, I agree to receive such treatment assuming that, despite being aimed at optimizing my health, they could adversely cause unwanted reactions and consequences.

In ...., on .... of 2---

Athlete's signature: Mr./Mrs./Ms.: .... Therapist's signature: ....

Inspection/physical exploration

It is essential to perform a physical examination or inspection of the athlete before treatment. This process should begin during the interview, observing the athlete's gestures when entering the consultation, how they walk, sit, etc., and also how they flex, undress, etc. During the appointment you should observe if the indicted painful or uncomfortable area presents an abnormal appearance. It is necessary to have a good "clinical eye;" that is, the ability to visually detect and afterward relate your observations as you explain to the athlete how you plan to treat them.

Initial observation

A more detailed visual examination requires correct placement of the masseur: in front of the different axes of symmetry of the various corporeal planes (front, anterior and posterior, right and left profiles, and transversal plane). This way a detailed and careful inspection of the posture at rest in the three spatial dimensions can be carried out.

Inspection should start at the feet and move upward, or from the head moving downward. The observed area must be at eye level. The gestures that the athlete performs voluntarily, whether requested by the therapist or not, as well as involuntary and unconsciousness movement should be closely examined.

Compare both laterals and asymmetries

Thanks to this inspection, by comparing homonymous side structures, how weight loads, skeletal and muscular asymmetries are distributed, we can detect trunk curvatures in anterior, lateral and posterior view in a general assessment. A more detailed assessment can reveal inflammations, anomalous movement paths, scars, calluses, etc. Simple objects such as plumb bobs, four-sided mirrors, etc., help to enhance this examination.

The dominant eye

During inspection it is practical to establish which is the examiner's dominant eye.
Arms are outstretched in front of the examiner and a small circle is drawn with both hands.
An object is centered in the circle, with both eyes open. First one eye is closed and then the other.
The eye that perceives the object centered in the circle drawn by both hands is the dominant eye.
It is advisable to use the dominant eye to assess asymmetries, distribution of loads, body volumes, compensatory movements, etc.

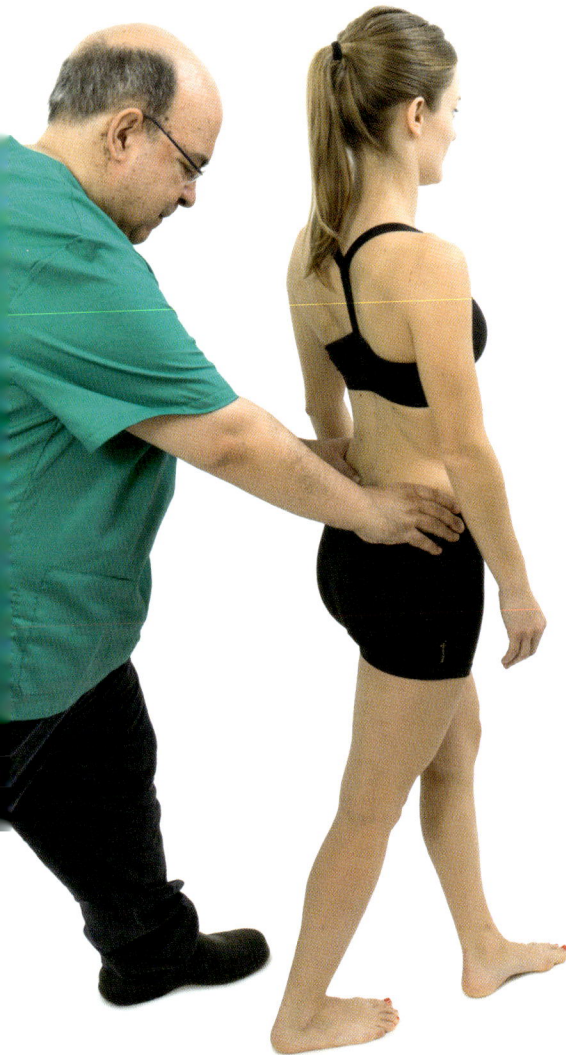

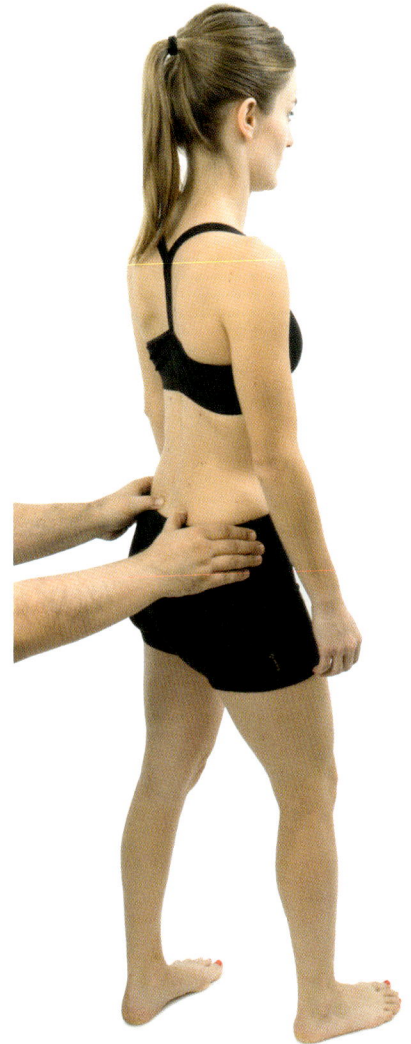

Exploring the lower back.
Dynamic tests provide information about body behavior when performing a movement. These tests are carried out on each side of the body to compare and establish differences.

The massage room and exploration

The inspection must be carried out in a calm, well-ventilated space and at a comfortable temperature. It should be fully lit and if possible be equipped with omni-directional lights, allowing you to detect problems. The athlete should be positioned so that their body structures are balanced and non-committal (as long as pain allows). If they must adopt a forced posture due to pain, you should avoid changing their posture (first observe, and then indicate and treat). The region to be examined should be free of clothing while respecting the athlete's modesty. You should look beyond the affected area (where the athlete is reporting some type of discomfort), observing both above and below that area. This area should be compared with the corresponding area on the healthy side.

Observing the obvious

Changes in the color of the skin may indicate a specific type of problem: A whitish color is a result of a lack of circulation; red, hyper-vascularization; cyanotic or violet, venous return problems; black, necrosis; dark and keratotic areas, excessive support or load; bulky, edematous-type areas, infiltration of cutaneous and subcutaneous tissue due to fluid retention, etc.

The appearance of the skin indicates the state in which the subcutaneous tissue is found: "orange peel," wrinkles, stretch marks, folds in articular areas, desquamation after long periods of immobilization, etc.

The secretion of the sweat and sebaceous glands gives the skin a dry, shiny, fatty appearance, according to its current activity.

Areas on the skin with wounds or sores may bulge, indicating that the healing process is underway. Its coloration also marks this process.

Bone or joint malformations can be seen from old or recent wounds. It is highly indicative of the pain-relieving postures that the athlete is obliged to adopt to relieve pain (knee and hip flexion, internal rotation of shoulder, Dessault's posture, etc.).

About pilosity: An increase in localized pilosity may indicate a vascular problem.

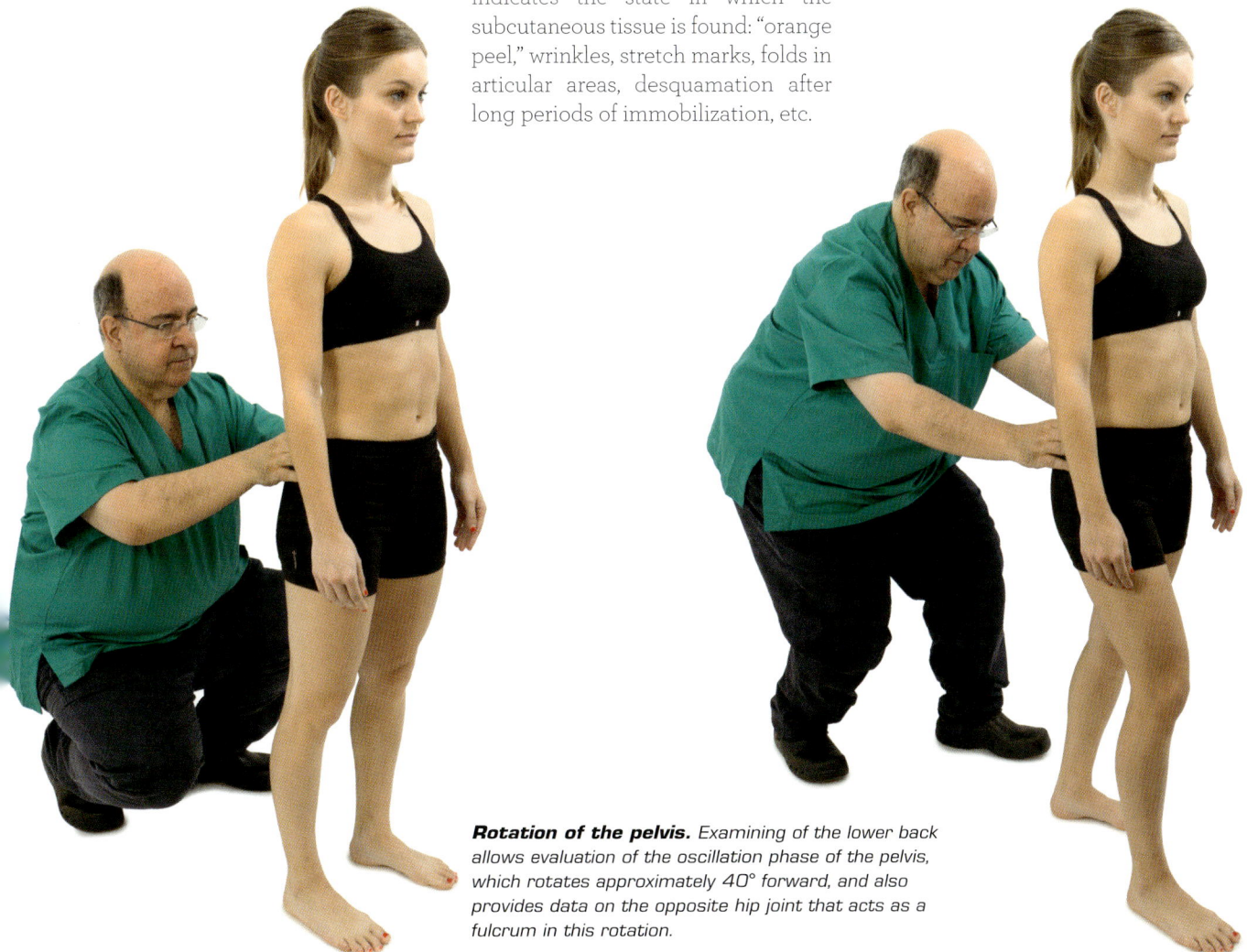

Rotation of the pelvis. *Examining of the lower back allows evaluation of the oscillation phase of the pelvis, which rotates approximately 40° forward, and also provides data on the opposite hip joint that acts as a fulcrum in this rotation.*

Understanding and interpreting body movement patterns

Movements are examined at this stage of the examination, for example in the lower back as it allows the assessment of the relationship between the scapular and pelvic girdles, the ability to swing the arms or the width of stride when walking. Global tests provide information on organizational level and the coordinating motor, and will always be comparative. They provide information on loss of joint range (amplitude) due to deficits in the joint, in strength or in active control of the movement.

Assessment of active mobility

Technical movements are evaluated: both those that cause discomfort, pain or disability and those that relieve it. This allows us to interpret, in the three planes of movement, which structures the movement uses in its execution and what the cause of the damage could be.

Active movements are evaluated in an analytical way, at each and every angle and direction that the joint articulation allows — bilaterally, if possible, to narrow down the causes of the problem.

As gravity influences movement, body placement can modify the athlete's responses when the different examinations are performed. You can use objects during the tests: fitballs, couches, benches, medical balloons, massage tools, etc.

The strength test

Subsequently, tests or resistance tests are performed on the different body segments that are involved in technical movements. These allow the masseuse to assess the contractile capacity (muscle and tendon) and the degree of strength or insufficiency.

Assessment of passive movement

Finally the global and specific movements should be passively reviewed again as the therapist arrives at the end of his examination of the joints. These tests provide information on the quality and quantity of the available range of motion and allow movements not actively evaluated (such as movement, stretching, traction, compression or any combination of them), which stress body structures. Crunches, cracks, projections, etc., whether associated or not with pain or discomfort, indicate an altered state of the joint and soft tissues.

Evaluation of the final movement stage

When a full-amplitude movement is made, the sensation of the tissues can be perceived at the end of the movement. This can be called a barrier, stop or resistance sensation and provides information about the joint's stability. It is necessary to establish exactly what tactile sensation is perceived at the end of the movement — it can be elastic, soft (soft tissue tension), firm (ligament or capsular) or hard (bone). The range of travel during passive movement is always somewhat higher than during active movement.

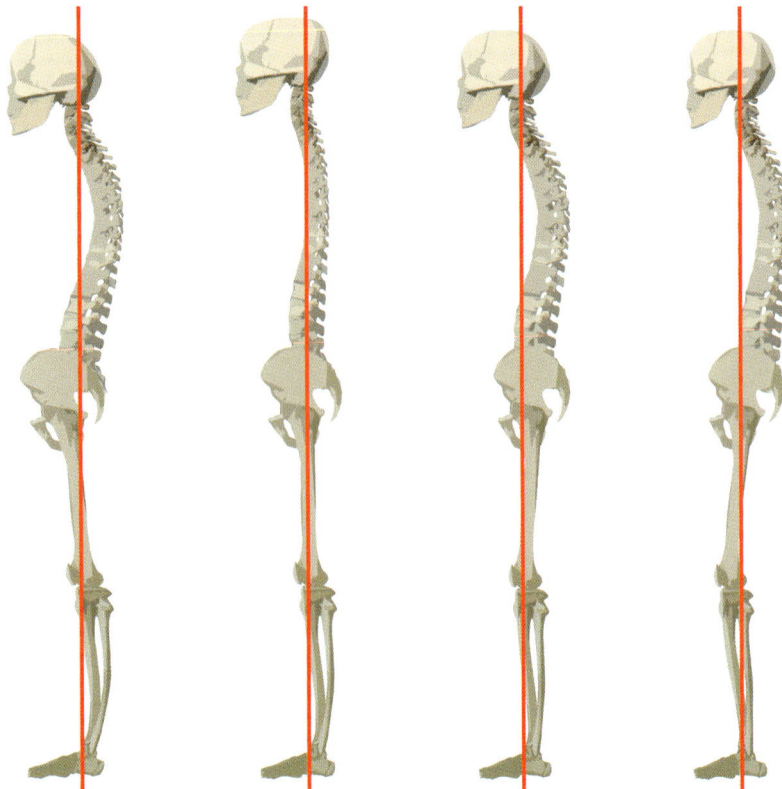

Lateral body alignment and alteration of physiological curves.

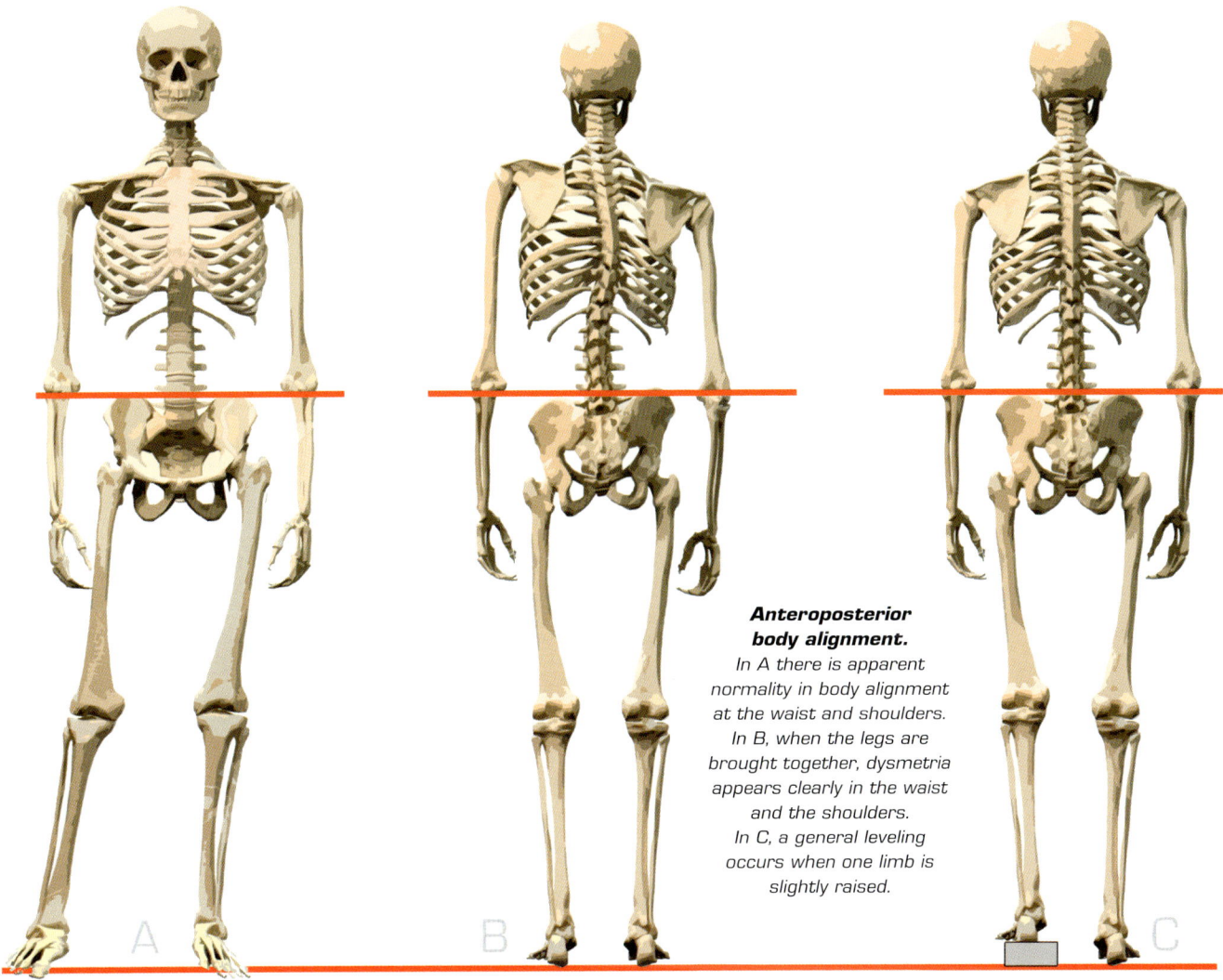

Anteroposterior body alignment.
In A there is apparent normality in body alignment at the waist and shoulders.
In B, when the legs are brought together, dysmetria appears clearly in the waist and the shoulders.
In C, a general leveling occurs when one limb is slightly raised.

Modification of impaired mobility

Pathological mobility implies a qualitative change in joint mobility and the final stop motion sensation. It is hard in bone alterations, hard-elastic in scars and non-elastic in blockages or contractures. The absence of movement indicates ankylosis, perhaps due to surgery or some other pathology (arthrosis, arthritis, malformation, etc.).

Movement may be very limited or limited. If active and passive movement is limited and painful in the same direction of movement, injury to non-contractile structures is suspected. If movement is limited or painful in the opposite direction, there is a lesion in the contractile structures. Finally, if passive movement is limited in different directions this indicates a capsular lesion and will be specific for each joint.

Techniques that enhance massage

Massage techniques should be combined with many of the maneuvers that are used to examine movement ability. The massage is enriched if the areas on which it is performed are put in tension or shortening, while the masseur uses a set of maneuvers ranging from traction to compression to forced movement, or any combination he deems necessary.

For example, to perform a friction massage on a knee ligament, lateral movement of the knee can be forced slightly to better access the ligament when there is tension. Tendons become more palpable if the athlete is asked to contract a particular muscle group, whether in tension or shortened.

In addition, these actions improve structural touch/perception by becoming more ostensible, allowing to therapist to differentiate if the structure that they can feel corresponds to a tendon, ligament or any other structure.

Posture

The muscles are classified into two main functional groups: slow-contracting red muscles (oxidative metabolism) and fast-contracting white muscles (non-oxidative metabolism).

Postural and dynamic fiber functions

Although most muscles contain both types of fibers, some perform a markedly more postural function and are "redder." Their metabolism allows them to constantly contract almost without experiencing fatigue. Their basic function is to stabilize and support the skeleton.

"Whiter" muscles are designed for dynamism. They are responsible for sudden contractions and rapid movements. Their metabolic behavior and anaerobic reactions make them fatigue quickly.

Response to stressful movement

When red muscles experience stress dysfunctions of whatever type (trauma, sustained postures, repeated movements), they tend to shorten and shrink. In contrast, when white muscles suffer from any kind of stress, they weaken and become inhibited and atrophied.

These dysfunctions can cause uneven posture, with groups of hypertonic muscles shortened while their counterparts are inhibited and weakened. These symptoms may be responsible, primarily or secondarily, for multiple pathologies (decreased range of motion, trigger points, predilection to repeat injuries, joint wear, etc.).

Cross-syndromes

A characteristic pattern is that of the "upper and lower cross-syndromes," described by Janda. The upper cross-syndrome is characterized by a myofascial retraction or diagonal shortening of the muscles and fascial posterior tissue of the neck and shoulder (trapezius and levator scapula) and anterior chest (pectoral). The antagonistic muscles above are weakened (inhibited reciprocally). They conjoin on a diagonal opposite to their anterior counterpart that encompasses the anterior neck muscles (deep neck flexors) and the interscapular (rhomboids and serratus).

Lower cross-syndrome also presents two diagonals. The shortened and hypertonic muscles join the lumbar and iliopsoas extensors. The diagonal of the inhibited and weakened muscles groups the abdominals and the gluteus maximus.

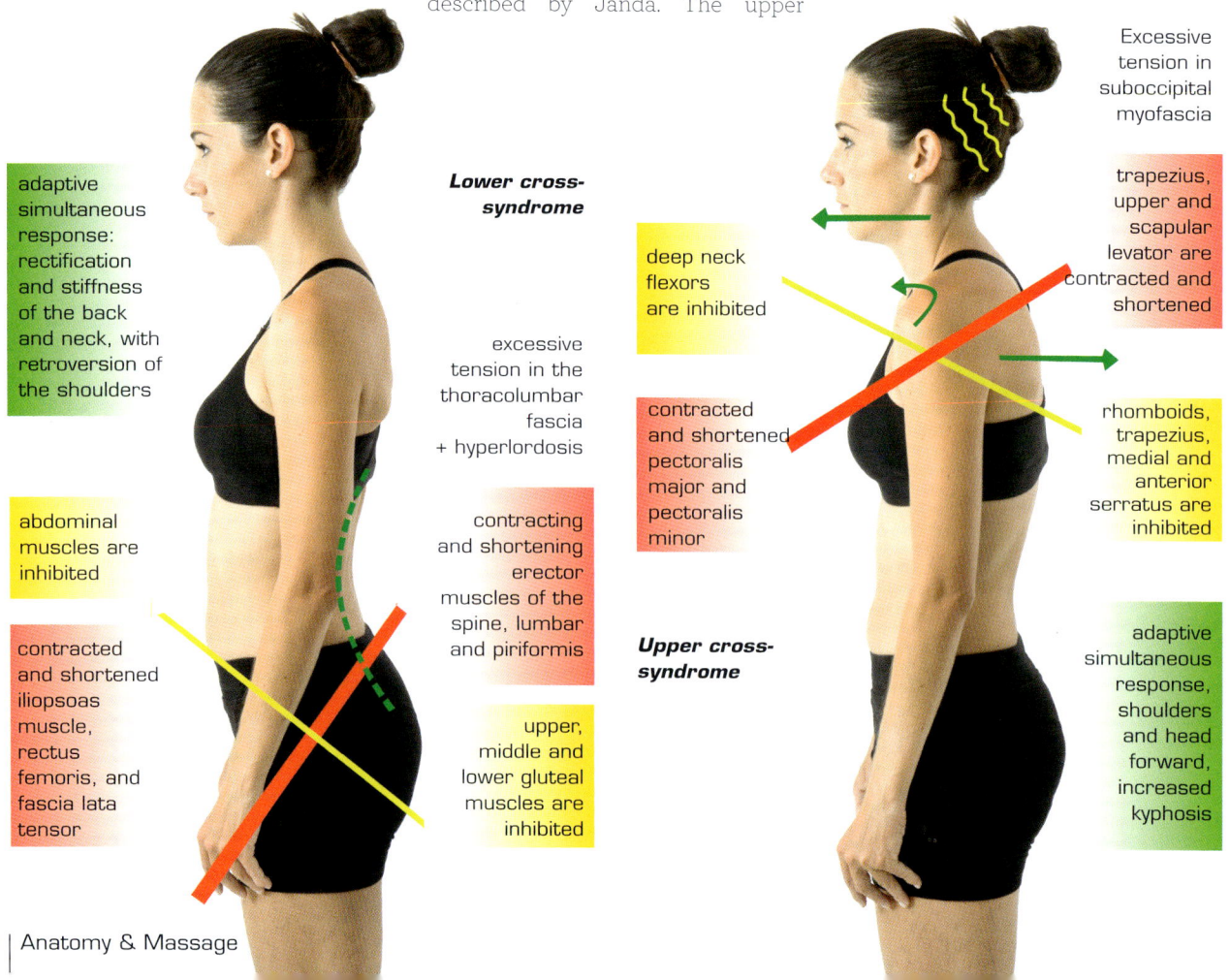

adaptive simultaneous response: rectification and stiffness of the back and neck, with retroversion of the shoulders

abdominal muscles are inhibited

contracted and shortened iliopsoas muscle, rectus femoris, and fascia lata tensor

Lower cross-syndrome

excessive tension in the thoracolumbar fascia + hyperlordosis

contracting and shortening erector muscles of the spine, lumbar and piriformis

upper, middle and lower gluteal muscles are inhibited

Upper cross-syndrome

deep neck flexors are inhibited

contracted and shortened pectoralis major and pectoralis minor

Excessive tension in suboccipital myofascia

trapezius, upper and scapular levator are contracted and shortened

rhomboids, trapezius, medial and anterior serratus are inhibited

adaptive simultaneous response, shoulders and head forward, increased kyphosis

Postural or tonic muscles shortened under tension:

- Plantar flexors: twins, soleus and posterior tibial
- Hip extensors: semimembranous, semitendinosus and biceps femoris
- Monoarticular hip adductors
- Piriformis
- Hip flexors: rectus femoris, iliopsoas and tensor fascia lata
- Lumbar square, erectors of the spine, rotators and multifidus
- Flexors of the upper extremity: pectoralis major (clavicular and sternal fibers), deltoid anterior position and long head of the biceps
- Lift of the scapula
- Upper trapezius fibers
- Sternocleidomastoid

Muscles that weaken under tension:

- Perineals
- Tibialis anterior
- Greater lateral and medial femoral quadriceps muscles
- Gluteus maximus, medius and minimus
- Abdominals: straight, oblique internal and external
- Serratus anterior
- Rhomboids
- Subscapularis
- Trapezius (middle and lower fibers)
- Upper-body extensors: posterior fibers of deltoid, teres major and latissimus dorsi
- Greater pectoralis (abdominal fibers)
- Long neck and head muscles

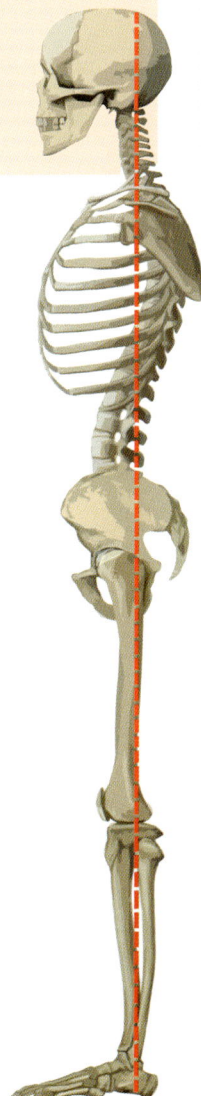

Restorative posture

An optimal recovery protocol initially seeks to achieve standard length and movement paths along its amplitude, based on standard amplitude tests. You can help the athlete by using tension while these tests are performed. Help the athlete to be more aware of when they make unconscious, compensatory movements, correcting the "traps" that the body falls into during a given movement and avoiding the contraction of weak muscles.

To strengthen or train a weakened or inhibited muscle, increase resistance, muscle mass and definition by increasing loads, repetitions and series based on individualized objectives. It is necessary to follow a regular tension program in order to maintain the different muscle groups in their resting lengths.

Massage and posture

Sports massage is an irreplaceable tool for the treatment of postural instability and asymmetries, as we will see in the chapters about massage in myofascial chains. It is possible to combine massage with tension to perform joint mobilizations, apply heat and stabilizing techniques, rehabilitating and reeducating altered posture patterns that are a source of problems for both the athlete and general public. It is therefore necessary to know where to perform a massage; that is why an initial assessment is so important. It tells the masseur which areas they must work on in order to obtain the desired benefits and results.

Ideal posture with load line.

Warning signs (red flags)

The term "red flags" refers to those signs or symptoms found during a scan that indicate something non-mechanical or non-musculoskeletal has been altered, and the athlete should be referred in order to further assess the status of their health.

Massage can be counterproductive at times. Therefore remain alert for:

Injuries or wounds. The repair-healing process of the tissues must be respected before manual handling. For example, fibrillar fractures should not be massaged without respecting this process. If this healing process is not respected, injuries may present major complications. Massage should only be applied when the scars have healed.

Inflammation. Tissue inflammation is associated with many different processes. It is present in all repair processes, whether aseptic (uninfected) or septic (infected) inflammation. Therefore the application of massage with the aim to normalize the tissues should be ruled out. The exception is chronic inflammation for which massage is very useful.

Red flags.

Pain threshold. Certain maneuvers can cause an amount of discomfort, but pain should not occur. Always ask if maneuvers cause pain and always respect the sensations the athlete is experiencing. Massage can be applied in areas alongside painful areas as this mechanism blocks pain perception (gate control).

Febrile syndrome and febrile. Fever is related to infectious processes. In this case a massage is not required.

Dermal pathologies. Massage is contraindicated in the case of skin conditions. In the case of contagious diseases, such as fungi, scrupulous hygiene must be maintained on both sides (use of disposable gloves, disposable gauze paper, antifungal products, etc.).

Varicose veins and circulatory problems. Strong massage should be avoided in cases of vascular fragility, phlebitis, thrombophlebitis, vascular inflammation, anticoagulant intake, after surgery, etc. In these cases great care is required when performing a massage. The most appropriate technique is manual lymphatic drainage.

Tumors. Massage must be avoided as it increases blood flow and therefore the proliferation of tumor cells.

In any pathology in which the pain does not improve or even increases and worsens with massage, its practice is contraindicated, and the athlete should be referred directly to the team doctor or director.

Legs with varicose veins

Points where direct and maintained pressure should not be applied

There is an extensive literature on the subject of indications and contraindications in massage, but here we only detail those that are essential to know about in order to avoid causing damage. Emphasis is placed on areas where constant and sustained pressure should not be applied.

Listed here are the areas where direct and maintained pressure should not be applied.

The nerves of the brachial plexus and the subclavian artery and vein are located **in the lateral triangle of the neck**. The common carotid artery and its bifurcation with the carotid sinus and its baroreceptors are also covered by the sternocleidomastoid muscle. Direct and sustained pressure on this area would cause blood pressure to fall and heart rate to decrease, as well as an altered response in the vagus nerve (in a posterolateral position relative to the common carotid artery), and therefore all organs receiving information from this nerve. The internal jugular vein can also be found in this area. Large neck vessels are located deep in the sternocleidomastoid muscle, with the exception of the external jugular vein, which diagonally crosses the surface of this muscle.

Axillary artery, brachial artery, axillary and branchial veins, cephalic vein and nerves corresponding to the brachial plexus.

In the medial aspect of the arm, next to the medial epicondyle of the humerus, is the ulnar nerve.

On the lateral side of the arm, next to the lateral epicondyle of the humerus is the radial nerve (on the elbow, in the middle section of the biceps brachii).

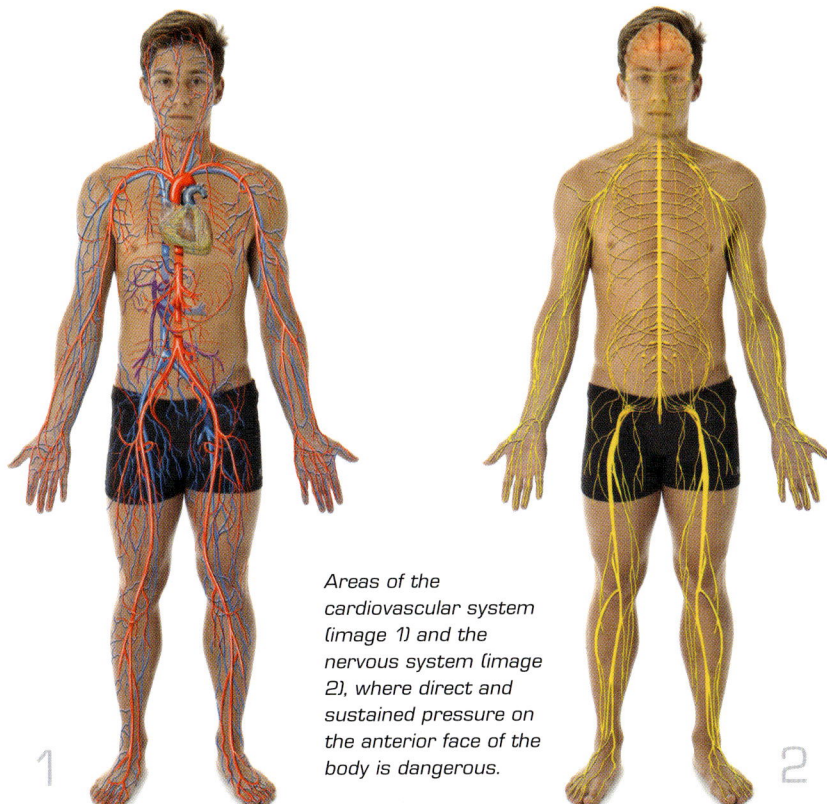

Nerves and blood vessel structures are very superficial; they are not sufficiently protected by muscle, connective or fatty tissues. Therefore, a direct and sustained pressure could trigger an altered response in these structures.

The nerves and veins that lead to the thyroid gland and vagus nerve are located **in the sternal notch and anterior part of the throat.**

Pressure on either side of the navel could damage the abdominal aorta.

The kidneys and spleen area **on the back next to the twelfth rib**. As these organs are suspended in adipose and connective tissue, strong, sustained pressure or percussive maneuvers applied with force **are prohibited** around them and their immediate area, as they could cause detachments.

Major sciatic notch. The sciatic nerve travels from the pelvis through the major sciatic foramen and is protected by the piriformis muscle.

The femoral triangle, which is located inside the pubis and medial to the sartorius. Within it we find the femoral artery (femoral pulse), the major saphenous vein and the femoral nerve.

Posterior of the knee. Here we find the popliteal fossa or hollow, through which the popliteal artery and vein and the tibial nerve pass.

Areas of the cardiovascular system (image 1) and the nervous system (image 2), where direct and sustained pressure on the anterior face of the body is dangerous.

1

2

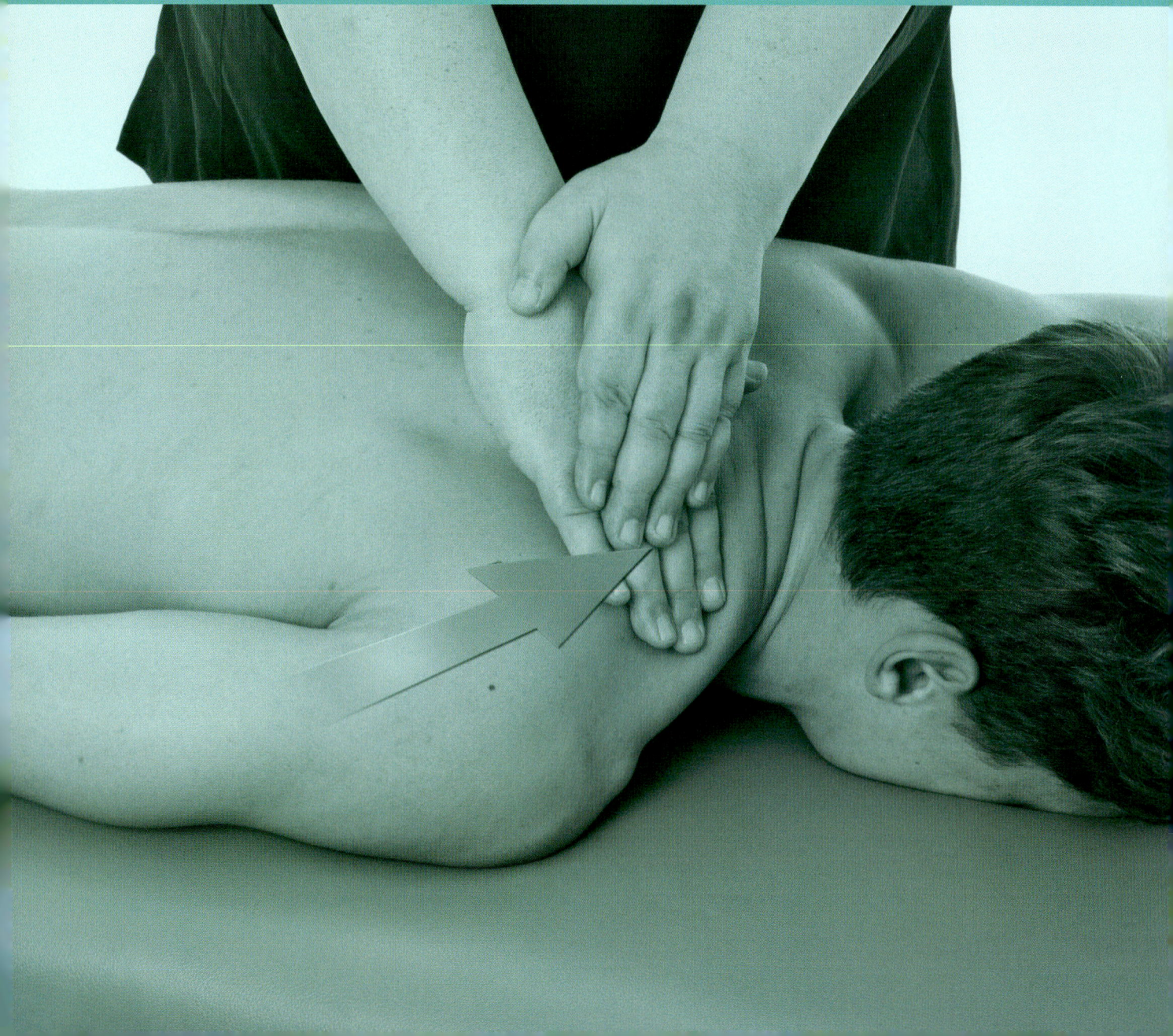

TECHNIQUES

2

This chapter details a variety of manual massage maneuvers: different classifications and most commonly used bodywork techniques in classic massage, deep-tissue massage and sports massage, as well as their characteristics, effects, indicators, contraindications and form of application.

Massage techniques employ a variety of technical movements called maneuvers. The objectives of these maneuvers are determined by the indication and time of training or competition during which the athlete is treated.

A variety of techniques is used in order to adjust the therapist's hands to the different areas and anatomical planes to which massage is applied and adapted, depending on the position of the person receiving the massage.

Maneuvers are described in terms that relate to general techniques: rubbing, friction, kneading, pressure or tweezers, etc. Specific techniques (pressure, pressure and drag, pressure with friction, or transverse friction, etc.), or bodywork techniques (compression, joint mobilization and stretching) are also included.

Gentle rubbing

Gentle rubbing is a pleasant way to initiate contact with the athlete. Gentle rubs can be applied in different ways depending on the rhythm, degree of pressure, direction or method in which they are applied. It is a technique that links to other maneuvers, for example during friction, to advance to a more distal zone. Massage that aims to drain or relax is used to finish treatment and is one of the three types of rubs used in sports massage: surface, soft and deep.

Gentle rubbing can also be used to produce a venous renewal effect, intermittently applying strokes to prevent bruising.

Effects of massage

- To act reflexively and mechanically, integrating both actions in numerous applications.

- To operate superficially at the start; skin massage stimulates the nerve endings.

- To provoke a reflex response, inducing deep and general muscle relaxation.

- On an emotional level it provides a relaxing effect, providing tranquility and calm.

- To improve venous return.

- To relieve interstitial pressure.

- To decrease heart and respiratory rates through the central nervous system (relaxation).

- Reduce pain sensitivity.

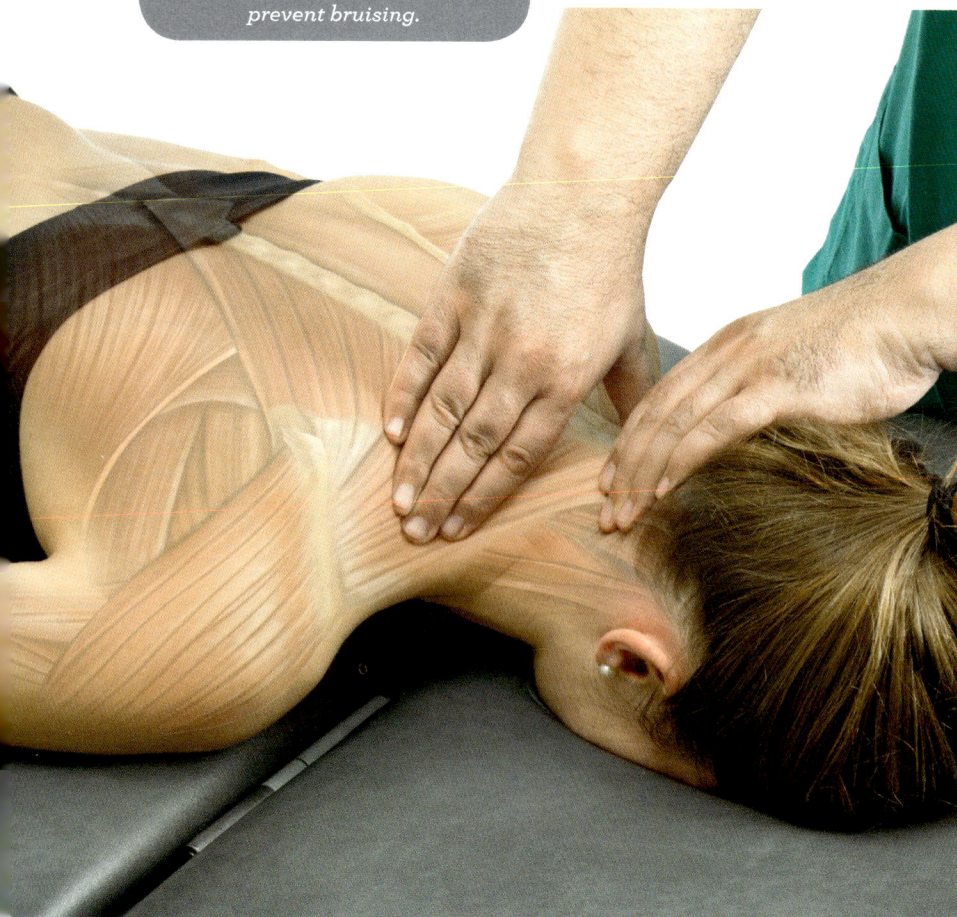

Friction maneuvers of the abdomen, or in the trapezius and cervicals applied with a slow and smooth regular rhythm promote a venous renewal effect. This effect is very useful for preparing the region for deeper maneuvering and also avoids bruising.

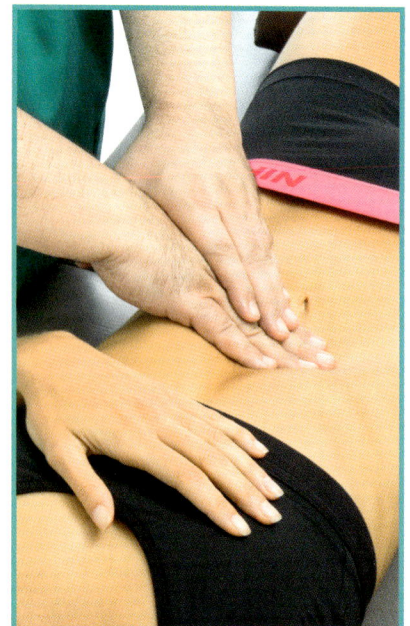

Mode of application

This consists of sliding the palm of the hand around the contour of the area to be treated, causing mobilization of the underlying tissues. It can be applied with both hands alternately (first one hand and then the other) or simultaneously and should cover large areas. The maneuvers must be performed smoothly and at a slow and regular pace. The palm of the hand is used, exerting a slight pressure and being careful not to penetrate tissues. The change in coloration of the nail plate, from pink to reddish, indicates the degree of pressure that is being applied with the fingers.

If the purpose of massage is to stimulate blood and superficial lymphatic flow, it is applied in a centripetal direction, following venous and lymphatic circulation (distal to proximal).

Indications

This is used as "first contact" at the beginning of a massage. It is the correct time to spread creams or oils and is also used to conclude treatment.

It contributes to initial assessment as it allows the masseuse to perceive the area's temperature and sensitivity and evaluate its level of elasticity and tissue tone. It simultaneously facilitates the circulation and mobilization of fluids in the tissues being massaged and reduces muscle tension, stiffness and hyperesthesia by relaxing and relieving the sensation of distress, contributing to pain relief. When used to treat the intestinal zone it collaborates mechanically and reflexively, potentiating peristaltic contractions.

Contraindications

◆ Avoid gentle rubbing near cutaneous alterations, open wounds, burns, or areas with abnormal sensitivity caused by allodynia, hyperalgesia or other disorders.

Sequence of smooth rubbing on both sides of the thigh by means of upward, semicircular compression movements.

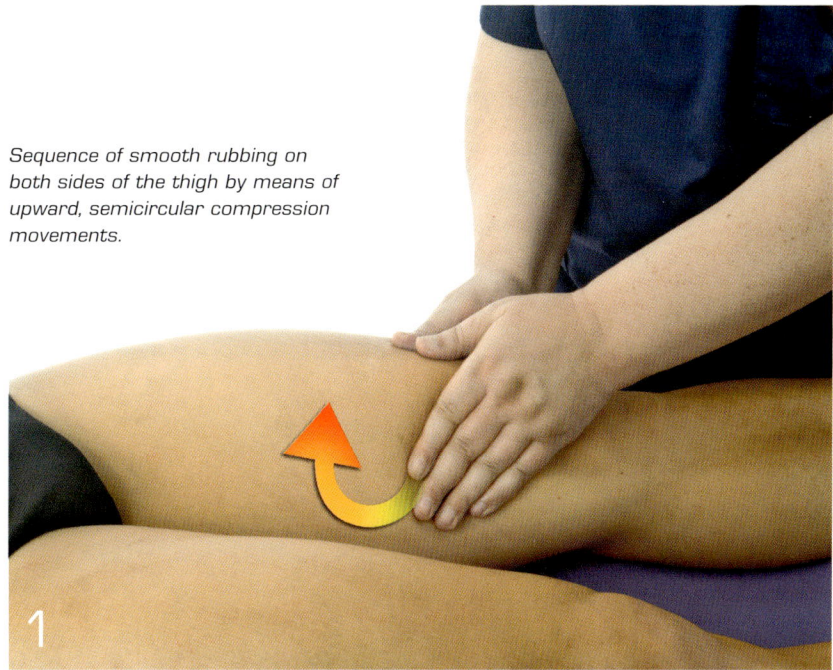

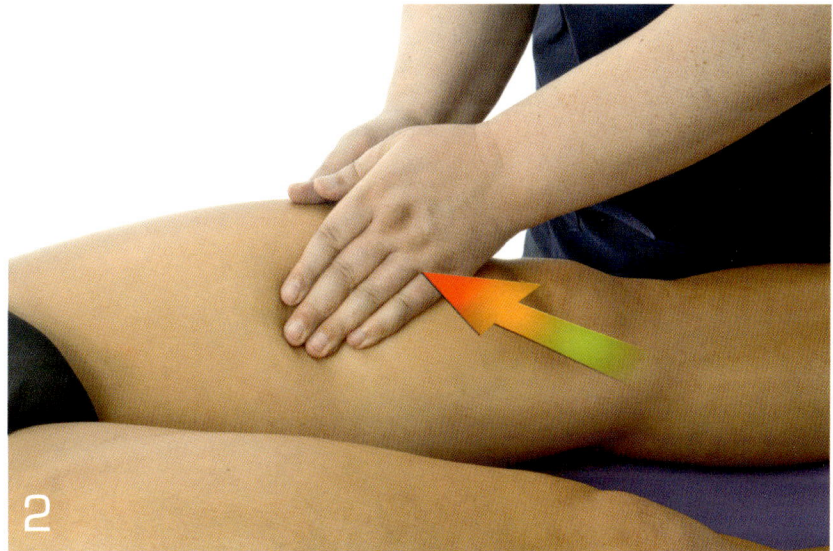

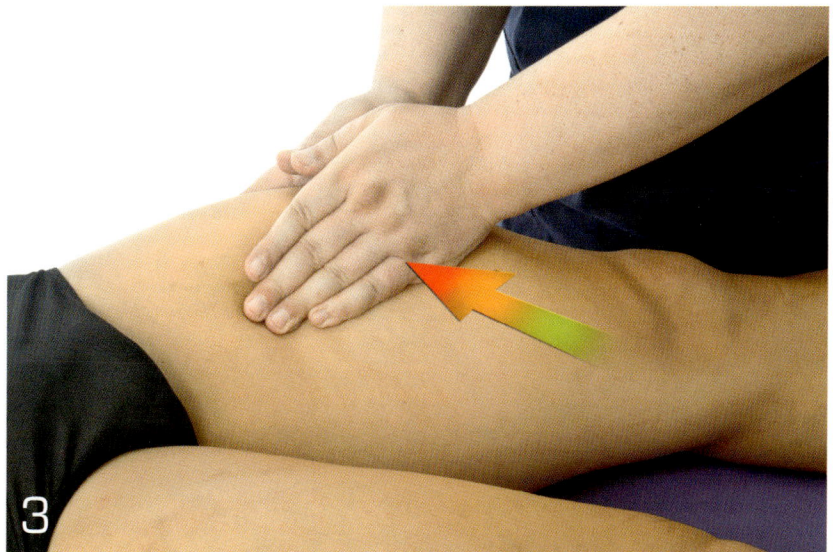

Palmar and ulnar friction

Pressure maneuvers and short-stroke rubbing combine to mobilize the skin plane covering structures located deeper inside the body. In sports it is very common to prepare muscles and joints before physical activity or after an injury to increase edema resorption.

Mode of application

This can be both ascending and descending, using strokes or circular maneuvers that are adapted to the extent, shape and anatomical contours of the area to be treated. Technique is chosen according to the purpose of treatment. It can be carried out with one hand, alternating between both hands, or one hand over the other (reinforced). The fingertips, palm of the hand, the ulnar edge or the knuckles can also be used, applying in an energetic, dynamic way before

Friction can be combined with gentle rubbing and passive joint mobilization maneuvers. The result is a "reverse massage," where the friction maneuver is performed by the osseo-articular structures of the person, which are mobilized by the static compression the therapist applies with the hands.

Effects of massage

- Allows vasodilation that causes local hyperemia.
- Increase the local skin and joint temperature (between 1 to 3 ° C).
- Activates muscle tone.
- Facilitates and enlarges joint movement.
- Depending on duration of application can serve to both stimulate and relax, producing an analgesic effect.
- Promotes desquamation of dead cells.
- Favors intestinal transit in the viscera.

Reinforced palmar friction is used, adapting the hand to contours with moderate-strong pressure and at a high rhythm in both ascending and descending directions.

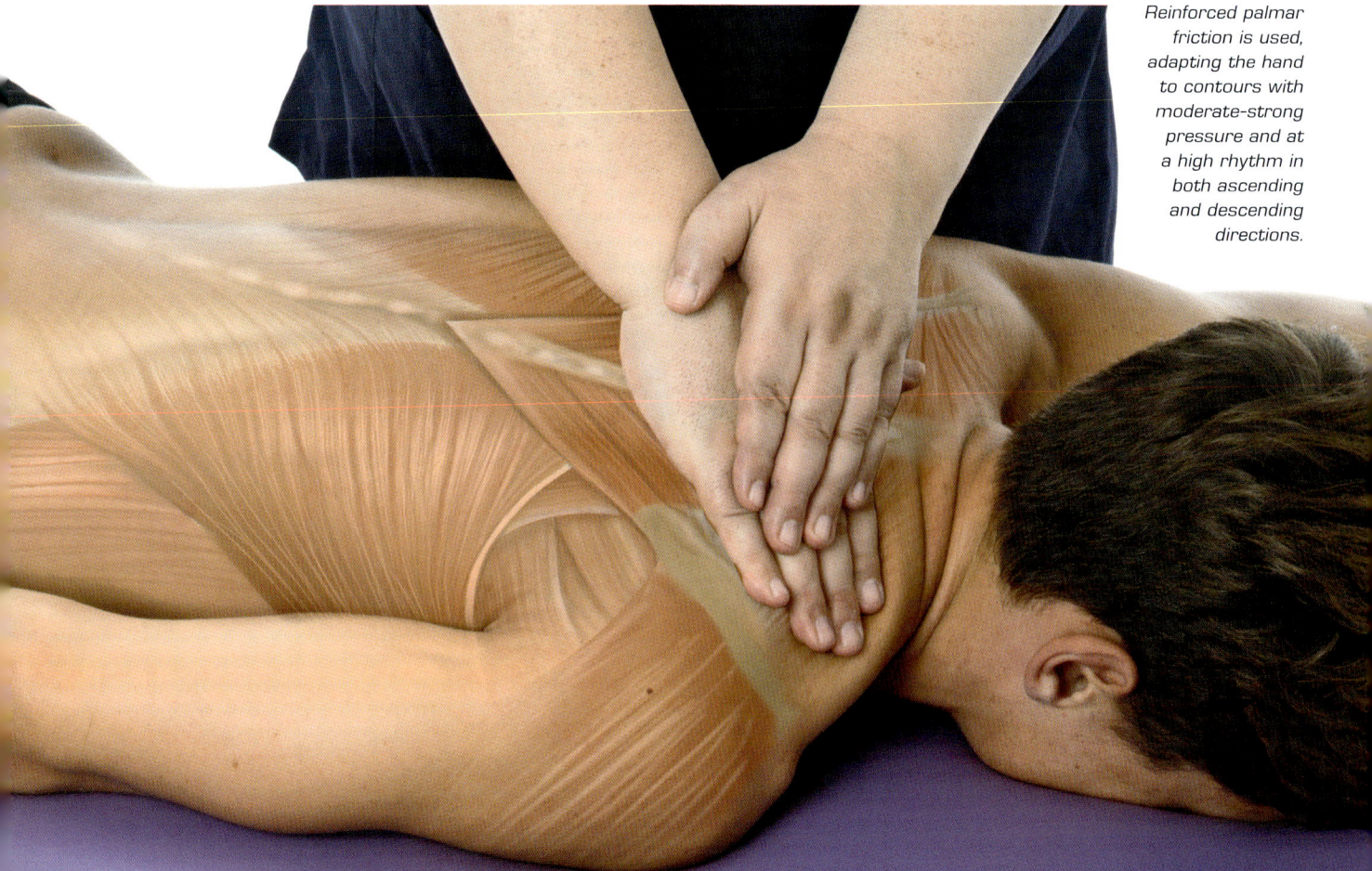

physical activity and, afterwards, to moderate it.

Pressure varies from mild to moderate-strong depending on the activity and the plane of tissue being treated. Friction movement does not start until the hand resting on the skin settles, little by little, forming a unit with it and establishing a fixed contact with the structure to be treated. Immediately work the tissue without losing any depth; rubbing movements are applied across the treated area.

Contraindications

◆ The same as the other techniques of sports massage. In addition there are contraindications specific to this maneuver, such as avoiding a fast rhythm after a strenuous physical workout as this could damage the hyperculture musculature or cause shearing movements, breaking a blood vessel. Circulatory problems and people with fragile capillaries or taking anticoagulant medications should be taken into account. This technique should not be used if the subject has varicose veins or suspected blood clots.

Indications

Contributes to the treatment of contractures, myogelosis, superficial adhesions and problems related to scarring. It helps to relieve the sensation of cold by providing heat to the treated area, making it a suitable pre-exercise activating technique. It also favors the reduction of pain in rheumatic processes or osteoarthritis. Its use is also indicated in situations of tiredness and fatigue due to its recuperative effect and is also a suitable technique for the treatment of stiffness and periarticular retraction, as well as pain sensitivity.

Sequence of reinforced palmar friction in the muscular grouping of scapular and cervical girdles. It is important to maintain homogenous pressure and slide. Practiced bilaterally.

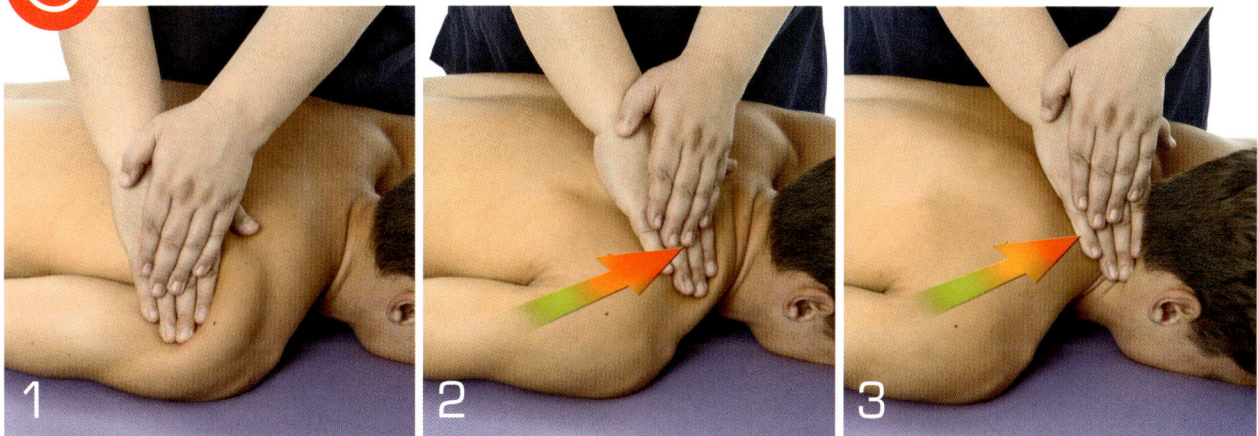

Reinforced ulnar friction. To facilitate perception of the active hand, the supporting hand is placed on the forearm when it should be supported at the tenar zone. Indications are the same as in the palmar friction.

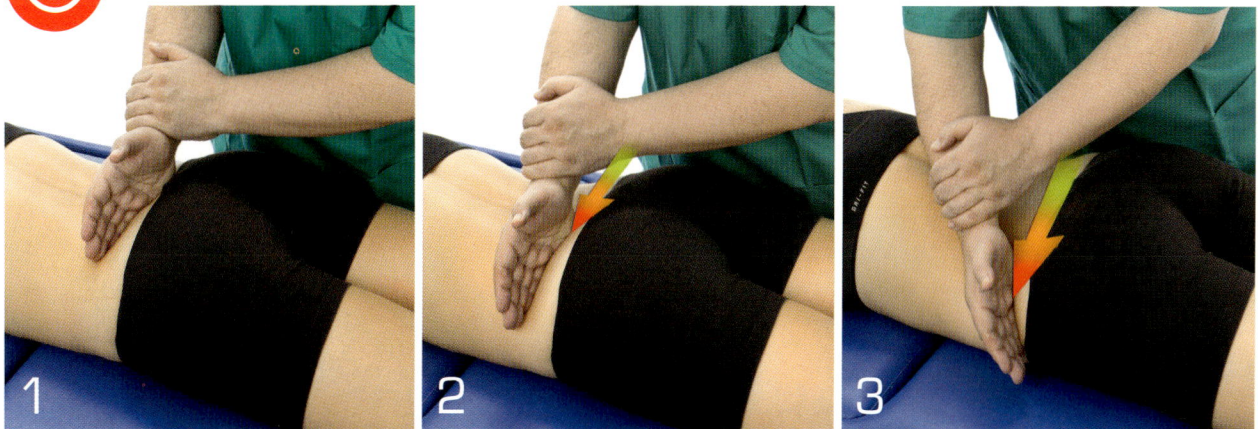

Kneading with fingers, tenar and pump

This maneuver combines pressure, displacement and elevation movements, mobilizing the tissues and moving their planes in a transversal way. It is very much appreciated by athletes because, depending on the desired effect, both superficial and deep muscles can be reached.

Mode of application
Digital kneading is the best known technique in the sports arena. It is

> The digital **kneading maneuver** is used at the beginning of the treatment and has a dual function: it allows the exploration of the area (massage is a form of continuous exploration), then afterwards treat the first layer of tissue in which the excess tension is found.

applied by making small clockwise twists with the fingers and thumb; however the palm of the hand does not make any contact with the tissue. This facilitates movement and allows mobilization of the underlying skin and tissues. If both hands can be used, alternate.

Types of kneading
The maneuver is designated as kneading plus the term that identifies the treated and the area of the hand used in massage. In addition to digital kneading, the following can be used:

■ Kneading with the **thumb**, using small circles in localized areas for only a short time, avoiding injury.

■ **Bimanual** kneading, using the entire hand and fingers. This affords the treatment of large muscles such as those of the lumbar and gluteal region (just like **knuckle** kneading), but utilizes the interphalangeal

joints.

■ Kneading with the **tenar zone**, which allows a squeezing movement on the tissue.

■ Finally the **pumping palmar**, which is very useful in case of pain and overload after exercise.

Criteria for use
All maneuvers can be carried out with one hand or with a reinforced hand. This is where the other hand is placed on top of the treatment hand to increase pressure and avoid fatigue.

Indications
These rubbings improve circulation, reduce tension or contractures, allow skin adhesions to function, are useful for relieving fatigue, treating DOMS, facilitating rest in sleep disorders and relaxation in cases of stress, and help to decrease or eliminate sensitivity to pain.

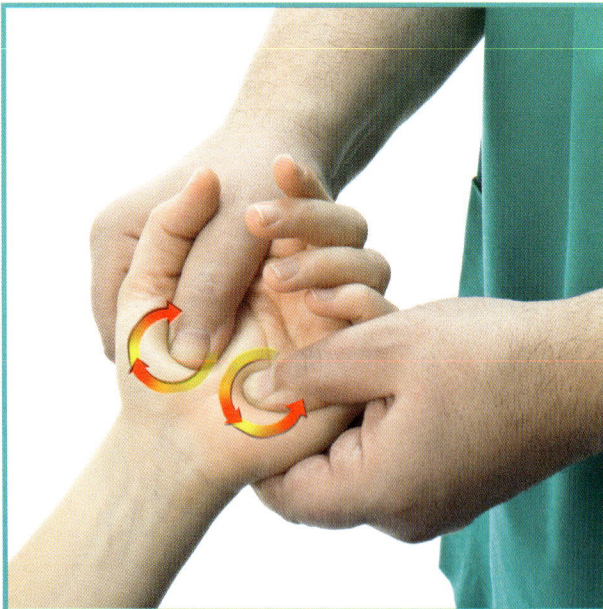

Hand kneading with thumbs. The therapist holds the back of the athlete's hand with the fingers of both hands and uses the tips of the thumbs to deeply knead the tenar and hypothenar eminences, progressively and alternately

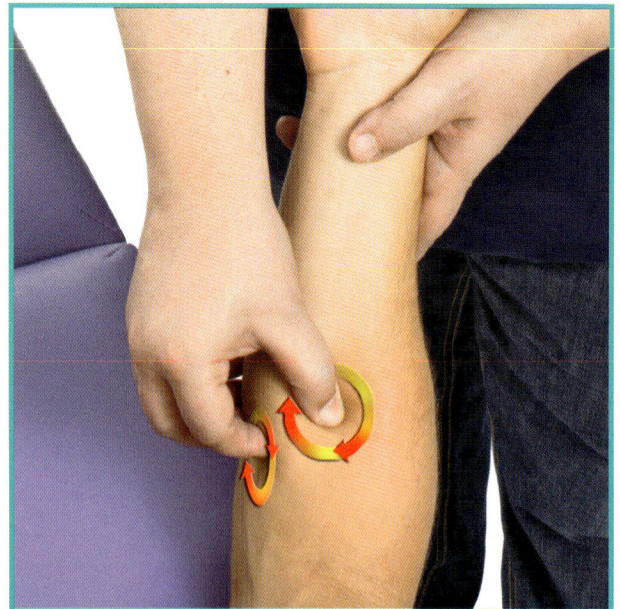

Digital kneading on the forearm. Can be done with the arm resting on a couch and the forearm semi-flexed and pronated, or with a flexion of approximately 90°. One hand holds the arm while the other kneads or both knead. Treats congestion and over stress.

Effects of massage

- Activate and improve blood and lymphatic circulation, increasing vasodilation and oxygenation of tissues.

- Peel the skin from underlying tissues and remove adipose tissue; help eliminate metabolic waste substances.

- Favor general relaxation.

- Very useful for treating contractures.

- Prepare the muscle for prolonged and more intense workouts.

Contraindications

◆ These maneuvers are applied to control excess tension after intense effort, but not to eliminate it during a single session; massage should not be painful to avoid a rebound effect. Should not be used on nerve plexuses or lymph nodes, in the case of varicose veins or a suspected blood clot (thrombus).

Palmar pumping/kneading of the thigh. *The athlete is supine, legs stretched out. The thigh is treated to relieve congestion after intense exertion. The palms of the hands work in unison, but the thumbs are not used. Similar to gentle rubbing, pumping functions with a higher degree of pressure.*

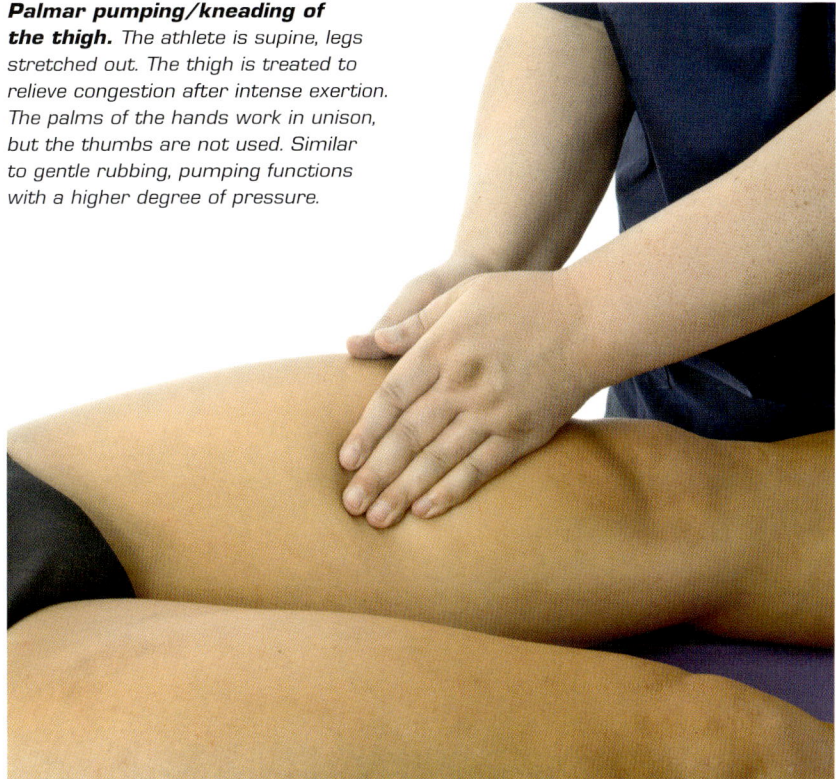

Tenar kneading of the arm. *With the athlete's arm flexed and raised, the forearm is held to access the biceps brachii. A compression movement is performed on the musculature with the tenar and hypothenar eminences of the hand, facilitating kneading in their deeper planes.*

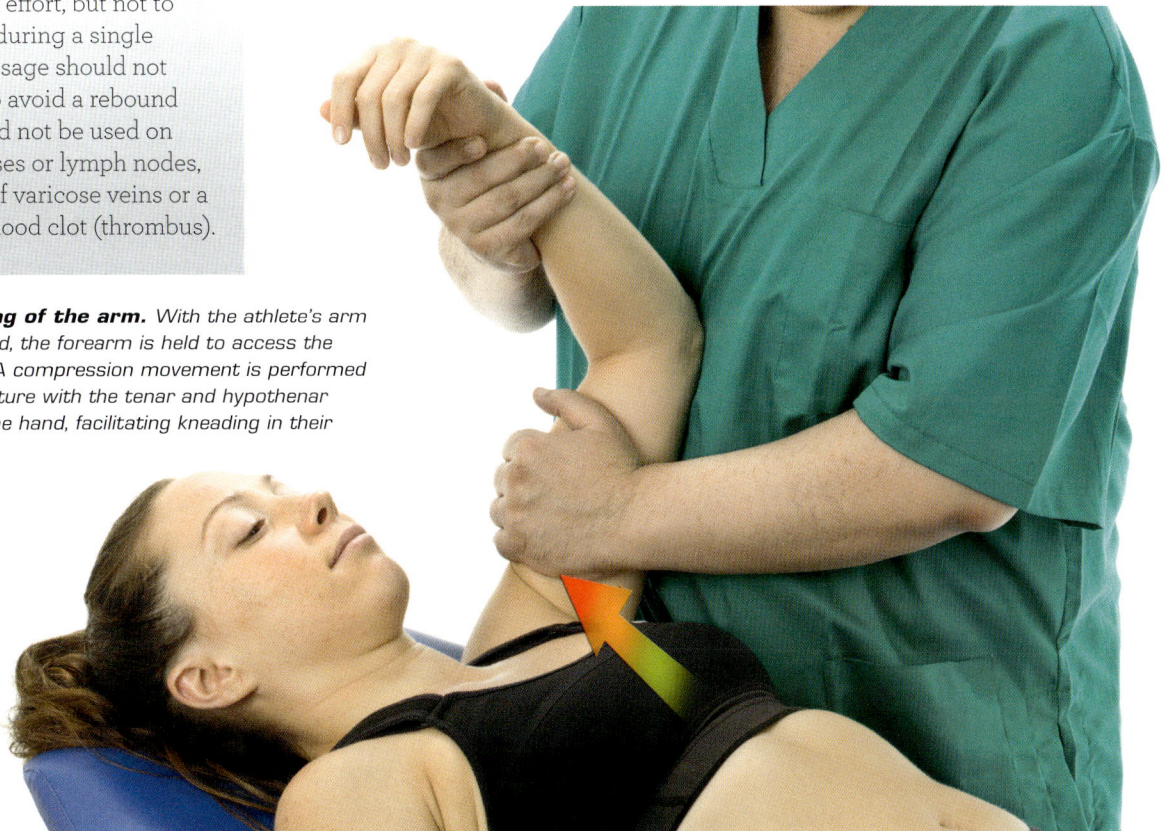

Clamps or cutaneous bearings

Consists of a movement that lifts and displaces the skin fold of the superficial fascia and kneads it gently. The maneuver is known as a roller or skin roller. During its practice the density, quality and painful sensitivity of the skin fold bearing can be assessed. In this case it is performed more slowly than other treatments and is called "Kibler's fold test."

Mode of application and types of clamps

The roller clamp is used in two ways, depending on the desired effect. One is to grasp a fold of the skin between the thumbs and the tips of the other fingers of both hands, positioned side by side and in ulnar deviation. An elevation and displacement is made in the form of a small wave that is maintained until there is no discomfort or pain.

Another way to apply this maneuver is to hold the area between the thumb and the index finger, simultaneously effecting a traction movement (sustained or rhythmic) or torsion while at the same time moving the bearing in the underlying planes in different directions.

Variations of elasticity and skin resistance are evaluated before using the roller clamp maneuver. In this case, proceed very slowly and without the use of oils or creams.

Face. *Roller clamp maneuver of the eyebrows, checking texture and movement and assessing pain sensitivity. This sequence shows the shape of the fold and is the formula that is used in any body area: First it is held and then raised and moved.*

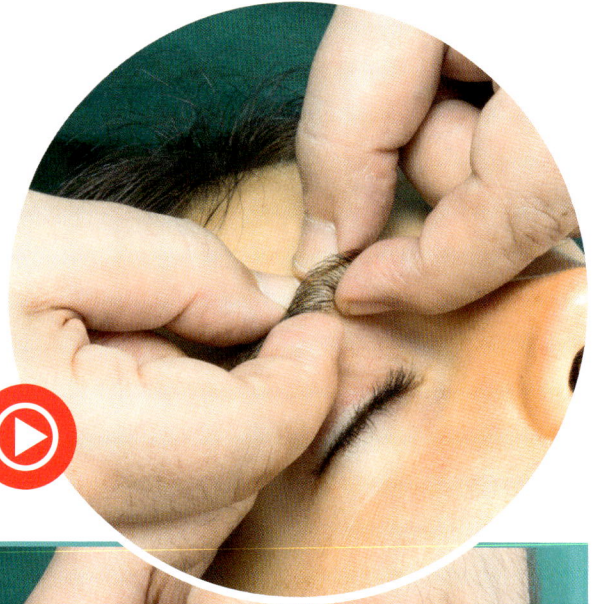

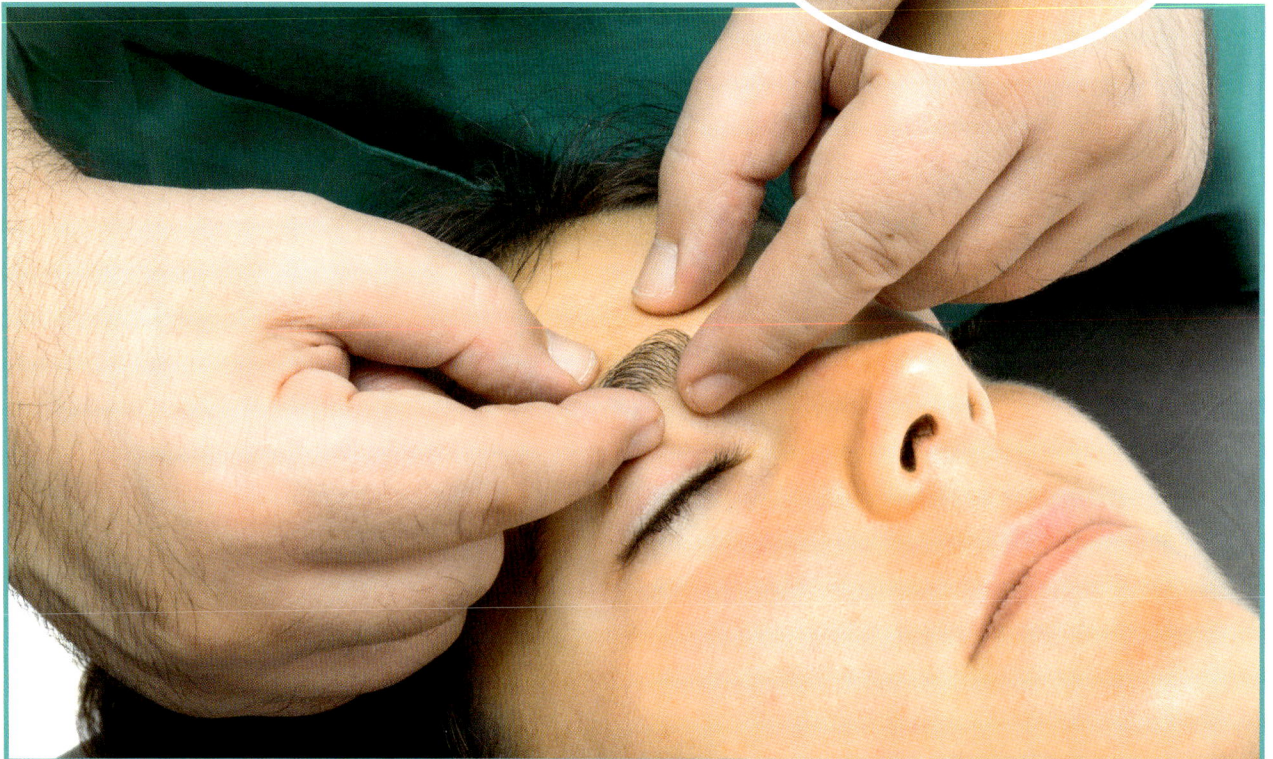

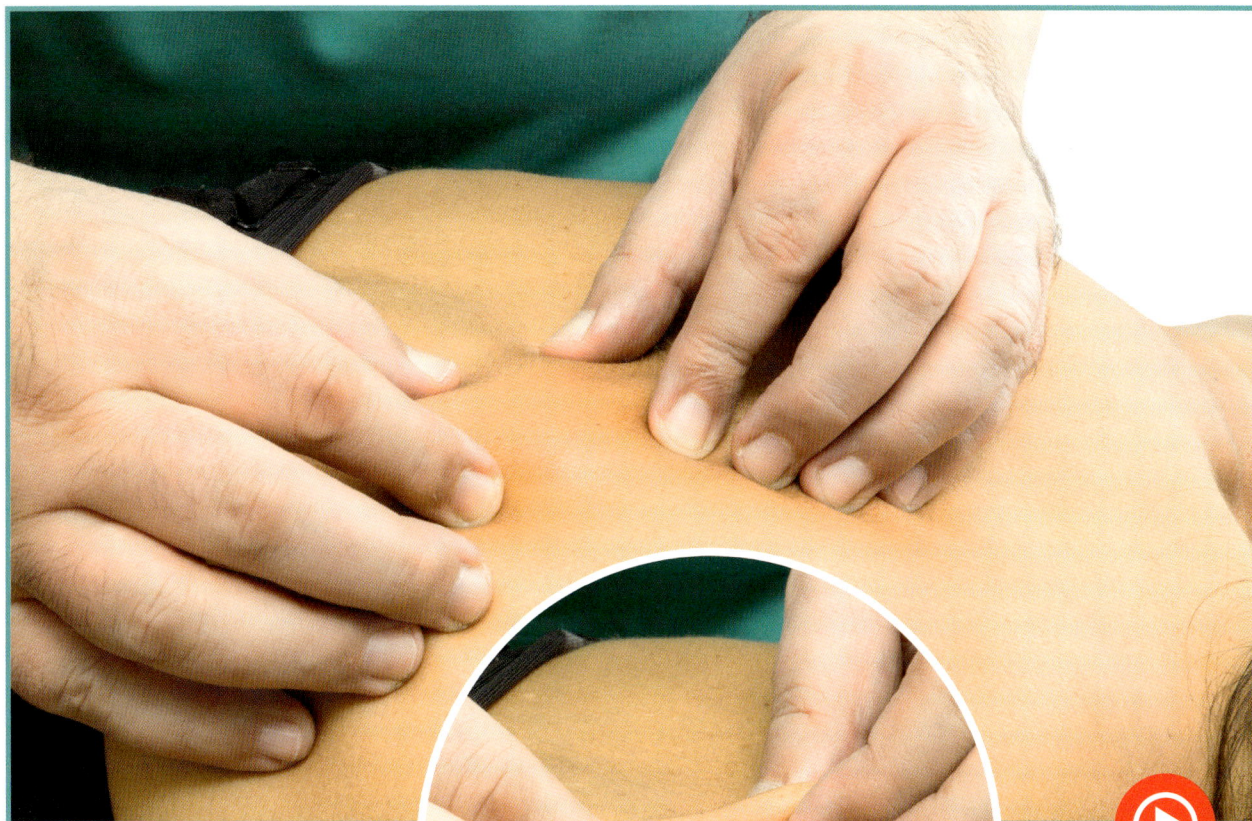

Back. *The rotated clamp maneuver in the dorsal area acts on the underlying fascia. The fold may vary in thickness depending on the area where the displacement was made, and must be pleasing and tolerable to the skin bearing.*

Criteria for use

In both cases, tissue is maintained at a comfortable tension that allows the fabric to roll. It can be applied over a localized or a wider area, and in both cases maneuvers can be carried out with or without displacement. To make the folding movement more comfortable the therapist can use the other fingertips to support the "heel" of the hand.

Indications

This versatile maneuver allows the exploration and treatment of excess tension between the cutaneous tissue and superficial fascia or contractures. It produces good results in cases of skin adhesions and retractable scars, and acts by mobilizing excess fatty tissue.

Effects of massage

■ This technique results in a considerable circulatory reaction and increases the temperature of the skin, activating cellular metabolism.

■ Decreases muscle tension.

■ Reduces pain sensitivity and is especially useful for peeling the connective tissue of the skin, as it facilitates the stretching and release of adhesions.

Contraindications

◆ Not to be employed in cold temperatures or if it causes an increase in pain sensitivity. Also avoid in cases of capillary fragility or hypersensitive skin, or when there are acute inflammations. Avoid using this technique quickly or carelessly.

TECHNIQUES

2

Applying pressure

The area to be treated is engaged and compression is exerted gradually and smoothly. It is a static and maintained maneuver that is applied over a short period of time.

Mode of application

The hand should be loose and relaxed to facilitate palpation. It should not move, so this is best done without creams or oils. A firm and steady intensity is required to obtain uniform effects. In large areas the direction of the pressure exerted is perpendicular, whereas in smaller areas it is preferable to use an oblique pressure on the surface of the athlete's body.

Types of pressure

Depending on the surface to be treated and the athlete's muscular tone, different devices may be used — such as the thumb, palm, fist (supporting the proximal phalanges, or the "heel" of the hand with the fingers semi-flexed), the forearm (both the fleshy part and the ulna) or the elbow. This maneuver can be reinforced with the other hand.

Criteria for use

The passage from one maneuver to another is gradual as changes are perceived under the compression area and according to respiratory rate. Apply pressure during exhalation. If it is a one-handed maneuver (sedative effect), a rhythm of 30 seconds to 1 minute is recommended. But in field sports that require a faster pace, from 4 to 6 seconds per movement. Immediate pre-competition treatment (activation) is done even more quickly, from 1 to 2 seconds per movement.

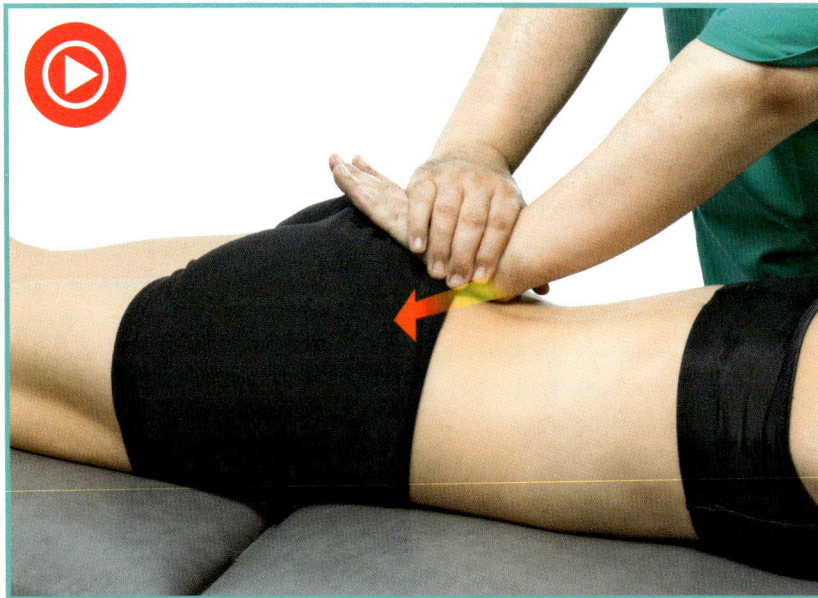

Palm of the hand. *Reinforced direct and maintained palmar pressure on the sacral zone. This maneuver, in addition to its own benefits on the sacrum, relaxes excess tension in the lower back.*

This maneuver is highly recommended in hypertonic states after intense exercise or in anxiety disorders, since the pressure exerted acts to dissipate excess tension and brings relaxation. Compression should be moderated in cases of osteoporosis.

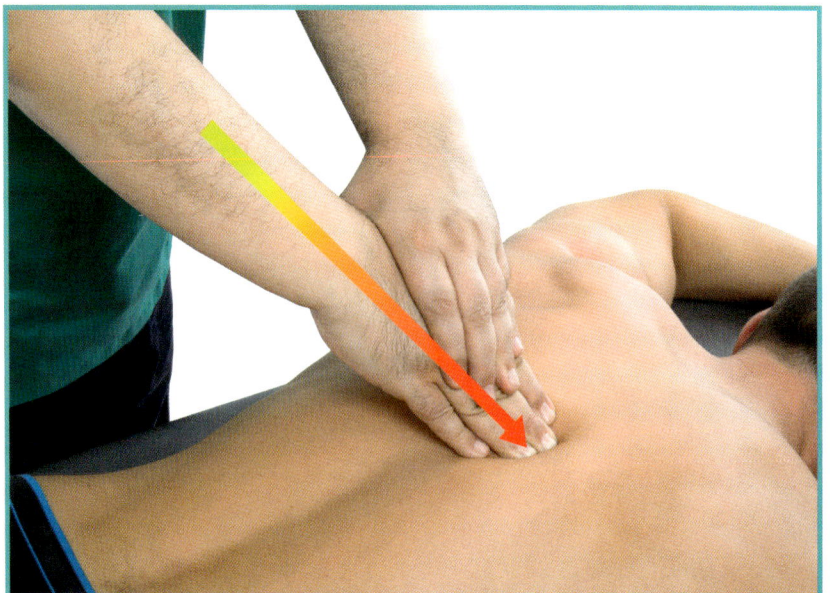

Digital. *Apply pressure with the tips of the four fingers on the paravertebral muscles of the back. Do not increase the pressure until you perceive a decrease in tension under the fingers.*

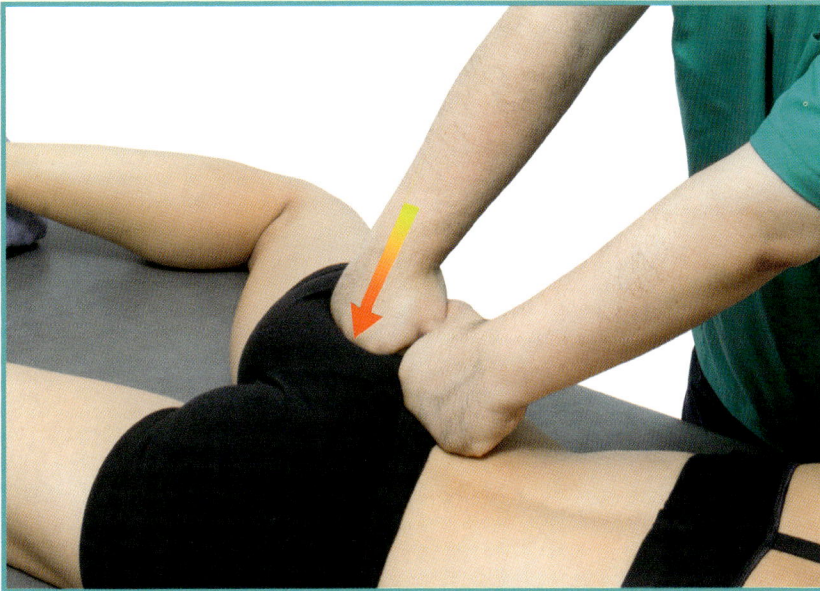

Indications

Useful for treating circulatory disorders as it compresses and encourages venous return. Applied in the extremities by a pressure sequence from distal to proximal. It inhibits excess muscle tone and decreases contractures. It facilitates relaxation in states of anxiety before or after an event.

Effects of massage

- Intermittent or discontinuous pressure promotes circulation by a pumping action.

- Relaxes excess muscle tension and has a sedative and analgesic action. Can activate or relax the athlete, depending on the rhythm used.

Fist. Oblique pressure on the muscular rounded piriformis. It is important to perceive how the deep-level tissue is stretched and excess tension decreases.

Contraindications

◆ Direct pressure should be avoided on a damaged joint area or if underlying soft tissues are damaged. Neither should be used if there is a suggestion of any spine pathology in the thoracic zone or if any of the ribs seem fragile.

Elbow. Static and inhibitory pressure on the lateral border of the sacrum, at the insertion of the piriformis.

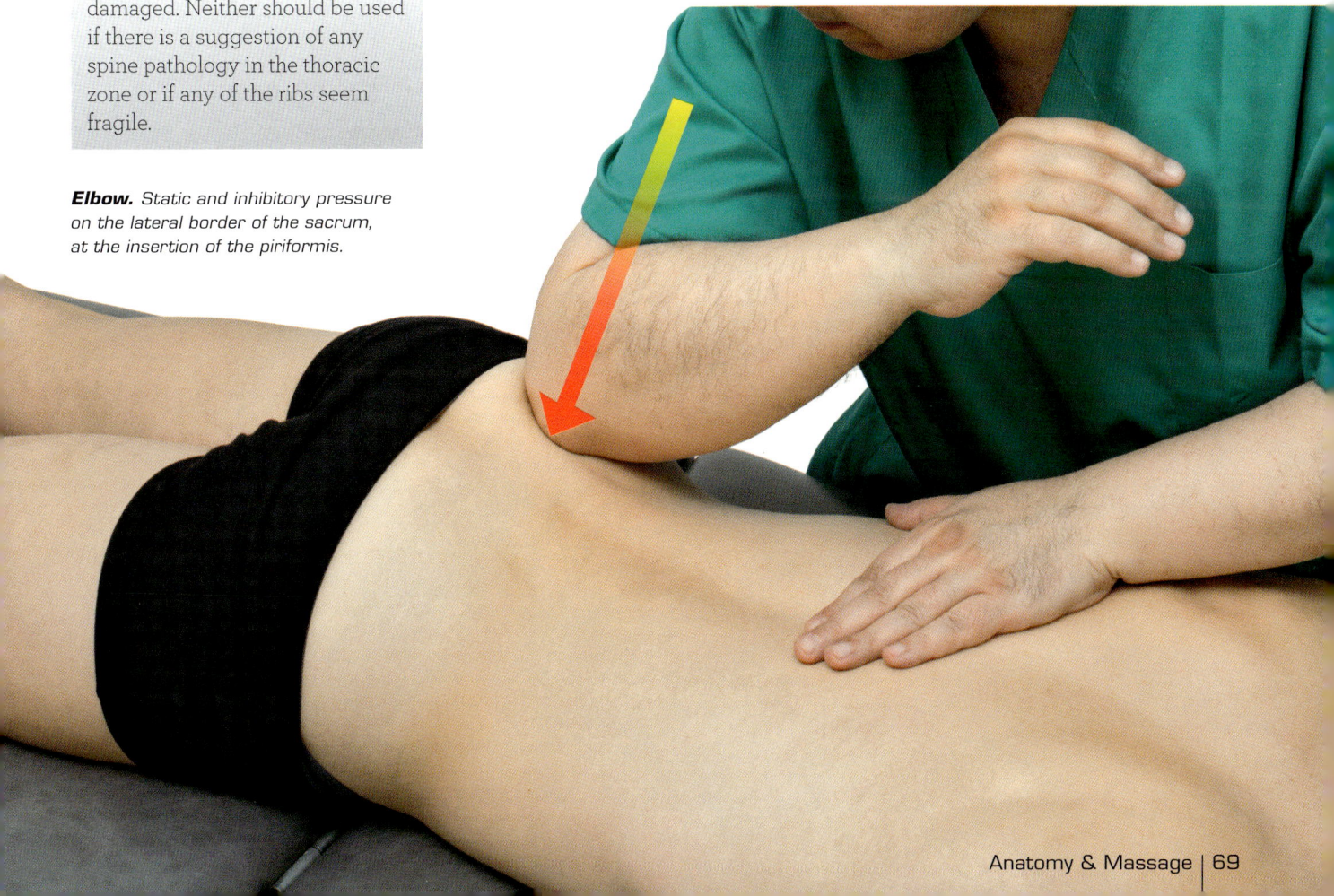

Pressure and dragging

Aiming to fulfill therapeutic objectives to treat connective (deep plane) tissue, several authors have created systems specific to their treatment methods. Deep tissue responds to heat and the forces of compression and elongation. From this great variety of manipulative techniques we have opted for those that perform pressure and drag functions, as drag technique heats the myofascial tissue through friction, and pulls at it when combined with pressure and drag.

Mode of application

The following procedure will allow access to myofascial tissue. Make sure the athlete is relaxed. Rest your hands softly on the area to be treated. Observe their respiration. Allow time for your hand to reach the anatomical plane where the restriction is located. Once you are there, work to recuperate the plane. When you feel that the excess tension has subsided, slowly withdraw. This formula is repeated in zones where you located the restriction with exploration.

Types of pressure and drag

You can use different devices when using the knuckles (middle phalanges of the second and third fingers plus support from the ball of the thumb), the fist (fingers do not close against the palm but are extended and the thumb relaxed), the forearm (fleshy area, or ulna) or elbow.

Criteria for use

In order to not compress structures the pressure and drag technique is applied to the tissue obliquely. Although it is the arms and hands that apply force, it is generated with the whole body, combined with body positioning.

Indications

A good tool to augment tissue tone, after immobilization, in case of restrictions between the myofascial planes, in deficient postures, if there is limitation of range of movement and when there is sensitivity or pain.

Lumbar area. *Short stroke technique using support of the middle phalanges of the second and third finger of the hand plus the thumb, which contributes to force of the drag and gives stability to the wrist, but does not exert pressure.*

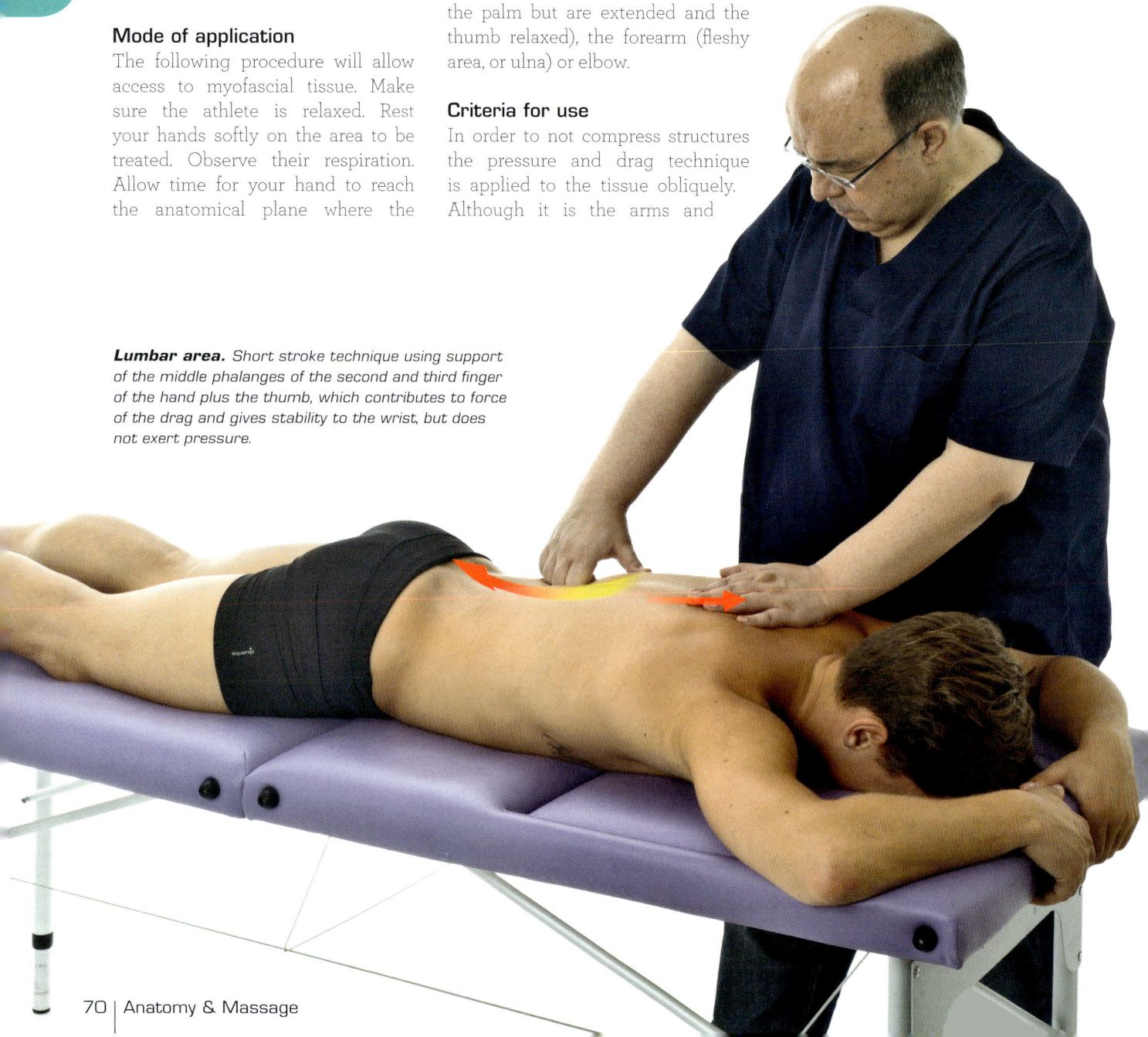

2

Lumbar and gluteal area. The athlete is accommodated in the lateral decubitus position, one leg with knee and hip in flexion, and the other leg extended on the couch to stretch the tissue. The therapist is positioned behind the athlete, holding the shoulder with one hand while slightly pre-tensing the tissue, while the other hand works to pressure and drag the tense tissue.

Effects of massage

- Allows the therapist to manipulate structures and increase flexibility, improving posture and fluidity of movement.
- Deep-tissue massage helps to mitigate pain of myofascial origin.

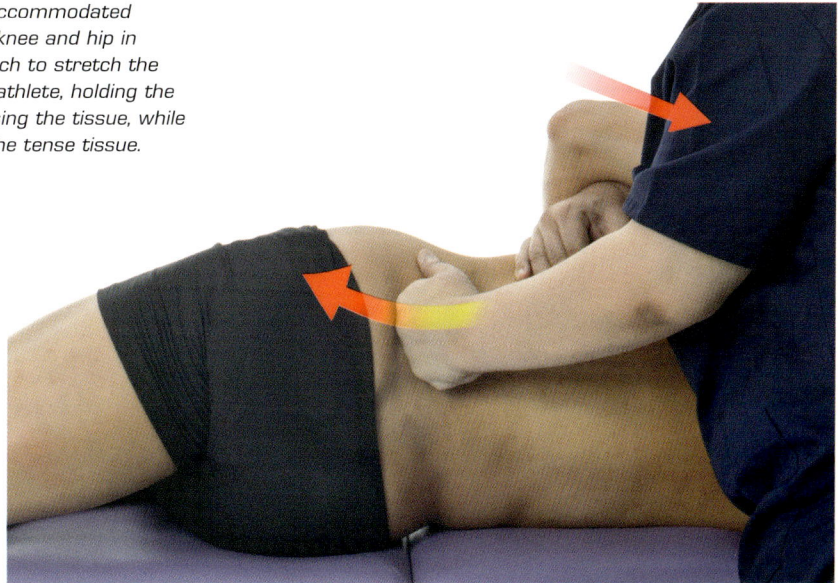

Contraindications

◆ If altered vegetative reactions (increased sweating, sudden changes in the color of the skin, etc.) are observed, it is advisable to stop the massage for a moment to allow the athlete time to react, stabilize and integrate changes. The usual massage contraindications apply, as described in general pressure techniques, pages 68-69.

Hamstrings and ischium. This position allows for assisted work on the back of the leg. When applying the pressure and drag technique, the athlete will display a small resistance in the hand, while flexing the hip a few degrees more will facilitate access to the ischium.

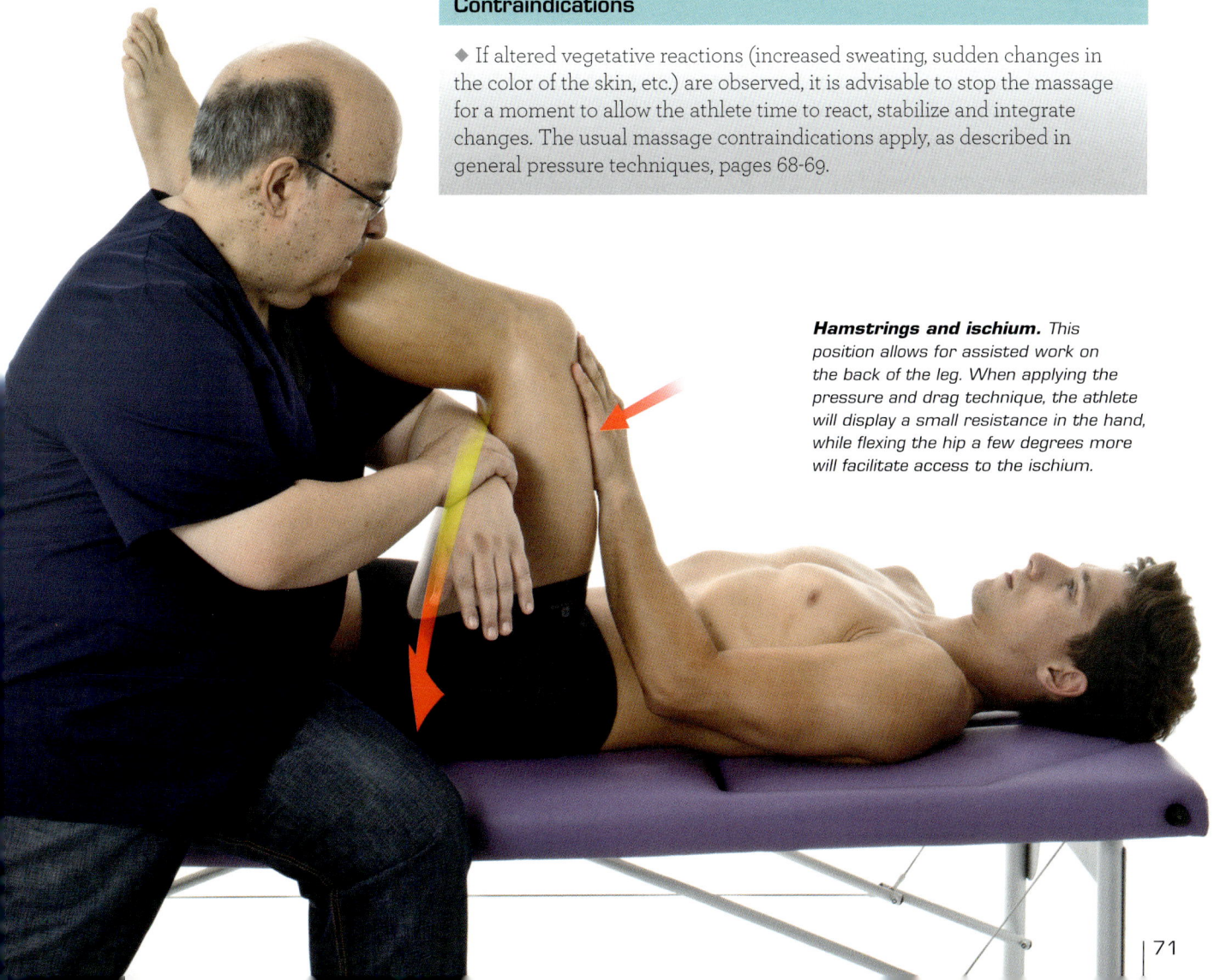

Joint traction

Traction is a technique in the field of sports massage which exerts forces of longitudinal traction to articular elements. Its objective is to act on the two articular surfaces to separate them while respecting their physiology, hence its name: "articular traction techniques."

Mode of application

The athlete's tolerance to the application of the technique must be taken into account, as well as the resistance of the myofascial tissue and the capsuloligamental elements before application.

Carried out using active fixation, with care taken to not press vessels or nerves, and is directed at a single joint. The therapist should not apply force roughly or use only the arms, but should make good use of their body weight to avoid fatigue, or even damage or injury.

Criteria for use

The aforementioned techniques focus on limbs with two objectives. One: to create a decompressive effect in the joint, and two: to result in its decoupling. The forces used for decompression are used to reduce pressure without separating joint surfaces. Conversely those used for decoupling are of greater intensity, facilitating the elongation of the treated structure and, with it, the physical separation of articular surfaces.

Articular traction techniques can be applied generally or globally (e.g., the entire limb), or partially over a segment (e.g., the ankle).

Effects of massage

- Reduces contact and pressure between the articular surfaces, which discharge and diminish pain of articular origin.

- It promotes the development, preservation and nutrition of the articular cartilage, helping to maintain the physiological and mechanical properties of the ligaments and the articular capsule.

- It sets groundwork for other techniques, such as joint mobilization (pp. 74-75).

Articular tractions are carried out in the vertebral column at high speed along a small area, applying impulse and denominated manipulations and must be supervised by a doctor. It must be stressed that poor use of this massage can create joint instability.

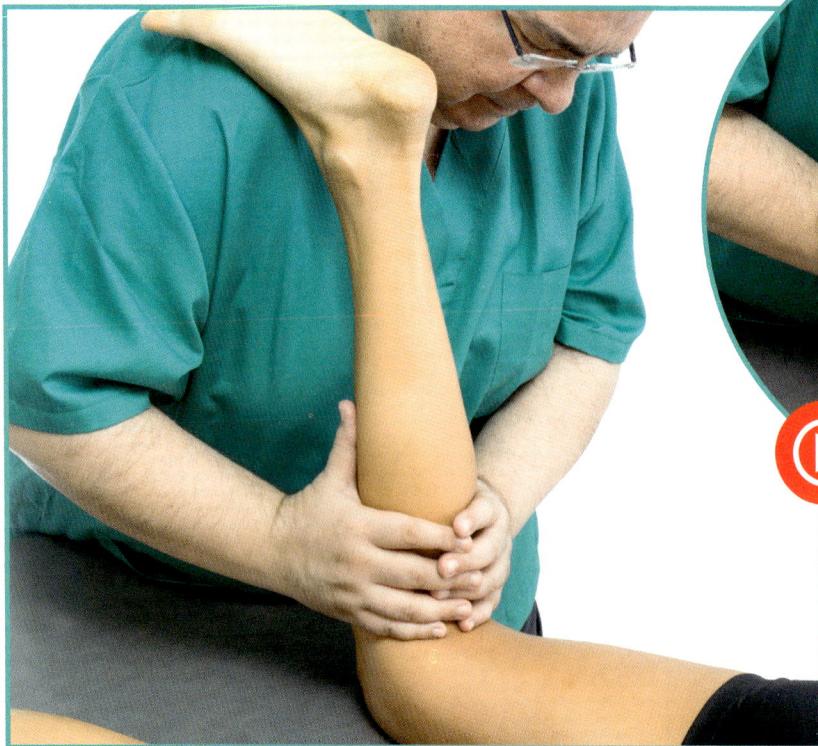

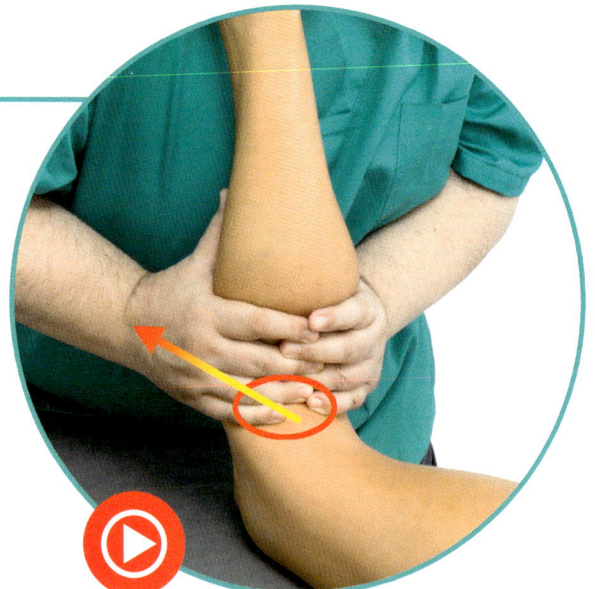

Knee. *With the athlete prone they are asked to flex at the knee, resting their foot on the therapist's shoulder. The calf area is held with both hands close to the joint line and caudal traction is performed on the knee. Several cycles of joint traction are required.*

Types of traction

Traction can be instrumental by using gravity and manuals. The instrumentals use pulleys, harnesses or electromechanical devices. Those that use gravity employ sloping planes. Manuals are only created by the forces of manual traction associated with a displacement.

Manual traction can be performed intermittently or continuously for a short duration. The tensile forces are applied progressively, maintaining traction for a few seconds and then relaxing, repeating the process again after a few seconds of rest.

Indications

Indicated when there is joint stiffness, retractions of the joint capsule and ligaments or by an excess of myofascial tension. Very useful to release small adhesions or as an aid in the treatment of muscular contractures. They can treat joint compressions of nerve roots and compaction of the joints (ankle, knee) after intense exertion or competition.

Contraindications

◆ To be avoided in cases of synovial effusion, established cases of ankylosis, acutely inflamed joints, recent soft tissue injuries, osteoporosis, dislocations and hyper-mobile joints.

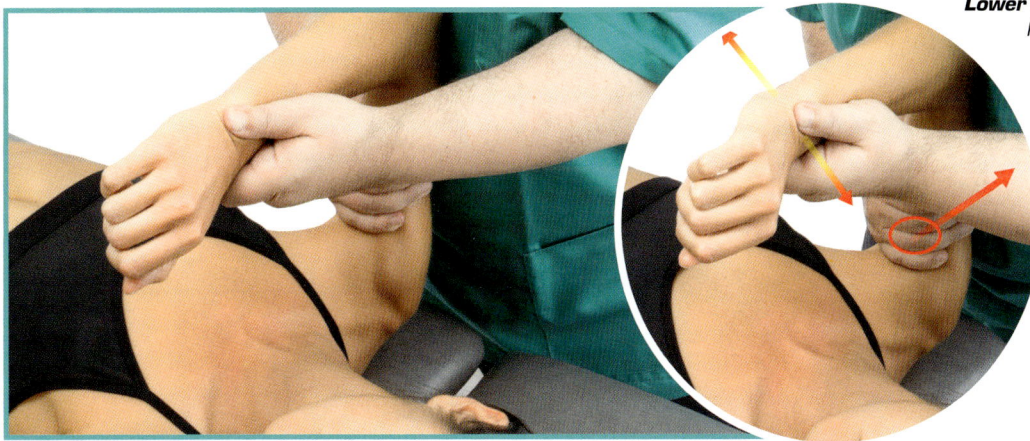

Lower thoracic elastification. Hold the athlete's arm still, away from the body. One hand is placed over the elbow fold, while the other hand rests on the lower ribs. Expiration is accompanied with a drop in pressure within the ribcage, while adduction of the arm is increased. Several traction cycles are performed. Upper and middle traction can be achieved by changing hand supports.

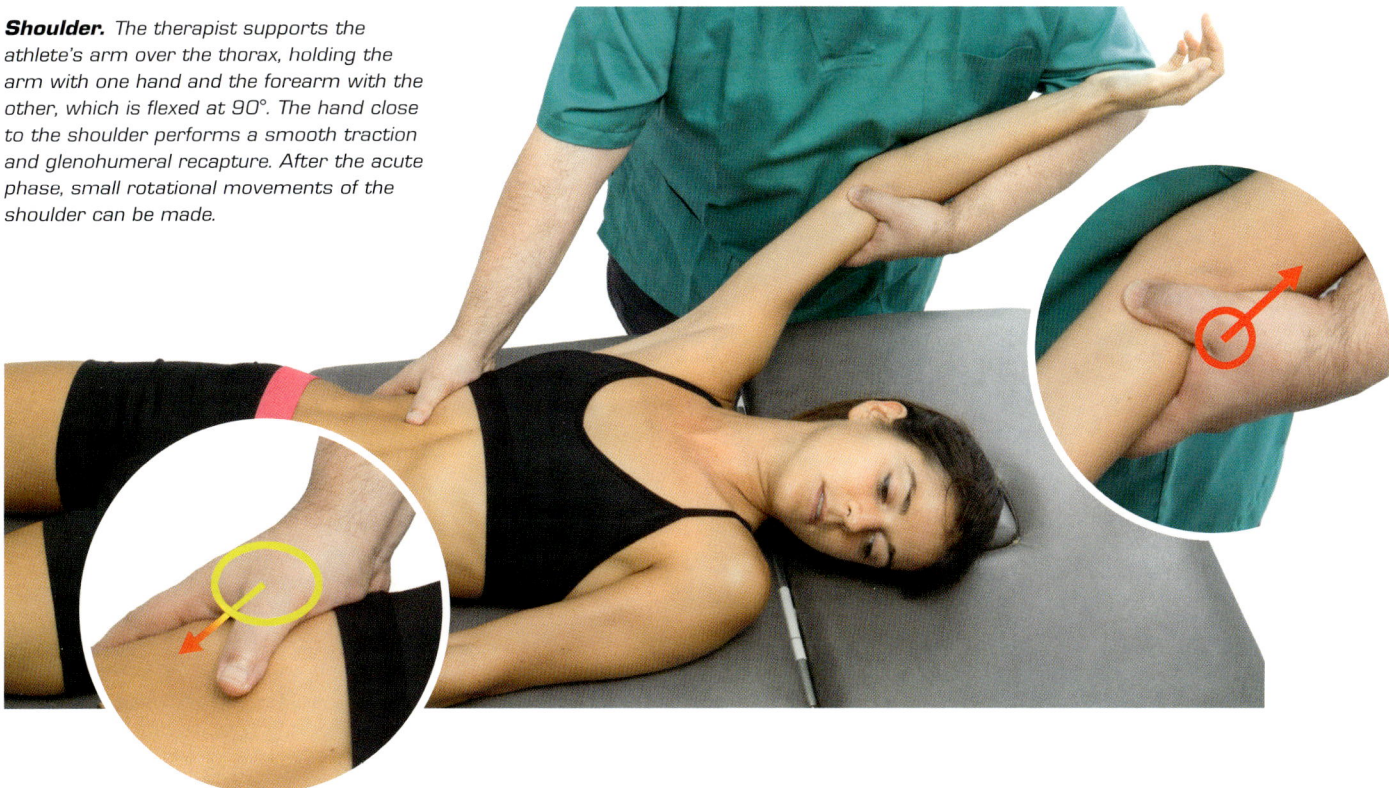

Shoulder. The therapist supports the athlete's arm over the thorax, holding the arm with one hand and the forearm with the other, which is flexed at 90°. The hand close to the shoulder performs a smooth traction and glenohumeral recapture. After the acute phase, small rotational movements of the shoulder can be made.

Joint mobilization

A joint is formed by a set of components that are joined to allow movement. Joint mobilization is the action by which a joint moves in its maximum arc or range of motion, or ROM.

This technique should be applied after investigating the athlete's condition, as carefully as possible, without forcing or exceeding the pain threshold in order to avoid the risk of a contracture or even injury. It is applied without a final return of the joint and also mobilizes the problem joint; for example, flexion along with extension, one after the other.

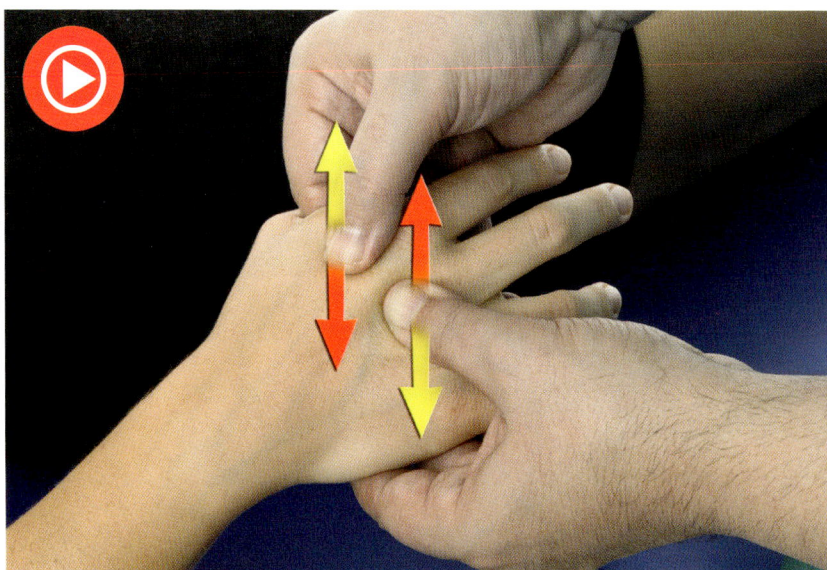

Mode of application

Articular mobilization is achieved by holding the articular components with one hand, while the other hand supports and applies mobility to the other. This movement is repeated, providing progressive amplitude at low speed and persisting in the direction in which movement is limited. At the same time, slight traction can be added or maintained.

Joint mobilization of metacarpals. The athlete stretches her arm out on the couch. The therapist uses both hands to grasp the athlete's hand and slowly manipulates the dorsal-palmar and palmar-dorsal movements of each of the metacarpals. This mobilization facilitates pressure function after a long period of immobility.

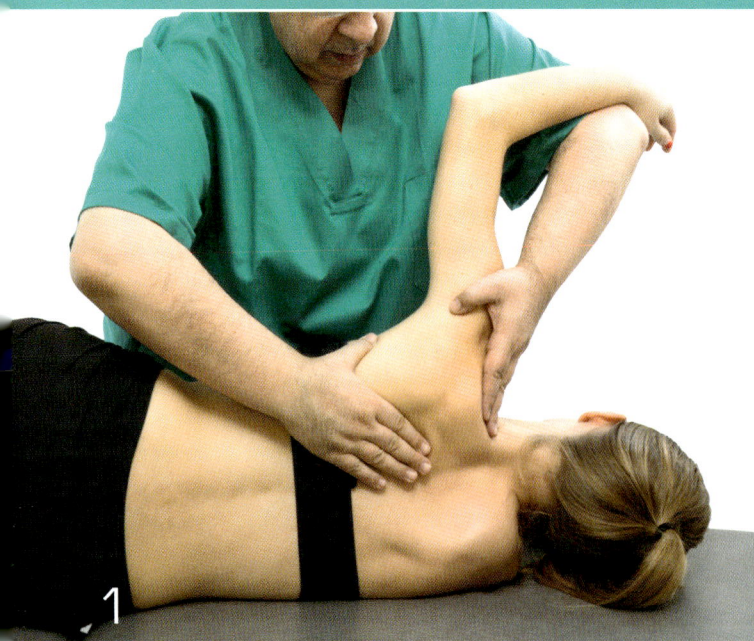

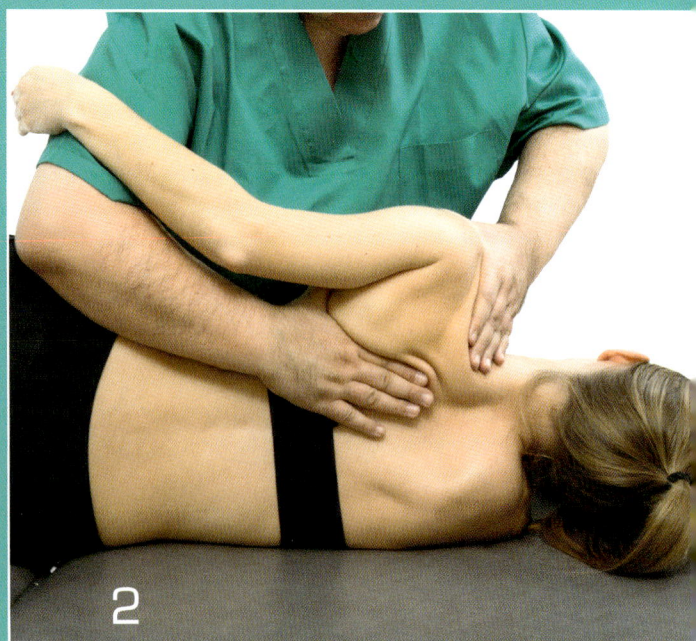

Passive mobilization of scapular girdle. The scapula is mobilized in a cranial or ascending, caudal or descending direction with the athlete lying on one side, (1). Internal and external tilting or swaying movements, and approach and separation (2) movement of the costal grid scapula.

Types of joint mobilization

These can generally be classified as active and passive joint mobilization. Passive mobilization is performed by the therapist without help from the athlete. Active mobilization is performed by the athlete. Although there are multiple combinations, this section only describes passive mobilization. The articular traction technique (pages 72-73) is also considered a type of mobilization.

Each joint has its own degree of mobility and is therefore assessed in isolation and compared to its mirror joint on the other side of the body. In addition, such mobility depends on multiple factors, such as age and gender, as well as the existence of injuries, scarring, hypo-mobility, etc.

Effects of massage

■ The benefits of mobilization allow the therapist to reach many organic structures beyond the joints themselves.

■ They provide a pumping effect that increases circulation and mechanical capacity of muscles.

■ In cases where injury has limited function of the joint, repetition of this movement until the joint has assumed its previous capacity helps to restore the motor image. This recovery of movement boosts the morale of the person being treated, who will see the potential for improvement during these exercises.

Criteria for use

This technique is performed where there is a need to increase the range of joint mobility and to cease and prevent joint stiffness. It also stretches a muscle or myofascial group, decreasing contractures.

Indications

Used in cases of circulatory deficit, alterations or limitations in joints and myofascial movements, capsulo-ligamentary retractions, postural problems, post-traumatic and subacute postoperative hyper-mobility, and adhesions.

Contraindications

◆ Not be performed in cases of acute inflammation, recent bone injury or bone fragility, in areas where recent soft tissue injuries or very old prostheses are located (in addition to the general contraindications of massage techniques).

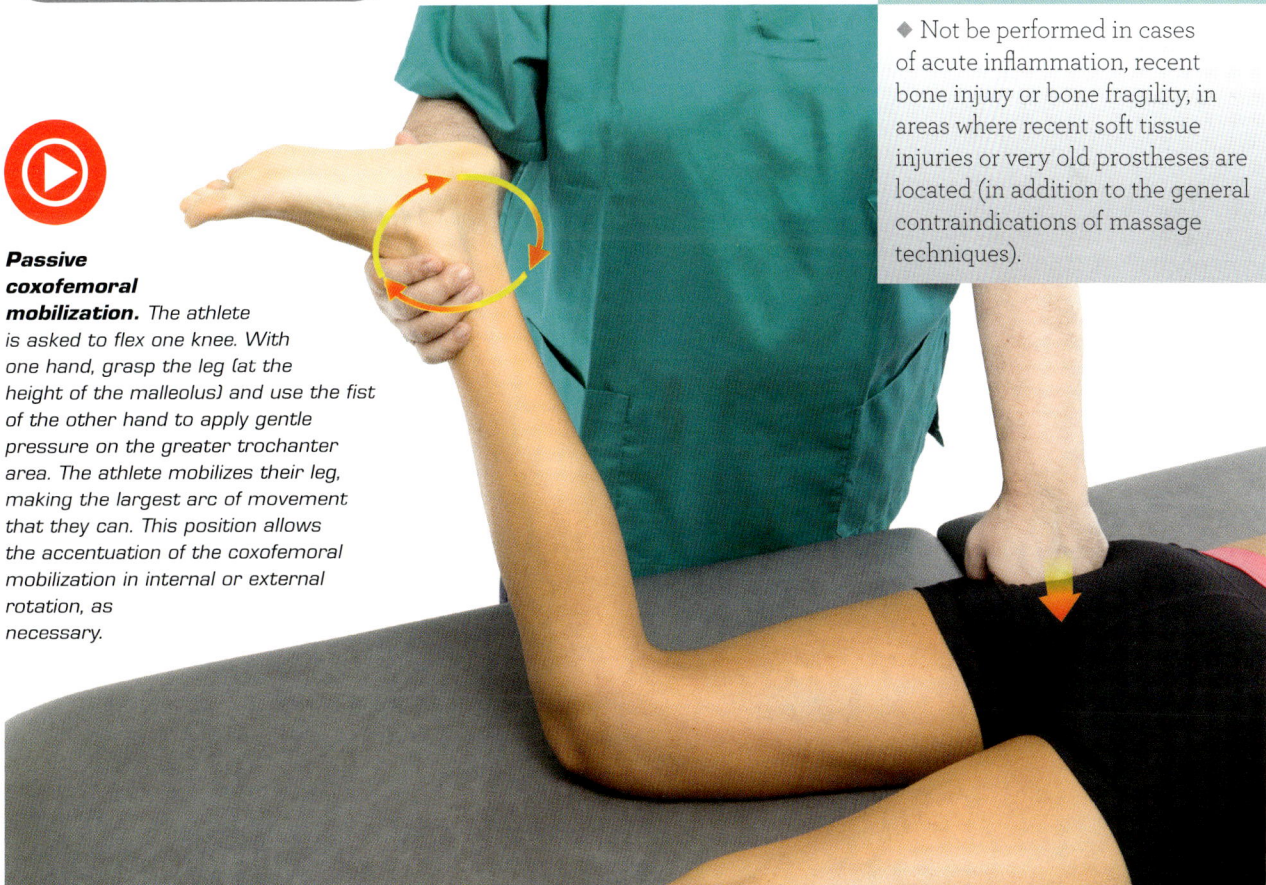

Passive coxofemoral mobilization. The athlete is asked to flex one knee. With one hand, grasp the leg (at the height of the malleolus) and use the fist of the other hand to apply gentle pressure on the greater trochanter area. The athlete mobilizes their leg, making the largest arc of movement that they can. This position allows the accentuation of the coxofemoral mobilization in internal or external rotation, as necessary.

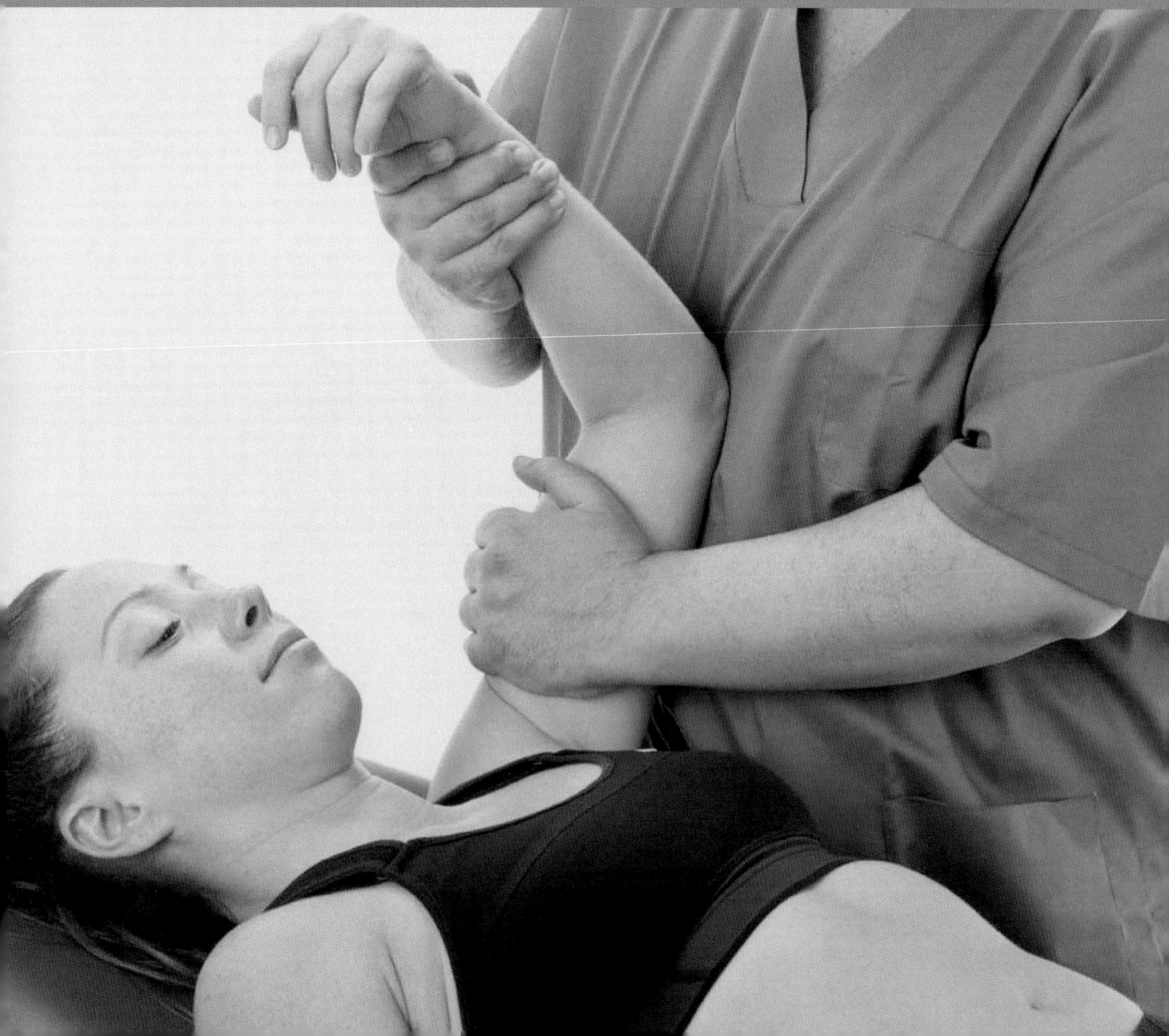

MYOFASCIAL CHAINS

3

Techniques for the physical treatment of the athlete, especially those of sports massage, evolve parallel to the systemic approach of anatomy and physiology. Its development has been driven by both studies in health sciences and the reorientation of technical skills. This global vision moves away from typical massage that is applied only to the muscles in order to encourage an association between them and any connective tissue. The need to understand the dynamics of global systems — in this case, the locomotor system — links the muscular, connective and nervous tissues through a formula, described as a neuro-myo-fascial chain or simply "myofascial chains." These chains make it possible to understand, analyze and treat postural behavior in a more comprehensive way, viewing organic compensations and human movement from an integral perspective.

The concept of global myofasciae

The term "myofascial" is a theoretical construct that guides the student or the practitioner toward a systemic, overall view of anatomy and physiology. It is a descriptive formula that synthesizes the dependence between two tissues that, although sharing a same embryonic origin (derived from the mesoderm), have always been studied separately: the connective tissue (the fascia) on the one hand and the muscular tissue on the other. This has led to the abandonment of the analytical approach (by segments) and the dissection of muscular activity in order to integrate it into a more effective system. The word "chain" denotes an object made up

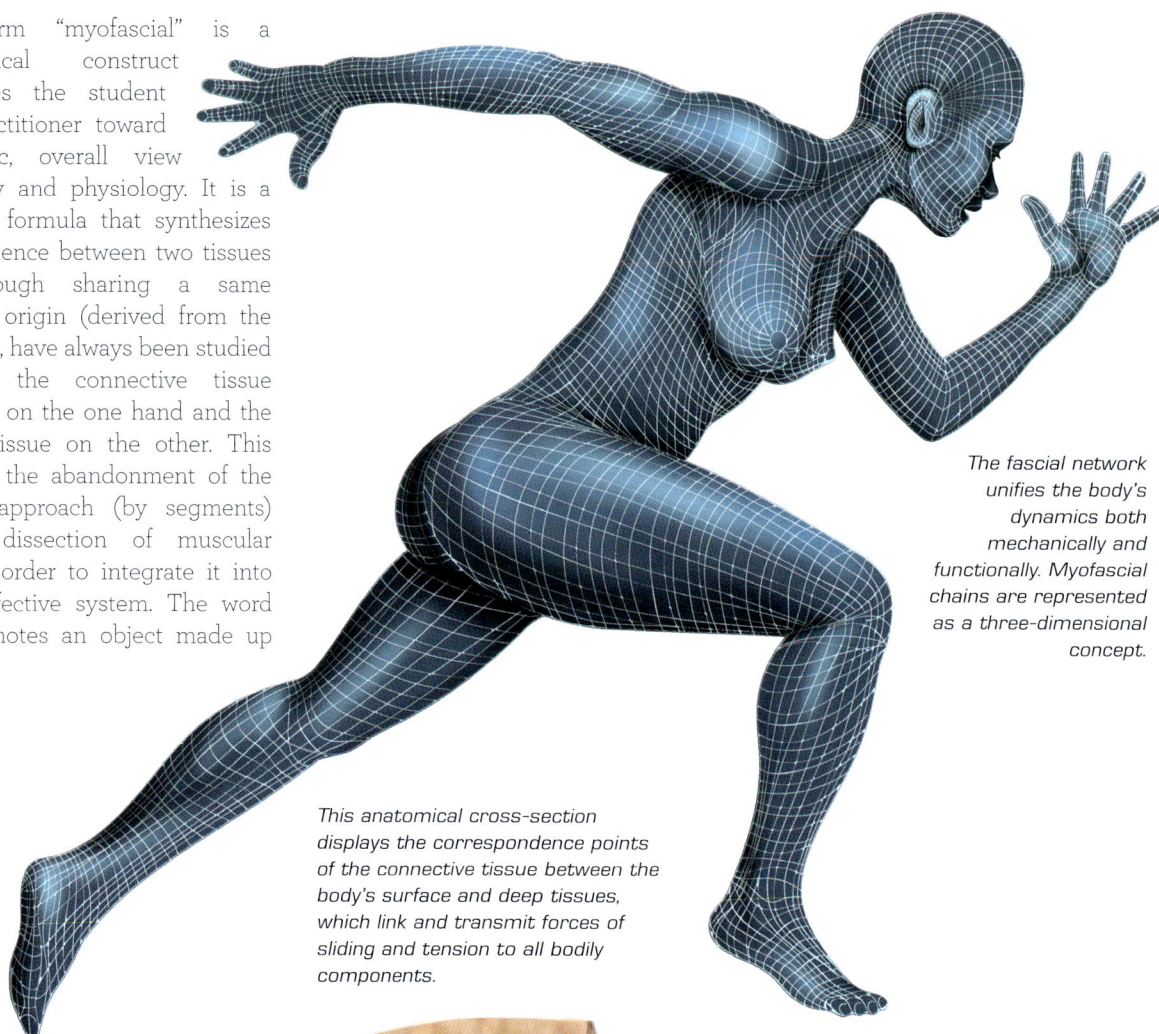

The fascial network unifies the body's dynamics both mechanically and functionally. Myofascial chains are represented as a three-dimensional concept.

This anatomical cross-section displays the correspondence points of the connective tissue between the body's surface and deep tissues, which link and transmit forces of sliding and tension to all bodily components.

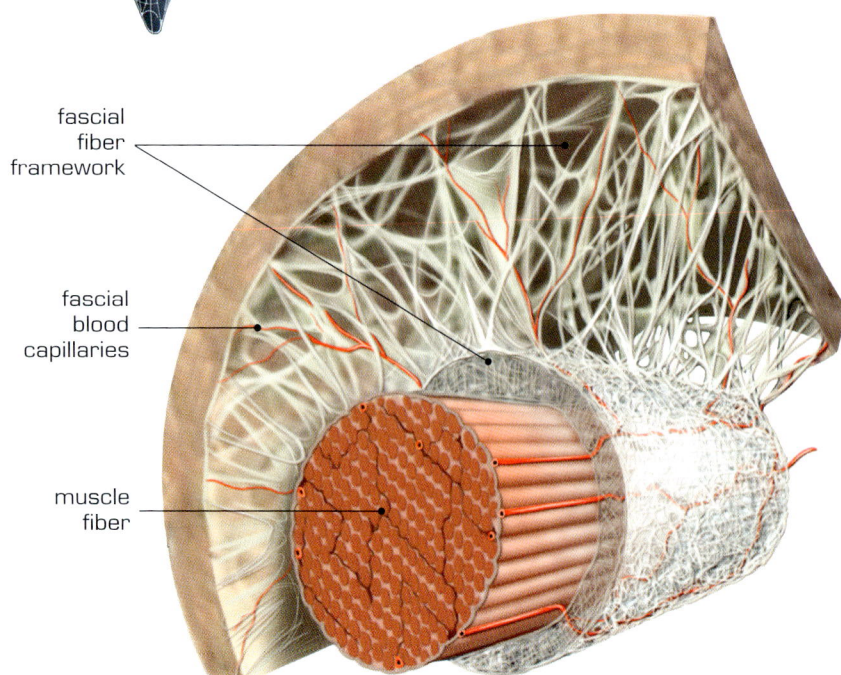

fascial fiber framework

fascial blood capillaries

muscle fiber

of interconnected links, and offers an idea of continuity that allows one to observe each of its elements without losing sight of the entire body as a whole. Thus the concept of global action is represented by myofascial chains, chain reactions that show how neuro-miofascial elements are associated with movement and put into play mechanisms involving different systems, and even the entire body. Therefore any action we carry out moves not just our musculature and joints, but also our veins, arteries, lymphatic system and nerve structures, viscera, etc.; all of this is thanks to the close relationship of all structures with the omnipresent fascia (see illustrations of the fascial network and the anatomical cross-section).

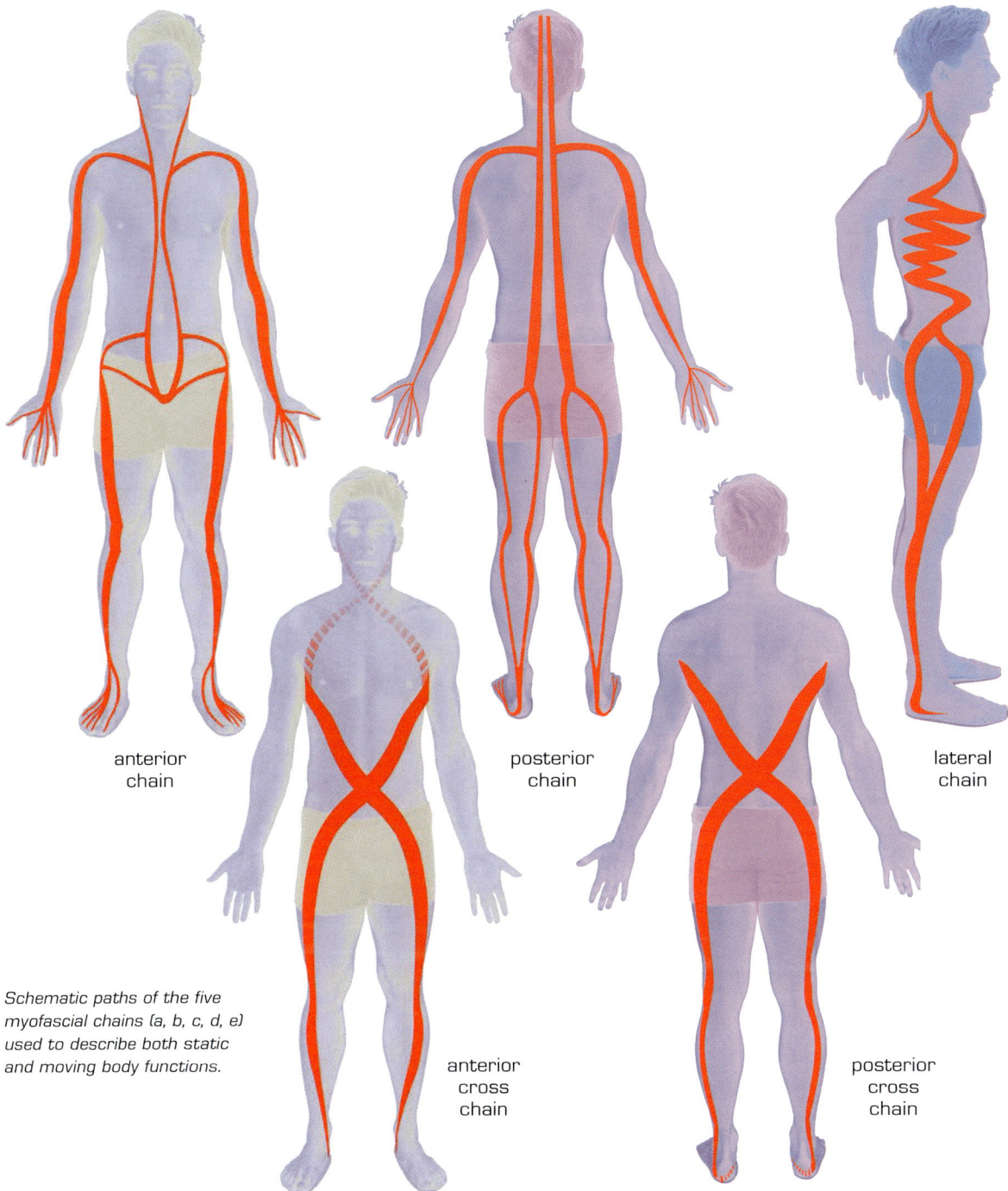

anterior
chain

posterior
chain

lateral
chain

anterior
cross
chain

posterior
cross
chain

Schematic paths of the five myofascial chains (a, b, c, d, e) used to describe both static and moving body functions.

Five "circuits" for body reading and massage treatment

The organization of myofascial chains shapes the idea of the body functioning as a unit; meshing of the fascial network provides a global overview of structure and function.

These chains are a useful resource for "reading" the body; they offer information about static and myofascial stress zones and help to locate shortening and compensation patterns. They explain many of the causes that dictate the execution of movement. At the same time they allow the therapist to draw up a massage treatment plan to treat pain of musculoskeletal origin.

This chapter employs a formula which encompasses five myofascial chains — anterior, posterior, anterior cross, posterior cross, and lateral — schematically represented in the illustrations. It reveals how to apply an individualized massage following these circuits, a treatment that must reach the affected structures and seek to restore the global and functional routes of movement without pain; this is an indication of success in manual massage.

The anterior myofascial chain

A flexion chain

The anterior chain connects the entire anterior front of the body, from the dorsum of the toes to the skull (temporal bone). Together with the posterior chain it is responsible for the stability of the body in the sagittal plane, which divides it into two equal segments — left and right. Therefore we must consider two anterior chains: one on the left and one on the right.

Its function is to prevent backward postural instability. Vertical displacements incite the regulatory activity of these chains, and they act together as a tensor.

Together with the posterior chain they are responsible for displacements in the sagittal plane, and although they perform opposite actions, they are related and coordinated.

Functions of the anterior chain

From its uppermost point in the skull, the anterior chain is charged with flexing the head, thorax and pelvis. It keeps the knees in extension and, thanks to abdominal myofascial tension, also fulfills a defensive function, protecting the viscera of the abdominal cavity.

The rectum in the abdomen raises the pubis or causes the sternum to descend in the direction of the navel, so we can choose either end as a fixed point.

This chain must be able to stretch completely in order to extend or hyperextend the trunk, flex the knees and perform plantar flexion of the ankles and toes.

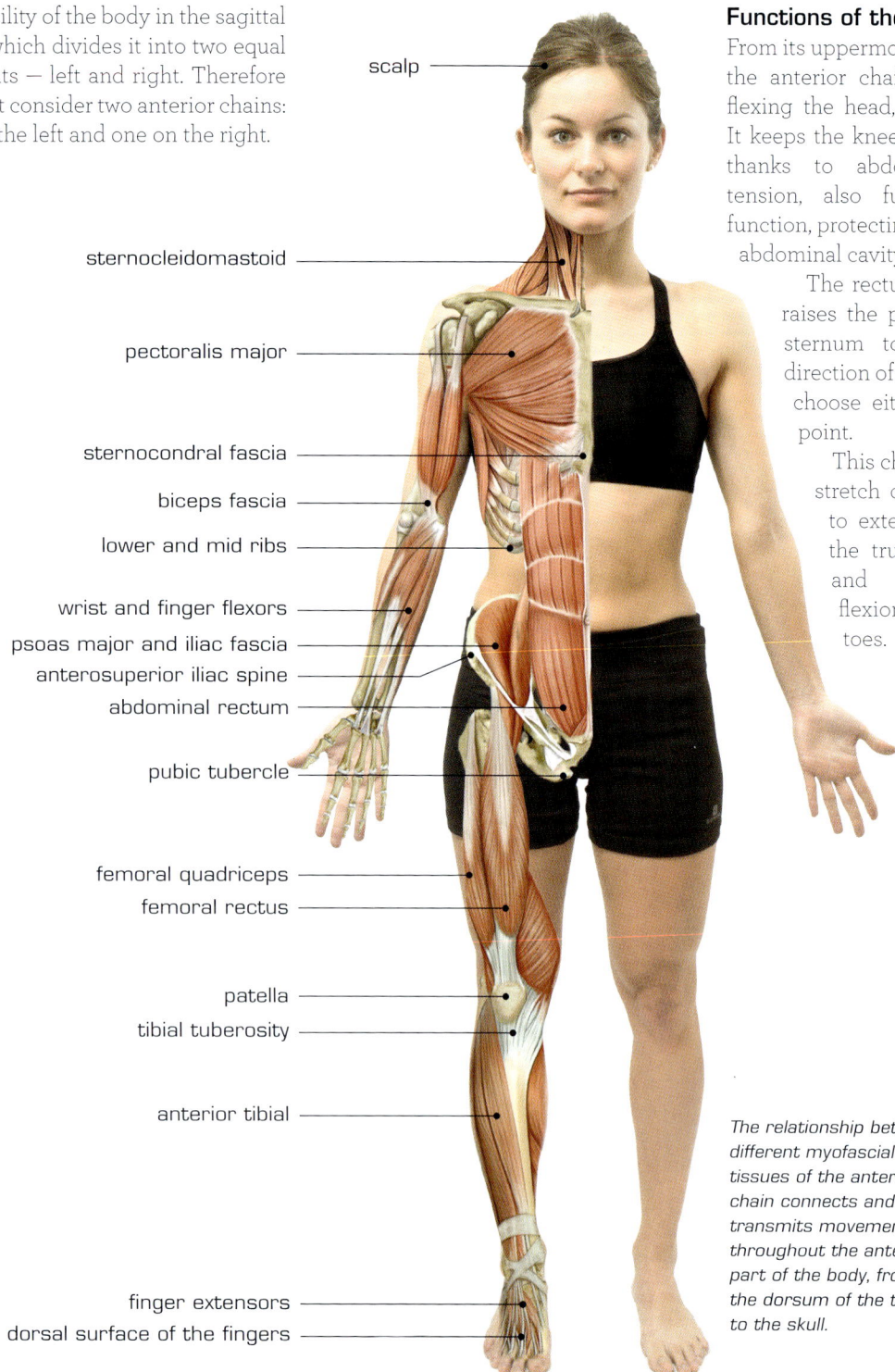

scalp

sternocleidomastoid

pectoralis major

sternocondral fascia

biceps fascia

lower and mid ribs

wrist and finger flexors

psoas major and iliac fascia

anterosuperior iliac spine

abdominal rectum

pubic tubercle

femoral quadriceps

femoral rectus

patella

tibial tuberosity

anterior tibial

The relationship between different myofascial tissues of the anterior chain connects and transmits movement throughout the anterior part of the body, from the dorsum of the toes to the skull.

finger extensors

dorsal surface of the fingers

The self-regulating activity of the anterior and posterior chains is observed during gait with each alternating step. The planting and takeoff of the foot or flexion-extension movements in the knees and hips are clear examples. Given its dynamic action, the previous chain shows a predominance for the motor phase (fast-contracting fibers).

The athlete and compensation in the anterior chain

Disorders of the anterior chain lead to excessive shortening or approximations between sections of the front of the body (especially from the head to the pelvis).

In disciplines such as boxing, a defensive position offers some degree of curvature of the spine and grouping of the upper limbs in front of the chest. This posture generates uncompensated work between the anterior and posterior myofascial structure of the trunk that also extends to striking techniques. This often leads to fatigue and imbalances between the anterior and posterior chains. Consequently, this attitude creates improper shortening and posture.

Some postures allow an extension of either the whole chain or one of its components, such as kneeling on the heels ("the Sphinx," trunk extended with elbows providing support) or "the bridge," flexing the neck and extending the toes and ankles. Holding this position will improve range of movement.

Detection of compensations

A first common shortening pattern in the anterior chain consists of loss of leveling of the scapula girdle. It is seen in individuals who maintain a posture with the neck and shoulders hunched forward, as well as an expiring (sagging) chest.

An excess of tension will be perceived during the palpation, due to the shortening of the myofascial tissue of the pectoral, sternocleidomastoid, upper trapezius and levator scapula muscles. The opposing (opposite) musculature will be weakened and flaccid in the deep flexor neck group and the interscapular group, as well as in the area of the spine erector muscles at cervical and thoracic levels.

A second shortening postural pattern is seen in the pelvic girdle, with pelvic anteversion, prominent abdomen, and hyperlordosis of the lumbar spine.

Excessive tension in the myofascial tissue, particularly in the hip flexors, iliopsoas, femoral rectus, fascia lata, short adductors, and extensor group of the trunk (lumbar spine erectors), will be noticeable to the touch. The tissues that are antagonistic to the aforementioned areas will also be weakened: abdominals and glutes. These situations constitute the upper and lower cross syndromes, respectively.

Reorganize constraints

Excessive tension in all of the components in this chain limits the body's straightening and extension movements, particularly when standing, in which the effort of the myofascial tissue moves from below (the feet) and then upwards.

Anterior chain in contraction. Joint action flexes the trunk and hips, increasing intra-abdominal pressure. She simultaneously performs dorsiflexion of the ankle and extension of the toes.

3

Lower body I: feet and legs

General Techniques

◆ Pressure strokes
◆ Longitudinal friction

Devices

◆ Knuckles
◆ Fingertips
◆ Elbow

First, assess mobility response

One way to start manual massage work on the anterior chain is to begin with an investigation of mobility in different joints of the foot, ankle and leg, assessing their behavior during movement.

Foot: evaluation plan

Depends on athlete's capacity for movement at presentation. Athlete lies on the couch face up (in supine position) with legs extended. One foot is held with both hands and light traction and afterwards pressure movements are applied in the direction of plantar flexion (around and down from the toes and ankle) in each and every one of the toe joints (phalanges), and in the ankle.

It is useful to assess the individual's ability to move. Compare both feet and then decide which areas require greater attention as you perform your massage work. Extensor tendons are also palpated on the dorsum of the toes and feet, on the retinaculum of the ankle neck, and always on both feet. Areas of induration, stiffness or lack of extension are assessed.

Following the model of the anterior chain, work on the lower limb begins at the back of the foot and, as if this were the side of a mountain, climbs up toward the hip.

Through traction and compression movements made with the foot and ankle, you can observe and assess their interrelation with the hip joint. Massaging the back of the feet frequently reveals a small fleshy mass on the outside of the instep, although not all people present it. If it exists, it corresponds to the shortening of the extensor muscle of the toes.

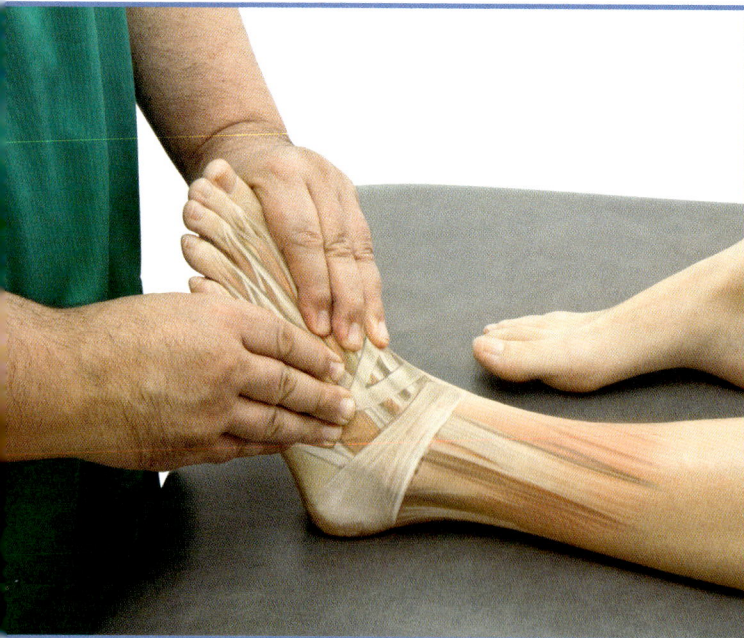

Massaging the back of the foot. *Applied while the athlete moves the toes and the back of the foot. It is important here to combine friction maneuvers with toe joint (interphalangeal) and foot dorsum (inter-metatarsal and Lisfranc and Chopart joints) movements. Very good results are obtained in individuals with claw toe and other alterations.*

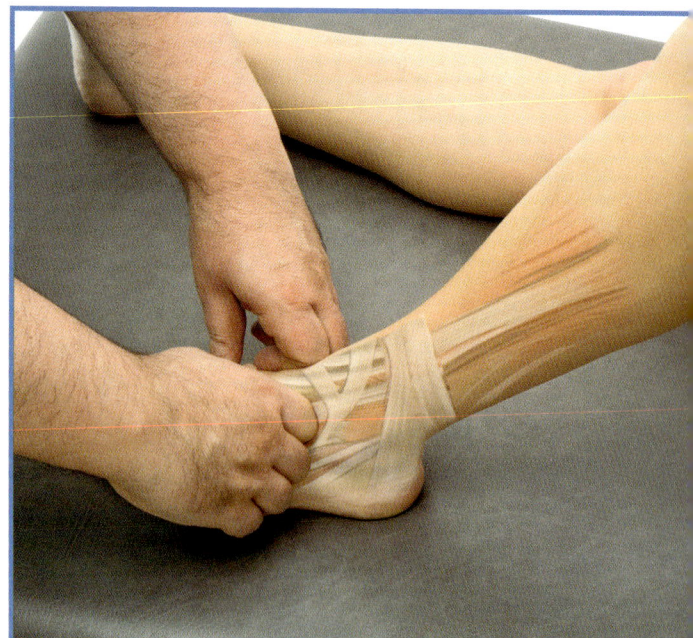

Massaging the retinaculum of the anterior part of the ankle. *Tension in this region limits the plantar flexion movements of the foot, such as standing on tiptoe. It is very important to massage this area after immobilization from fractures, sprains or surgical procedures, etc*

3

Leg: evaluation plan

Work now continues with the anterior part of the leg tendons (tibialis anterior, long toe extensors and third peroneus), which can be examined well through palpation. When the athlete lifts their ankle and brings the tip of the foot inwards (dorsiflexion and inversion), the muscle and tendon of the tibialis anterior can be perceived by touch.

Elevation (extension) of the toes will allow the therapist to note the long extensor of the big toe and also the other long toe extensors.

The tendon of the toe extensor disappears under the anterior tibial, ending under the protrusion below the knee (tuberosity of the tibia).

Anatomical precision

The interosseous membrane, intermuscular septa or connective tissue wall can be found between the anterior tibialis, located anterior to the medial malleolus (during dorsiflexion) and the peroneus (during plantar flexion), which are located behind the lateral malleolus.

This septum can be traced to its upper limit in front of the top of the fibula. It can be felt by moving the fingers upwards from the lateral malleolus. This area can be better explored if the individual makes alternate elevating and descending foot movements.

Massage techniques with active movement

Massage maneuvers are mostly carried out with the athlete at rest, and on some occasions it is the therapist who passively moves the athlete's joint while performing massage. But you can also work while the athlete

Dorsiflexion and eversion of the ankle allows the therapist to explore the muscle and the tendon of the third peroneum, as it is absent in same people.

makes a series of active movements, as this facilitates the observation of restrictions. The athlete can then be informed of which movement he should use, how to move and how far to move during treatment, allowing the athlete to perceive and integrate these "new" movements.

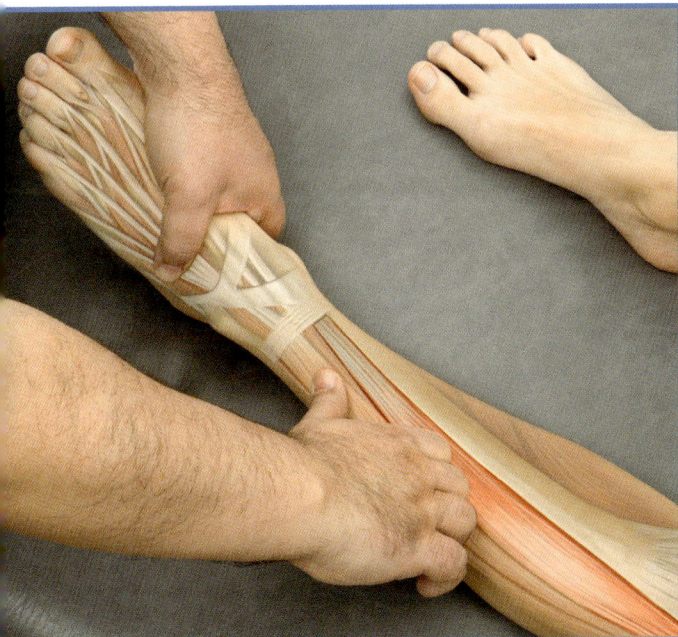

Massaging the front of the leg. *The athlete is asked to perform a plantar flexion to treat the anterior tibial. Pressure is maintained with the knuckles in the fleshy anterior compartment of the leg and the tibial shaft (shin) while the athlete maintains flexion of the ankle and toes.*

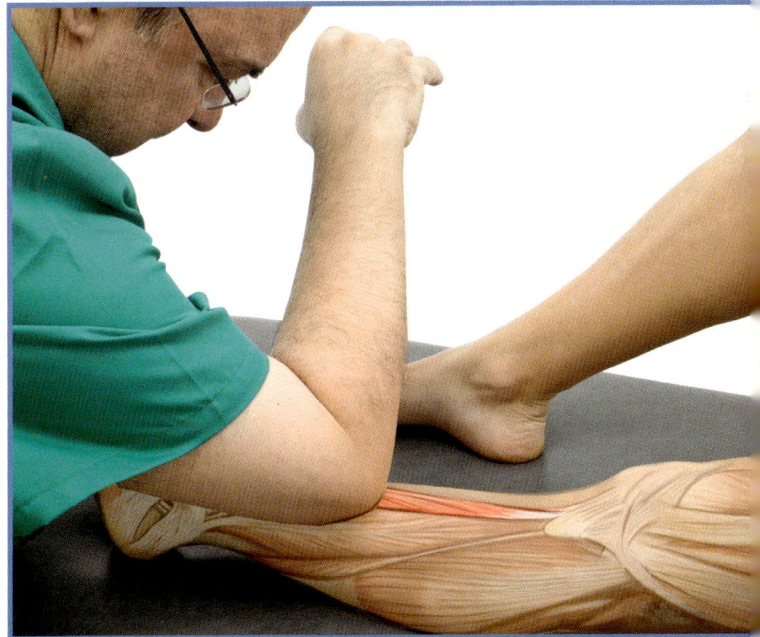

Massaging the leg. *Applied to athletes with bulky or very compact muscular mass by exerting sliding pressure with the elbow. Work carefully and intermittently to avoid causing damage.*

Lower body II: thighs and pelvis

General Techniques

◆ Pressure
◆ Static pressure with longitudinal friction
◆ Oscillatory movements

Devices

◆ Forearm
◆ Hand
◆ Fingertips

Areas in relation and translation

The close relationship of self-adjustment established by the anterior and posterior chains can also be observed in the myofascial tissue of the pelvis and thigh. These tissues are under great stress, combining to keep the body standing "still," walking, running or jumping.

Walking and myofascial chains

When walking, the force of the heel impacting on the floor is absorbed in part by soft tissue, in particular by the knee extensors and hip flexors. These act as a spring against the force of gravity, avoiding further joint flexion. Nevertheless, the body subsides during the impact caused by the heel-step.

As the step progresses the sole of the foot supports more load, until the take-off phase begins. At this stage only the front of the sole of the foot is in contact with the ground. The body experiences a gentle lift. These slight movements of descent and rise in the body transmit elastic energy to the corresponding chains (posterior chain for the previous stage of the step, and vice versa).

When a limb is advanced (flexion of the thigh), the pelvic bone on that side (the coxa) rotates backwards in retroversion. The posterior chain is prepared in pre-tension for the coming step stages: those of load support and subsequent take-off.

The opposite happens at the other extreme, with the thigh (hip) in extension while the pelvic bone on that side turns forward in anteversion of the coxa. The anterior chain functions as a spring, effecting the previous phase of the step. This simultaneous alternating in the chain operation reveals the coordination required for any bodily movement.

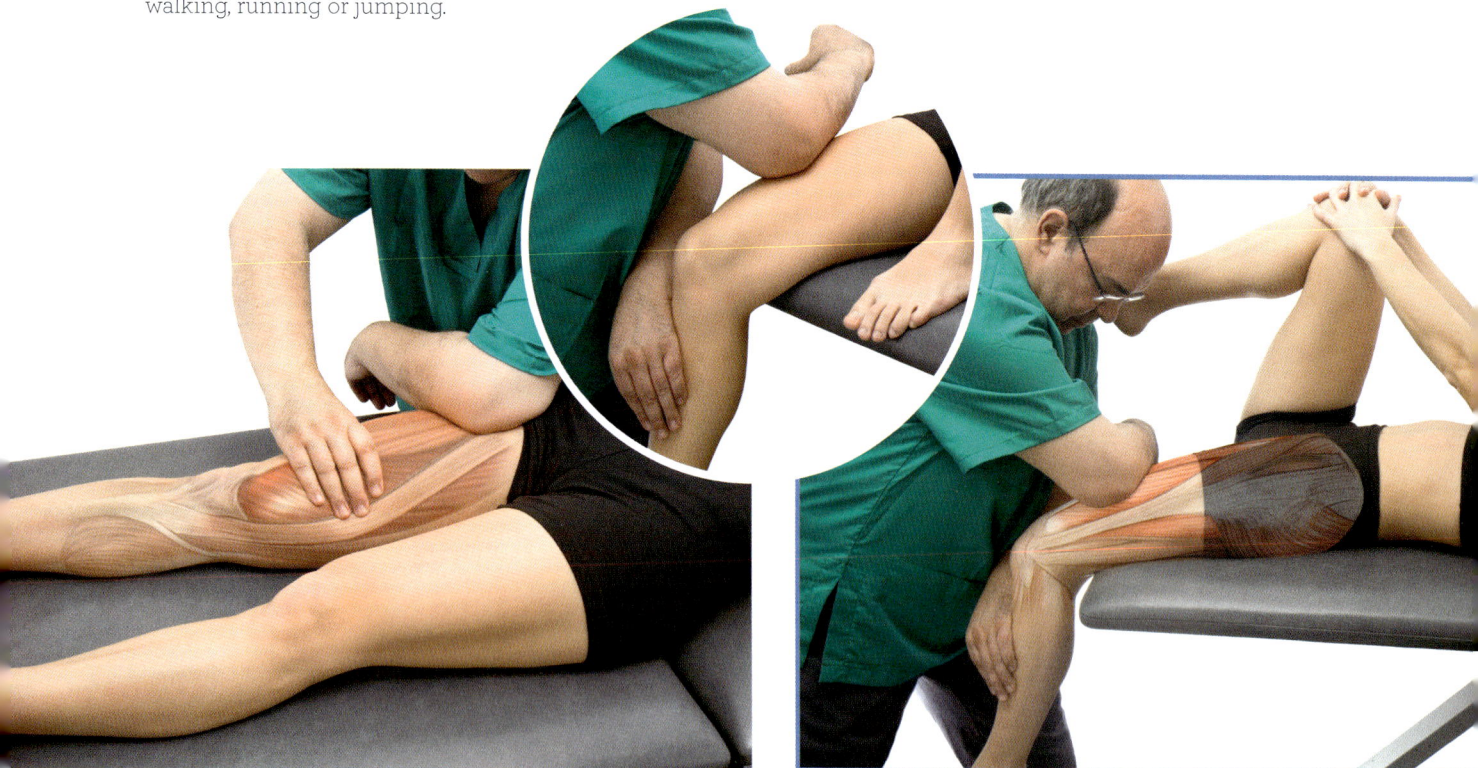

Massaging the front of the thigh. *Knee and hip should be extended. The therapist places his lower hand on the knee to perform thigh rotations, which can be applied with the knuckles or forearm. You can also simultaneously perform oscillatory movements on the thigh with the lower hand to facilitate distension.*

Massaging the front of the thigh. *With part of the thigh hanging from the couch and the hip extended, the knee is bent to stretch it. The patient rests the foot of the other leg on the therapist's shoulder. This technique is applied in an upward direction. With the other hand the therapist can perform oscillatory movements of the leg (internal and external rotations of the hip).*

The hip and psoas posture

One of the myofascial tissues that condition the position of the anterior chain is located in the hip area. Observing its stability allows us to obtain information about whether the posture is forced or comfortable, clearly visible by the positioning of the hip and the lumbar curvature. The Thomas test can be used, providing information on shortening of the hip flexors.

Hip flexors: evaluation plan

The athlete lies on their back (supine) with limbs well aligned. They are then asked to leave one leg hanging from the couch while the masseur holds the thigh of the other leg in contact with his chest in both hands.

Precautions

◆ It is advisable to not work on the psoas above the height of the navel as this can compromise renal structures.

◆ The inguinal ligament area should not be compressed, as the lateral femoral cutaneous nerve passes underneath. Try not to compress any region in which you perceive an arterial pulse.

◆ When applying this first technique it is advisable to avoid causing pain in the peritoneal structures and to keep the massaging fingers relaxed and free of tension. You must slowly move your fingers to the rhythm of the athlete's breathing, asking him to describe any sensations until you make full contact with the muscle. Only then can friction movements be applied, as far as the specified work plan permits.

■ If the hanging thigh is not kept parallel to and resting on the plane of the couch, we can conclude that the flexor group of the hip is tense and shortened.

■ If the knee does not flex more than 45° with the entire limb hanging freely off the couch, the quadriceps femoral rectus is tense and shortened.

For the Thomas test a hand is placed between the couch and the athlete's lower back in order to observe the position of the pelvis and the quality of the lumbar lordosis.

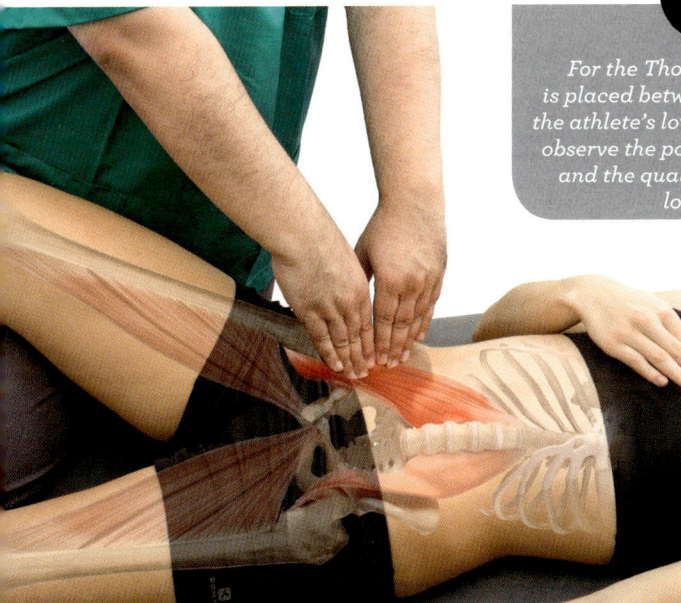

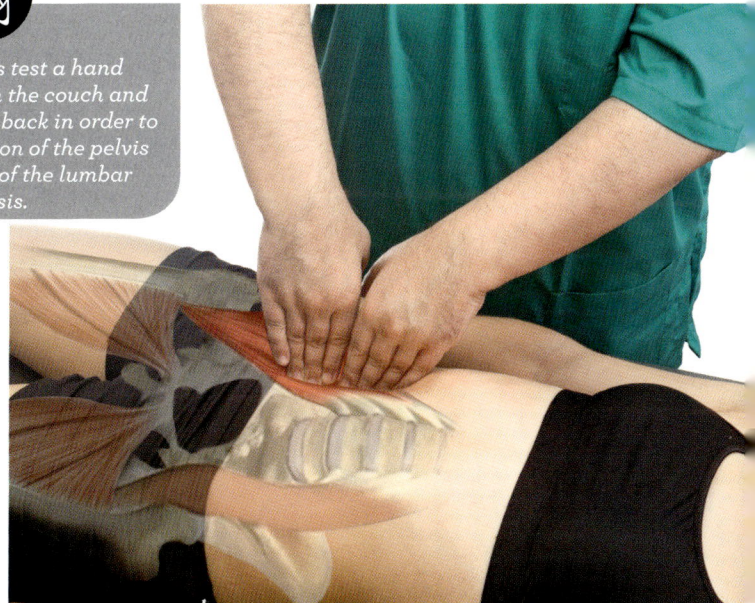

Massaging the major psoas. The iliac and psoas can be palpated above the inguinal ligament in the abdominal area. The athlete is supine, with knees bent and feet supported. If you need to improve muscle perception, ask them to lift the foot from that side off the couch. It will strain and you will be able to better assess the psoas. Maintain contact with the coxal bone or iliac, and slowly sink the fingers into the anterior superior iliac spine.

Massaging the myofascial iliac area. Legs are extended in the final phase of supine treatment of the muscle and its fascia. It is necessary to move the skin in a lateral direction toward the anterior superior iliac spine so that it will be possible to slowly sink the fingers into the iliac fossa.

Upper body I: abdomen and thorax

General Techniques

- ◆ Opposite sliding pressures
- ◆ Pressure
- ◆ Longitudinal friction

Devices

- ◆ Knuckles
- ◆ Fingertips

Connection area

The abdominal area is a communication space between the lower and upper trains, where the legs communicate with the trunk through important connections of myofascial tissue. It also contains an important visceral and nervous area, as well as blood and lymphatic vessels.

Abdomen: assessment plan

There are three distinguishable layers in the abdominal rectum: the most superficial fascia, which covers the front of the abdominal rectum, the muscle itself, and the deep fascial layer behind the fleshy mass. It is important to assess the degree of tension at all three levels.

If the abdominal rectum is flat, tension is located in the superficial layer and in the sinus of the muscle. If the muscle itself is prominent carefully evaluate muscle tone in order to ensure shortening of the deep fascial layer.

Precautions

- ◆ Compression of the abdominal area should be avoided in areas where the arterial pulse is detected, such as the descending aorta. This technique is more about approaching the area rather than compressing it.

In overweight people this deep-layer tension is responsible for respiratory restrictions and difficulty moving the abdomen backwards (pelvic retroversion and lumbar flexion). Spinal extension exercises (such as "Sphinx," etc.) provide the necessary elongation in cases of shortening in this region. Friction maneuvers and kneading may be added to stretching postures in order to combine the two principles, allowing tissues to be stretched taut.

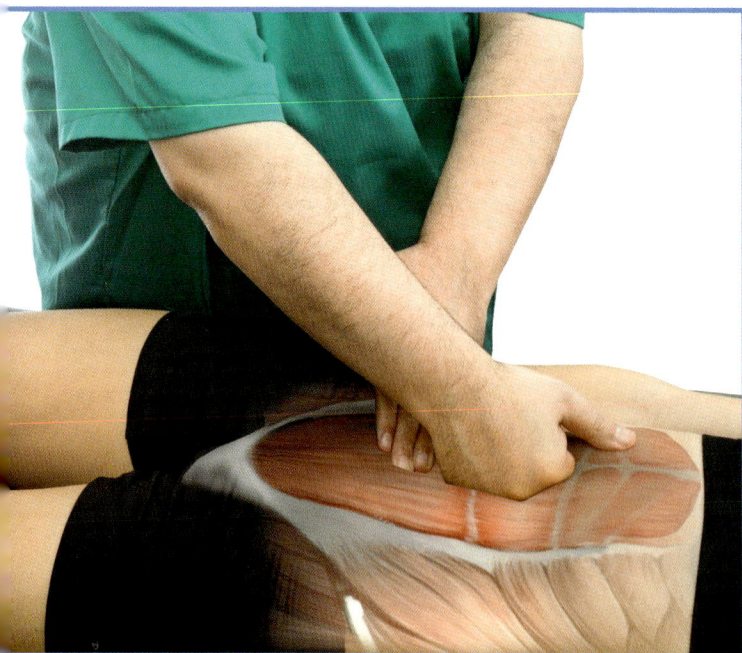

Massaging the abdominal rectum. *A descending friction is performed with the caudal hand, while the knuckles of the cranial hand exert upward sliding pressure. In this way opposing forces are applied, stretching the fascia and releasing muscle tension.*

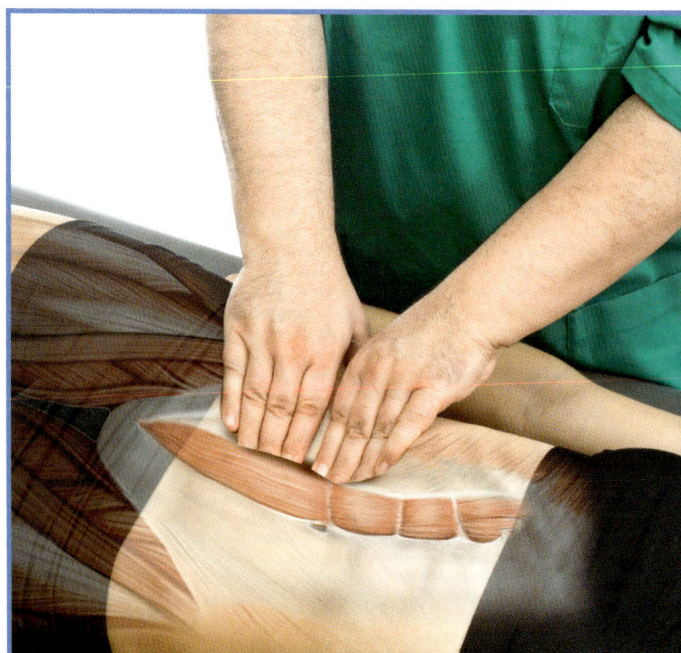

Massaging the abdominal rectum. *Each myofascial package or "tablet" is treated from the abdominal rectum area to its tendinous section at its insertion above the fifth rib. The back section is accessed from the lateral edge of the muscle, and it is lifted by pulling upwards from the fascial vein.*

The ribcage and respiration

The dynamics of respiratory costal movement must be assessed in this area. Costal displacement capability in the "manual water pump lever," or the upward and forward movement during inhalation, and downward and backward movement during exhalation, is always working on the upper ribs.

Thorax: evaluation plan

Respiratory dynamics in the lower ribs correspond to the "bucket-loop" movement in which the lower ribs expand outwards during inhalation and close during exhalation.

Tactile exploration allows the therapist to determine which zones present more movement restrictions, helping to focus the massage.

The abdominal rectum ends above the costal arch at the fifth rib. It can be moved with extended fingers or with the heel of the hand in the direction of the athlete's head. In addition, although the muscle ends there, the chain continues through the sternal area, covering the sternum itself and the tissues that run on the costochondral joints between the sternum and the insertions of the pectoralis major.

Massage techniques with active movement

In this area the movement performed by the athlete to assist massage action is directly related to the breathing movements and linked to volume changes in the rib cage or abdomen (they are difficult to appreciate in the photographs). Once the areas with rigidity and restricted movement

Precautions

◆ If the individual presents with tension or congenital deformity of the ribcage (such as a sunken chest in the area of the sternum) or a protruding chest (in the shape of a boat keel), hand displacement may be more difficult. Tissue should be accessed sideways and from the outside toward the midline. Breathing should be observed.

are located, the therapist adapts to the respiratory rhythm of the athlete, applying these techniques when the athlete exhales.

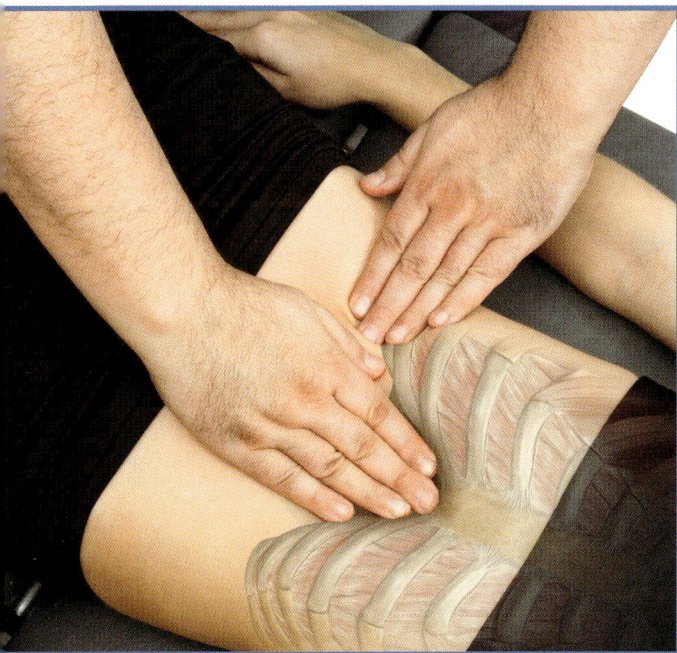

Massaging the chest and rib cage. *Massage around the ribs to release its anteroinferior section, continuing the maneuver through the sternal area.*

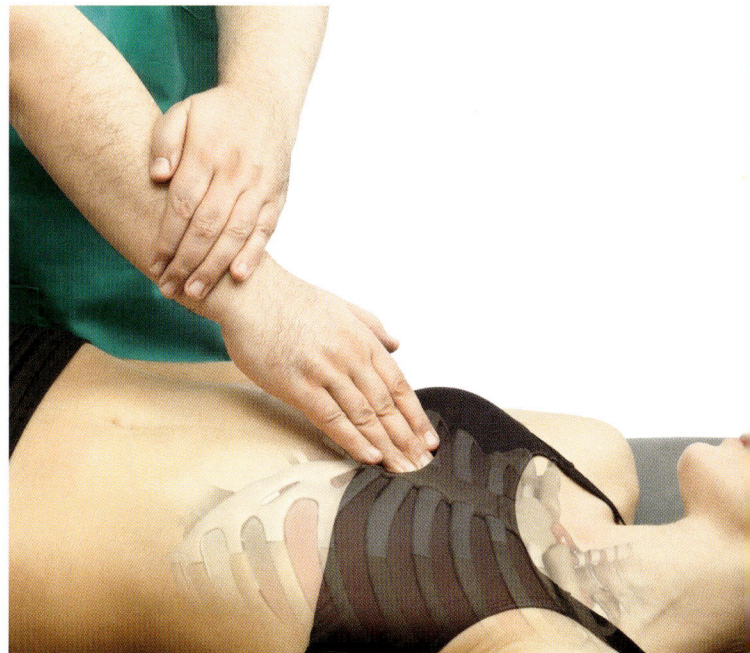

Massaging the thorax, sternum area. *Use friction on the sternum tissue and the lining of the sternocondral joints between the sternum and the medial border of the pectoralis major. This maneuver moves the tissue in an upwards (cranial) direction and laterally.*

Upper body II: head and neck

General Techniques

- Pressure
- Pin
- Kneading

Devices

- Fingertips
- Hands

Head above shoulders

The anterior chain is responsible for the flexion of the trunk during movement or when still, but the head produces hyperextension in the upper part of the neck. This is due to its biological makeup, as when inserted in the mastoid apophysis it causes flexion of the lower part of the neck (lower cervical). In contrast, it causes hyperextension in the upper part of the neck (high cervical).

Head and neck: evaluation plan

Observation of the athlete's general posture provides information about his verticality, and especially about how he perceives it. We must focus on the global image that it offers us; then we can assess where it oscillates. If the head tilts forward or falls backwards, how the head is seated and whether it is stable on the neck, its relationship with shoulders, etc.

The lateral view (in profile) also allows us to see if there is any type of alteration in the cervical curvatures (i.e., if the head or the shoulders are anteriorized). The quality of the amplitude of movement is then evaluated.

Observation of breathing gives an idea as to whether the up-and-down costal movement is good. This indicates a correct correlation between the thoracic and pelvic diaphragms, both controlled by the anterior chain.

Clavicles

It is also very important to assess the state of both clavicles, since they are closely related to everything that affects the neck region, the scapula, the upper limb and the upper half of the ribcage.

Precautions

- During treatment, the fascia in the lateral area of the neck is pushed with the fingertips, taking care not to affect the carotid artery. Any change in coloration on the face of the athlete should be addressed, or if there is a sensation of intracranial pressure.

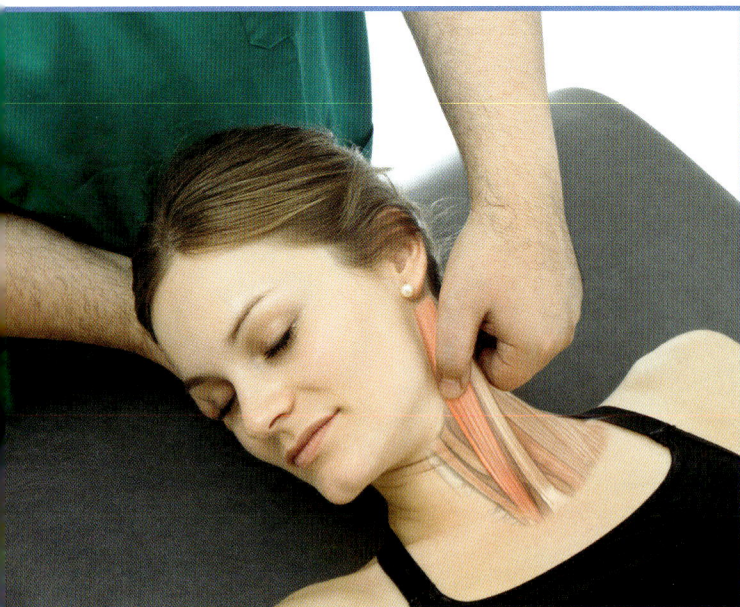

Massaging the anterolateral part of the neck.
For the sternocleidomastoid, the caudal hand raises the head slightly to better access the tissues, and downward friction is applied with the cranial hand. It should not be pressed on the neck itself; the direction of pressure will be anterior or posterior, but concentrated around the neck.

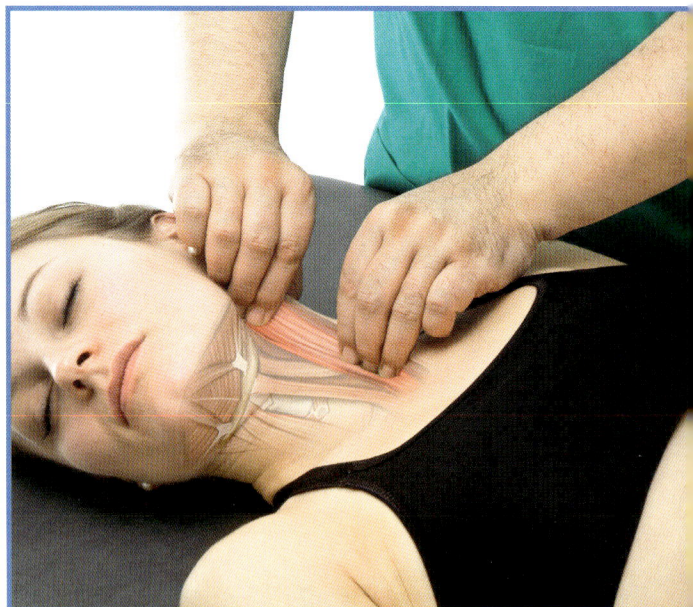

Massaging the sternocleidomastoid. *Performed in the anterolateral area of the neck using a clamp. The athlete's head is placed lateral to locate the muscle, which is accessed with the first two fingers of each hand, making clamp movements in a transverse form to detach them, and sliding the fingers along its length.*

The skull, central chain confluence

The fascia of both the neck and sternocleidomastoids are joined in a band of resistant fibrous tissue that connects the frontal and occipital areas of the scalp to the upper part of the skull.

Neck: evaluation plan

In cases where the athlete's posture is head forward, the loop created by both sternocleidomastoids in the form of a scarf along the posterior part of the skull (between the parietal bones and the occipital scale) means that the previous chain restriction will be perceptible to touch.

It is important to note that the anterior and posterior chains meet at the back of the skull.
A detailed examination of the scalp, from the occipital to the frontal, should be carried out in order to locate small, spindle-shaped fascicles that are extraordinarily tense and painful to the touch. They are treated by progressive fingertip pressure. Pressure is applied in the center of the nodule while the athlete confirms sensitivity for about one minute until the nodule loses the tension.

Any technique that relaxes the tension of these structures through friction, pumping, rubbing, traction of the scalp, etc., will ensure correct functioning of cranial dynamism. A critical region on which to perform massage techniques is the area of the asterion, located at point of the skull in which the occipital, temporal and parietal bones converge.

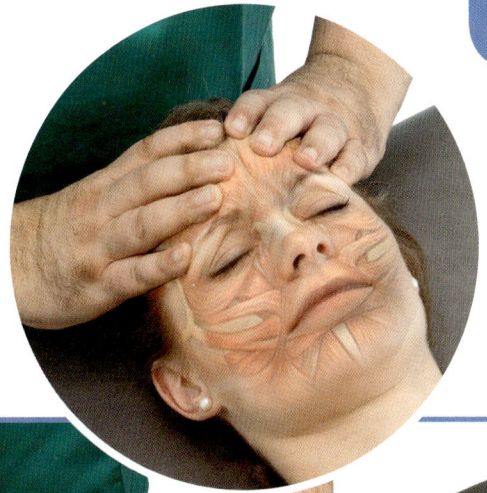

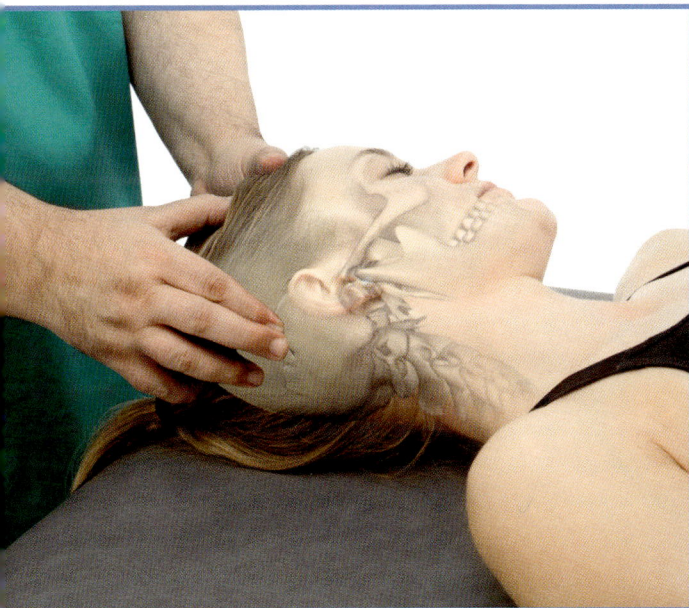

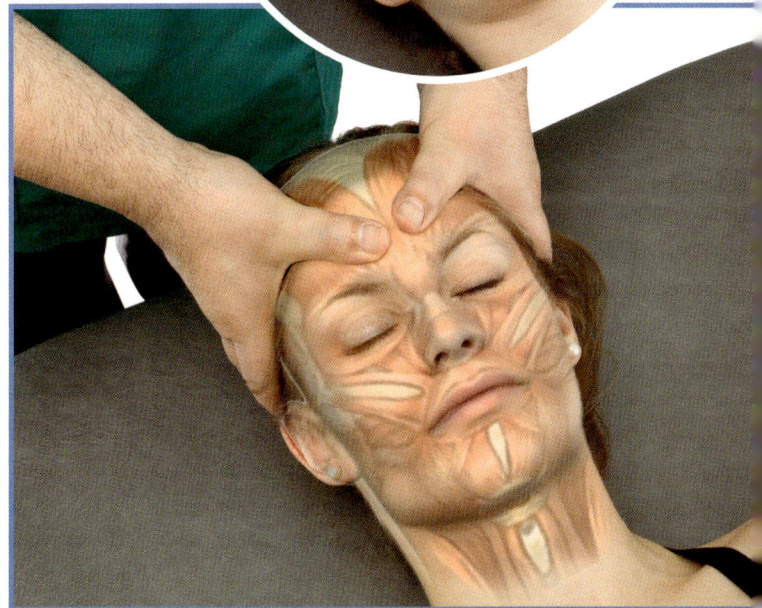

Massaging the skull area. *This area should be carefully treated, as pressure is applied to the skull at sensitive points particularly in the asterion, where the parietal, occipital and mastoid areas of the temporal bone converge. Different areas of the skull can be worked. Press the center of these points with the fingertips while questioning the athlete about his sensitivity.*

Massaging the skull and scalp area. *This area should be treated delicately. Place both hands with your fingers apart and apply a pumping motion, as if squeezing, carefully but firmly. This massage is very useful for treating headaches.*

Upper body III: torso, arms and hands

General Techniques
◆ Opposing slide pressure
◆ Pressure
◆ Longitudinal friction

Devices
◆ Knuckles
◆ Fingertips

Tissue of the torso
It is important to evaluate a possible shortening of the pectoralis major. During observation the therapist may note the characteristic posture of hunched shoulders.

Torso: evaluation plan
The patient lies on his back several centimeters from the top edge of the couch and the therapist asks him to raise his arms above his head and let them hang loosely. If the muscles are not shortened the arms should be able to hang until they reach the horizontal. If the back of the arm cannot contact and come to rest on the plane of the couch, we will have established that the pectoralis major is shortened and tense.

Another option, with the patient standing, is to hold his wrists and ask him to drop his body forward. If travel is limited and the athlete perceives a significant stretching sensation in addition to limited movement, the shortening of myofascial tissue is indicated.

Obviously, examining the state of tension of the anterior axillary border formed by the pectoral muscle by touch will conclude the evaluation of the state of the torso tissue.

Precautions

◆ This is an area very rich in blood vessels, lymphs and nerves which are located in a superficial plane (such as in the brachial plexus directly on the clavicle and the cephalic vein that passes through the deltopectoral groove, etc.).

◆ Consider that the myofascial component of the pectoralis major area originates at the latero-external border of the breasts and axillae and that the mammary glands are located on the superficial fascia of the pectoralis major. For this reason massage must be carried out slowly and delicately.

◆ A thin gauze should be used to protect the fingers and prevent nails from damaging the athlete's skin

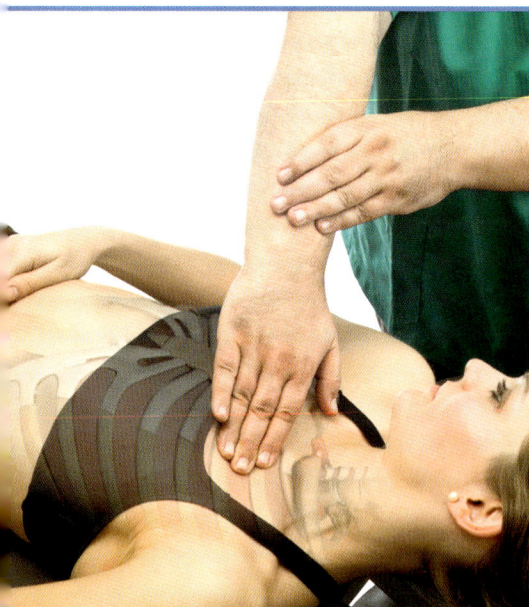

Massaging the pectoralis major and its fascia. *Sliding pressure is applied with the fingers and the palm of the caudal hand over the muscle. If the area is very tense, knuckles can be used.*

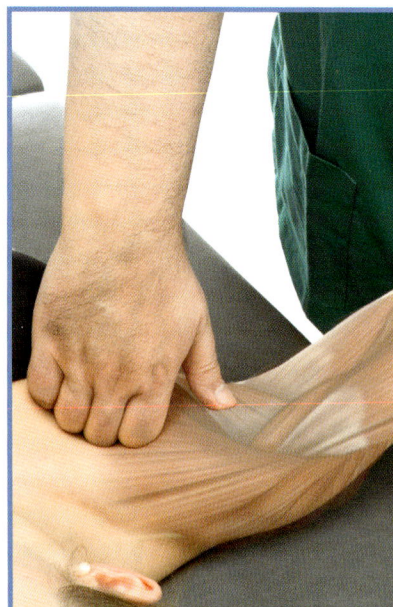

Massaging the pectoralis major. *It is necessary to make space. To do this, the cranial hand separates and holds the athlete's arm while the other exerts a sliding pressure with the knuckles on the taut muscle fibers.*

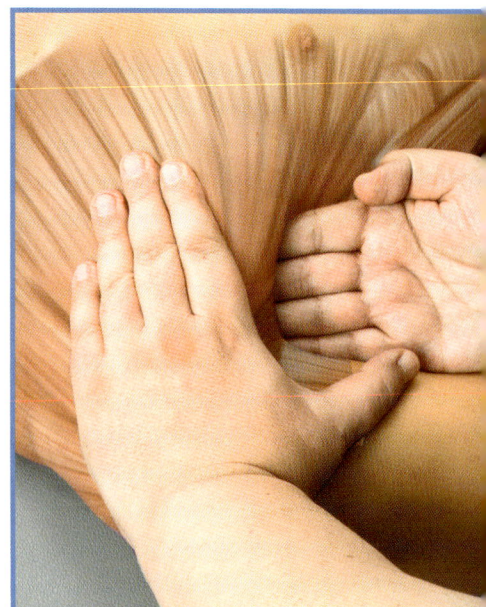

Massaging the posterior fibers of the pectoralis major. *Approaching this (supine) area can cause discomfort, so it is advisable to proceed slowly.*

Arms and hand control

The upper extremities link us to the outside world thanks to their great capacity for expression and the versatility of our hands. In particular the ability to combine different degrees of strength, flexibility, dexterity and control with more or less refinement characterize most sporting actions that involve the arm and hand.

The chain arrangement provides an anatomical and functional continuity that also extends to the upper extremities. The arm attaches to the trunk through connective tissue, which connects to the skull, cervical and neck as well as the chest via the clavicle and scapula (shoulder girdle), extending down to the abdomen.

The myofascial chain in the arm expands to the aponeurosis in the form of flattened tendons between the arm and the forearm until it ends in the palmar fascia. The release of tense myofascial tissue in the arm area facilitates movements at the shoulder girdle, as well as better control and management of the hand.

Precautions

◆ Care should be taken with the compression exerted on the anterior side of the forearm, especially in the area of the sulcus of the ulnar nerve, at the elbow and at the wrist, where the median nerve runs in the carpal tunnel.

◆ Avoid compressing the bony protrusion of the hooked bone in the palm of the hand, as that is where the ulnar nerve is located.

◆ During the application of arm massage it is necessary to take into account that in the anterior face (approximately to the middle of the humerus) and in its medial section (between the biceps and the triceps) great care must be taken when using compression on the structures of the neurovascular package, where the median and ulnar nerves are found, as well as the artery and the brachial veins.

Massaging the arm. *Transverse friction maneuvers are used over the biceps brachii, from the septum or intramuscular septum forward.*

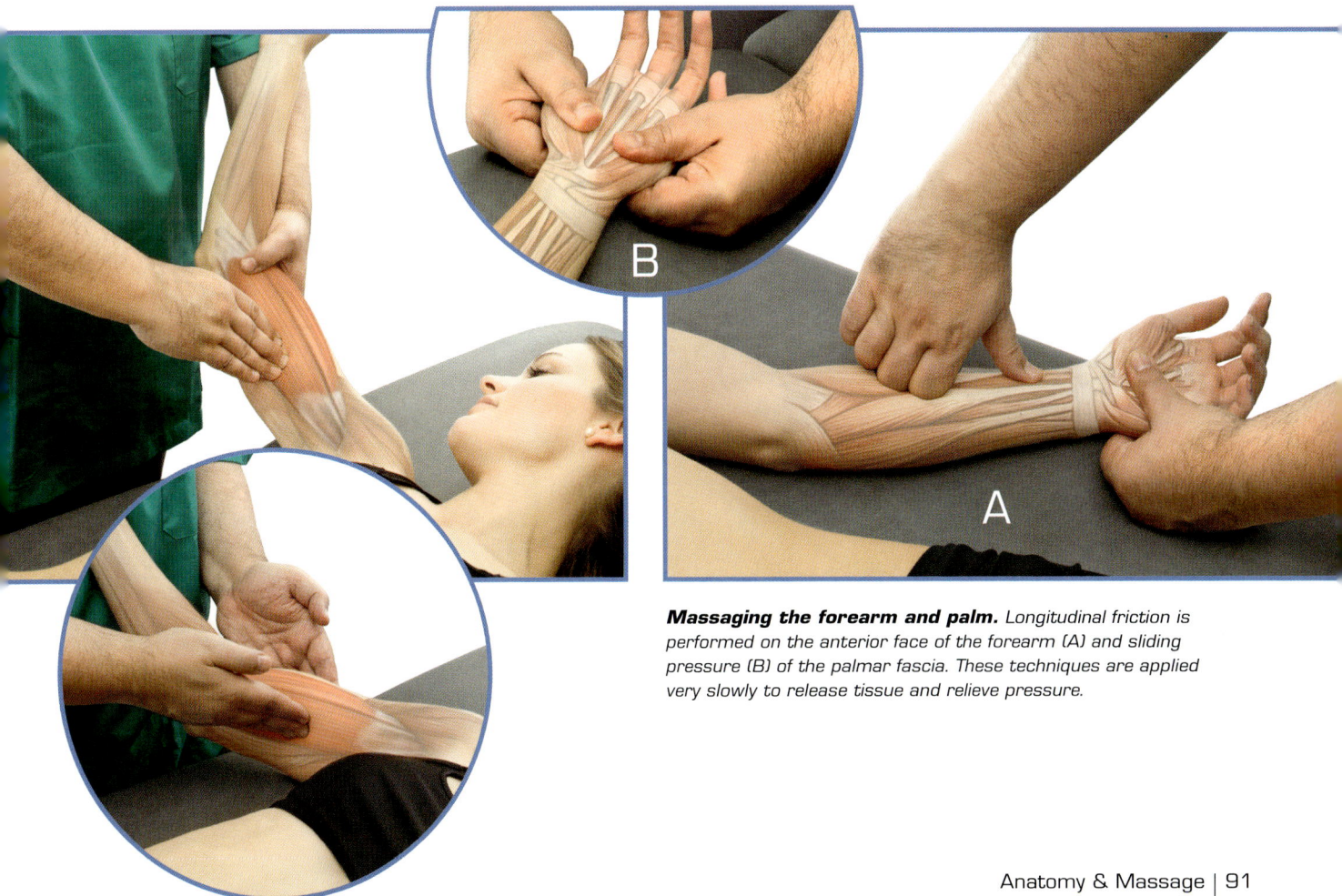

Massaging the forearm and palm. *Longitudinal friction is performed on the anterior face of the forearm (A) and sliding pressure (B) of the palmar fascia. These techniques are applied very slowly to release tissue and relieve pressure.*

Posterior myofascial chain

A chain for extension

The posterior chain functions as an extension. It straightens the body, avoiding postures in flexion. Its combined action prevents the person from falling abruptly forwards (falling headfirst).

It is responsible for linking tissues across the entire posterior surface when the knees are in extension. Conversely when the knees are in flexion, two blocks can be considered: the first from the sole of the toes to the knees, and the second from the knees to the frontal region.

The stretching sensation in the chain is different depending on whether the body flexes forward with the knees in extension or in flexion.

Functions
of the posterior chain

The posterior and anterior chains work together to balance displacements in the sagittal plane.

There are two posterior chains, one on the right and one on the left, which must be balanced, in turn, before other chains can act. The posterior chain presents a predominance of tonic and postural motor with slow contraction fibers and provides a more static action, ideal for controlling posture.

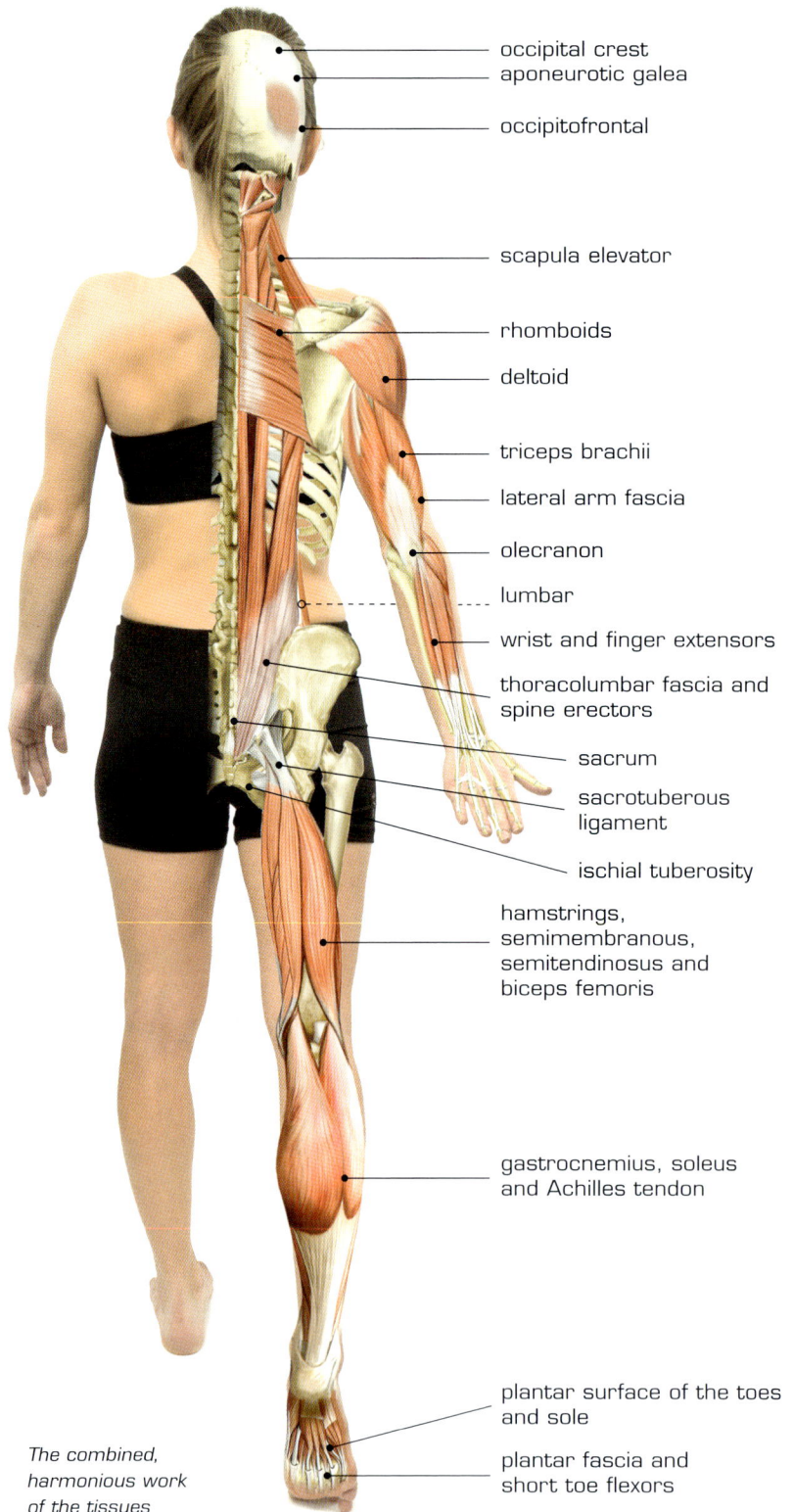

The muscle segments of the posterior chain are rich in red fibers, highly resistant to fatigue, and necessary to maintain a standing position throughout the day against the force of gravity. Hard fascial bands abound, such as the calcaneus or Achilles tendons, the tendons of the popliteal region, the sacrotuberous ligament and the complex structure of the thoracolumbar fascia.

- occipital crest
- aponeurotic galea
- occipitofrontal
- scapula elevator
- rhomboids
- deltoid
- triceps brachii
- lateral arm fascia
- olecranon
- lumbar
- wrist and finger extensors
- thoracolumbar fascia and spine erectors
- sacrum
- sacrotuberous ligament
- ischial tuberosity
- hamstrings, semimembranous, semitendinosus and biceps femoris
- gastrocnemius, soleus and Achilles tendon
- plantar surface of the toes and sole
- plantar fascia and short toe flexors

The combined, harmonious work of the tissues of the posterior myofascial chain facilitates movements of extension and body straightening, avoiding flexed and shrinking postures.

The athlete and shortening of the posterior chain

Shortening of the posterior chain involves hyperextension postures, for example in neck or spinal cord hyperlordosis. In athletes, knee hyperextension may be considered secondary to the aforementioned problems.

The use of techniques that correct the hypertonic state in the indicated curves (hyperlordosis of the neck and loins) would solve the problem. This problem is frequently observed in swimmers, whose sport's discipline demands constant work of the posterior chain, especially in these areas.

Detection of compensations

The first step when assessing the common shortening of the posterior chain lies in analyzing the behavior of the athlete when he is asked to flex forward and try to reach his toes with his hands. It is necessary to observe if it is able to perform this action. The contour of the back should be evaluated on both sides and the position and distance of the fingers relative to the ground measured.

In situations in which the individual is unable to achieve a proposed goal, we can conclude that the posterior chain has some type of shortening. It only remains for us to establish which segments of the chain present this shortening to a greater degree.

Precautions

◆ Reorganize limitations. Remember that it is a priority to stabilize the anterior and posterior chains before evaluating other chains.

Strategy for massage

To perform the treatment along paravertebral structures the position of the different spine segments must be taken into account.

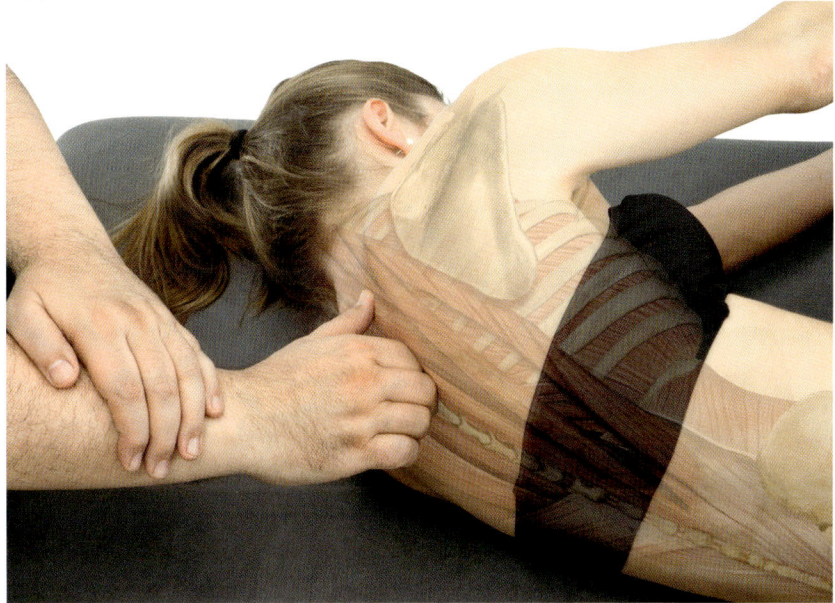

Massaging kyphic areas. *The kyphotic zones (posterior incurvation) make the spinous apophysis more prominent. They can be mobilized by indirect pressure, improving their flexibility and relieving excess tension.*

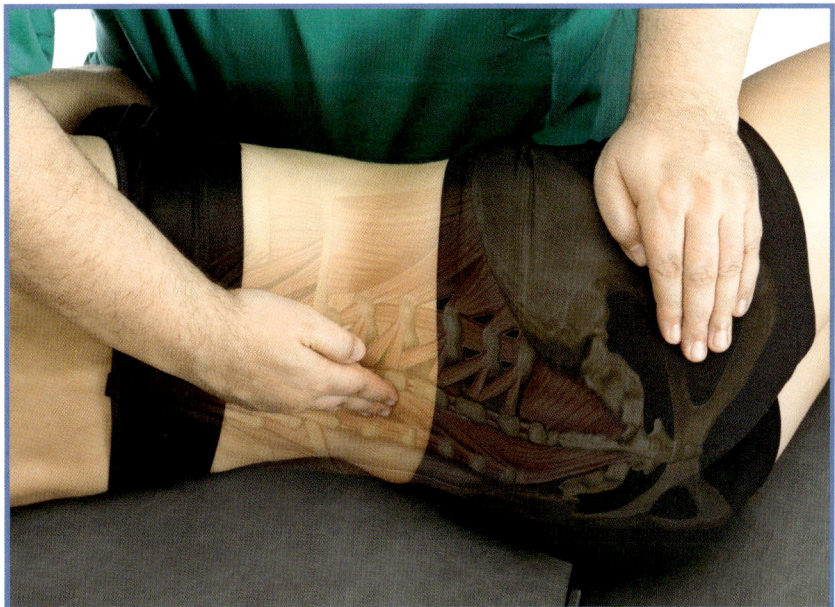

Massaging lordotic areas. *In forward curvatures (lordosis) the goal is the same: to improve movement ability and relieve excessive tension. To achieve this, maneuvers are performed that stretch the taut fabric and provide flexibility in the rigid areas.*

Lower body I: Feet and legs

General Techniques

- ◆ Friction pressure
- ◆ Transverse friction

Thread Tools

- ◆ Knuckles
- ◆ Thumbs
- ◆ Fingertips
- ◆ Elbow

First assess response to mobility

If during the previous flexion test you observe that the athlete's leg loses its vertical relation to the bearing surface of the sole of the foot moving backwards, it means that the leg muscles are involved in the shortening of the posterior chain.

Put another way, the tension exerted by the muscles and the fascia of the sole of the foot and the back of the leg when flexing the body forward make it impossible to keep the foot-and-leg axis at right angles, whereby the angle increases as it moves backwards.

The foot: evaluation plan

The state of the plantar fascia in the foot is evaluated, comparing its internal aspects (from the big toe to the heel) and external (from the fifth toe and the base of the fifth metatarsal to the heel).

In cases where the distance from the big toe to the heel decreases and cupping occurs in the inner arch, there may be a shortening of the medial border of the foot, which should be treated to adjust the tension.

If, on the contrary, the decrease appears in the distance between the fifth toe and the heel with internal arch collapse and slight hollowing of the external arch, the shortening of the plantar fascia is considered to be more external.

Precautions

- ◆ It is important to use very little lubricant so you can work with the fabric with drag and elongation maneuvers, avoiding a slipping sensation.

During a running stride, the impacts and pulls of the foot hitting the floor create tension in the tissue covering the foot vault, called the plantar fascia, and can cause it to inflame. This scenario, very common in runners, is called plantar fasciitis. In advanced cases an ossification (bone formation) may appear at the base of the heel bone, known as the "calcaneus spur."

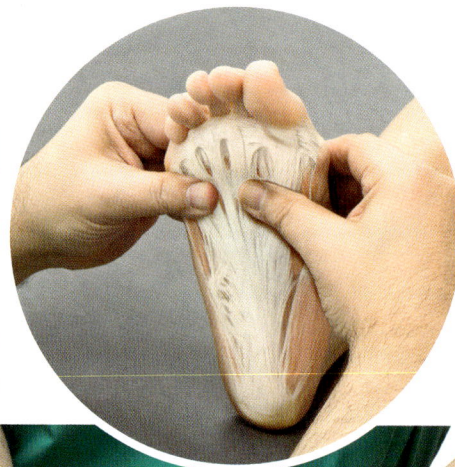

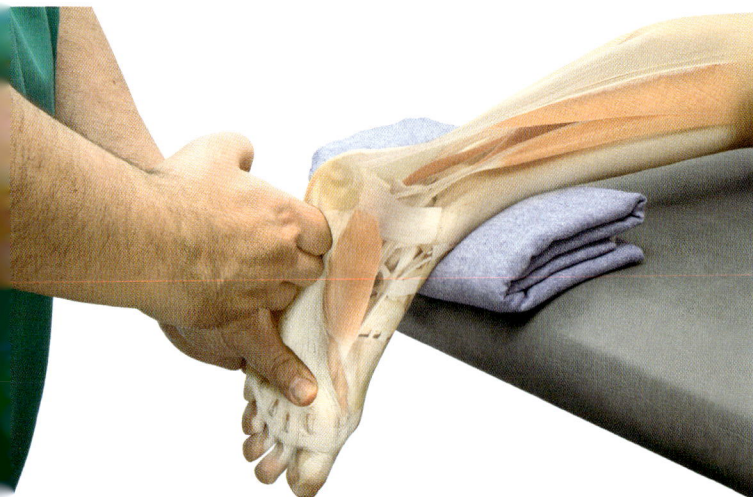

Sole of the foot. *Use friction with knuckles or second phalanges. The elbow may also be used transversely or longitudinally and from the calcaneus to the toes to flex the tissues. The foot is held with the other hand and flexion and extension movements are used to increase mobility.*

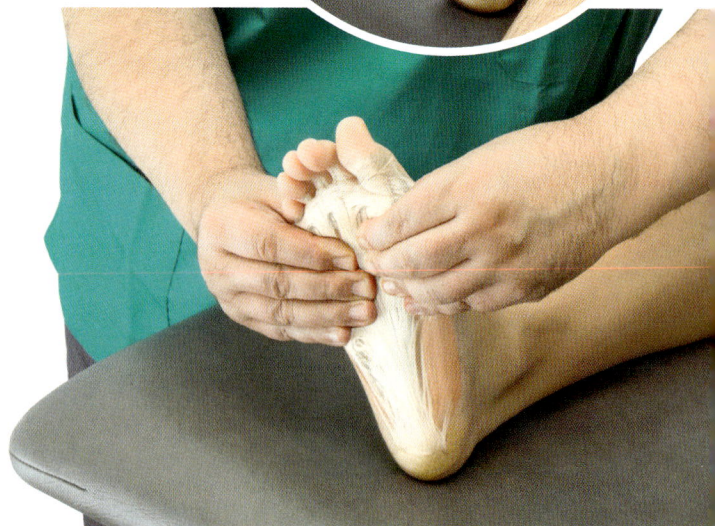

Sole and arch of the foot. *Work on the sole of the foot and its transverse arch. The foot is held with both hands and open movements are applied to the sole by means of a clamp with the medial joints of the fingers*

Evaluation Tactics

The Achilles tendon is a powerful and dense structure and is considered the extension and posterior focus (behind the ankle) of the plantar fascia. It is the tendon of the gastrocnemius and soleus muscles.

Attentive and comparative palpation of both tendons (right and left) allows us to establish their consistency and sensitivity. This determines the degree of stress or force to which both tendons are subjected.

The gastrocnemius crosses the joints of the ankle and the knee, acting functionally in both, so it is necessary to take this aspect into account when assessing its degree of tension.

Therefore during the elasticity test the ankle is flexed dorsally while the knee is kept in extension, since its motor actions are opposite. The ankle should be able to perform a 90° angle bend range. If discomfort or pain develops before the ankle reaches that position, we conclude that the muscles are shortened.

Precautions

◆ It is important to relax the ligaments of the back of the ankle, rubbing slowly and deeply from the malleolus toward the calcaneus. This massage technique is performed to help athletes with previous pelvic (pelvic anteversion) movements.

The soleus crosses only the ankle joint and is explored by placing the knee in flexion and assessing only dorsiflexibility of the ankle. If the latter does not reach 90° or there is pain or discomfort in the posterior aspect of the leg, the soleus is contracted.

Anatomical precision

The soleus is a deeper muscle than the gastrocnemius. It is a tonic and postural muscle. Its shortening should be taken into consideration in individuals who have recurvatum of the knees (knees move backwards) while standing. It is also one of the muscles that make it impossible to squat with feet flat on the ground.

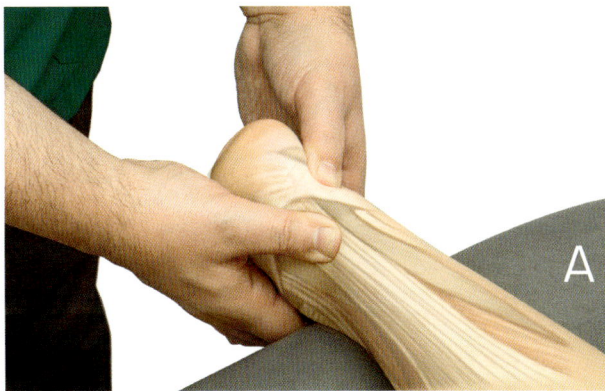

Achilles tendon. Massage with longitudinal or transverse friction, with the thumbs (A) or the knuckles (B). If the athlete can tolerate it he may be asked to flex and extend while the massage is applied.

Gastrocnemius zone (popularly known as twin muscle). The caudal hand fixes the leg and, with the fingertips of the cranial hand, the transversal friction of the tense muscles is performed (A).
● Transverse friction (B) may also be applied with both thumbs at once. It is very useful to perform this massage before stretching exercises in case of injury.
● The entire tendon can be rubbed with the elbow, from the middle of the leg to the heel.

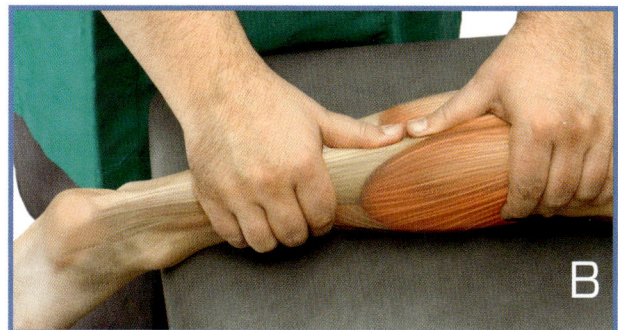

Lower body II: thighs and pelvis

General Techniques
◆ Oblique pressure friction
◆ Friction plus traction

Devices
◆ Forearm
◆ Thumbs
◆ Knuckles

Areas of relationship and translation

The pre-tensioning ability of the posterior chain musculotendinous elements during an event or the race determines the propulsive effect on thigh, leg and foot.

We can observe that athletes with a lack of knee extension are forced to increase movement of the ankle, the hip, the pelvis or beyond, in order to compensate.

In addition this leads to an increase in energy consumption and a greater extent of tissue overload, which fatigues the athlete.

Displacement and the posterior thigh group

If the athlete shows a shortening in the hip flexors (see Thomas test in the previous chapter), the hamstring test is performed with the unexamined leg in flexion and the foot resting on the couch or floor. This allows the pelvis to assume a neutral position. If there is no shortening of the hip flexors, the unexamined leg will remain stretched.

In both cases the knee of the examined leg is kept in extension and raised until the movement limit barrier is perceived. If the limb cannot be lifted up to 80° (reaching near vertical), there will be some degree of shortening of the hamstring muscle group.

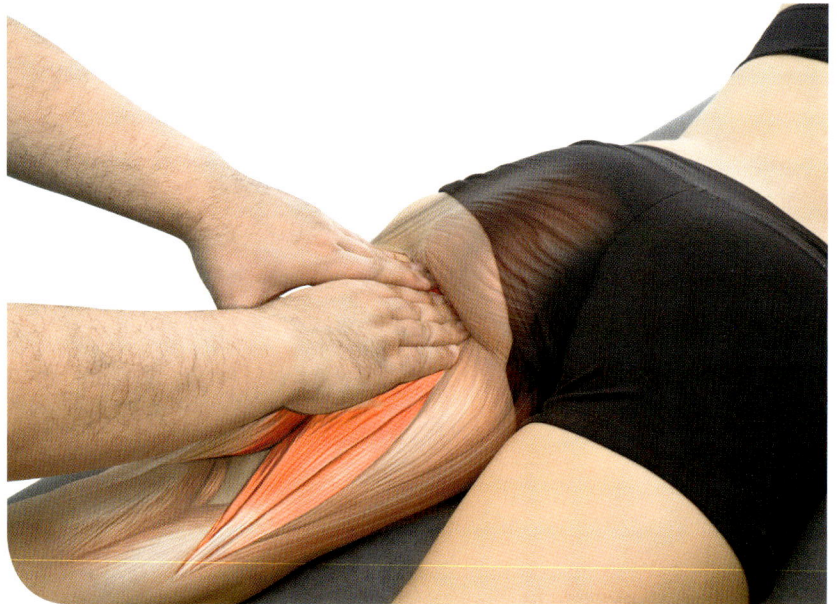

Ischium area. It is important to work the ischium area with sustained pressure against the ischium, the area of insertion of the posterior musculature of the thigh. It can be rubbed with the fingertips of both hands longitudinally, or with the index finger reinforced with the third finger transversely.

Hamstring area. Athlete in supine position, hip flexed and the leg over the shoulder of the masseur. The posterior musculature of the thigh is worked with the forearm under controlled stretching tension, using friction with pressure, on both the outer (biceps femoris) and the inner (semimembranous and semitendinosus) side of the thigh.

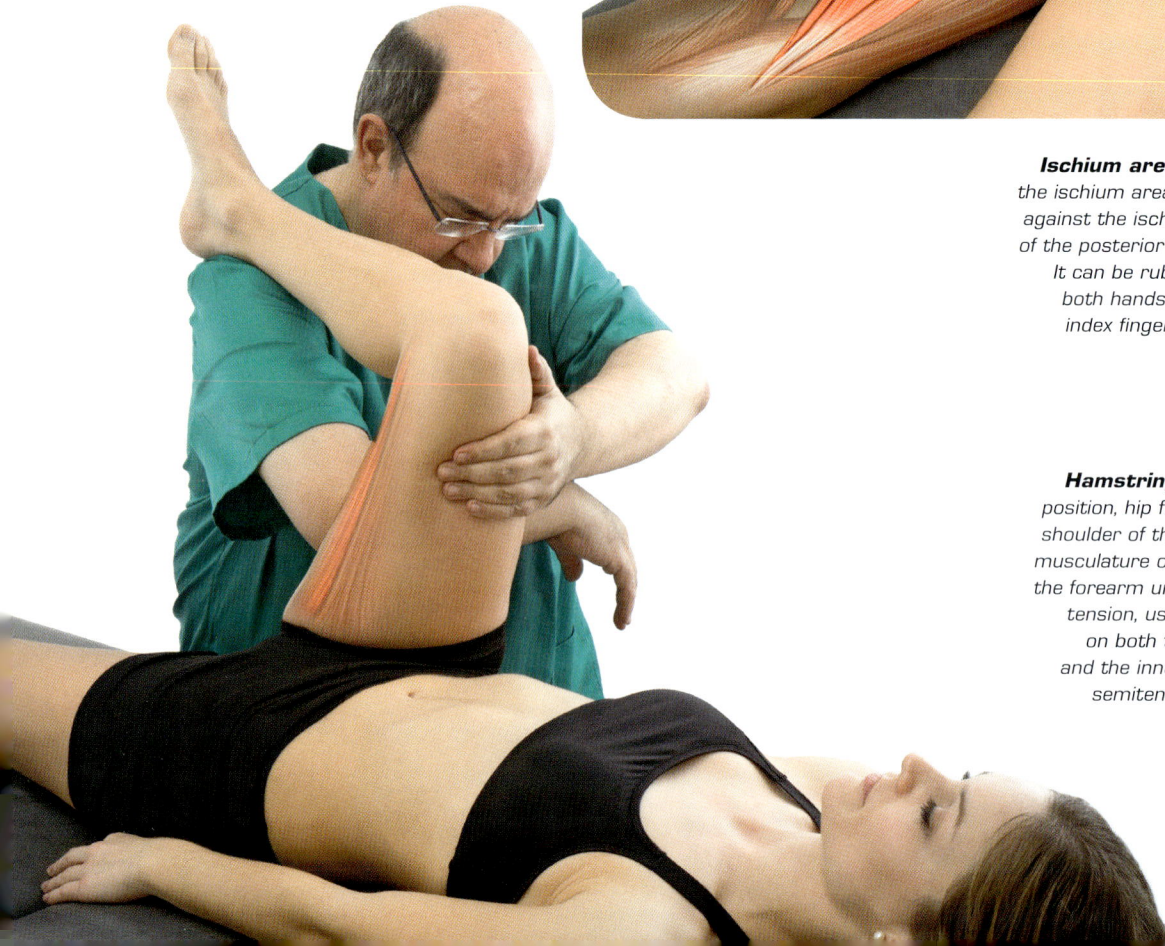

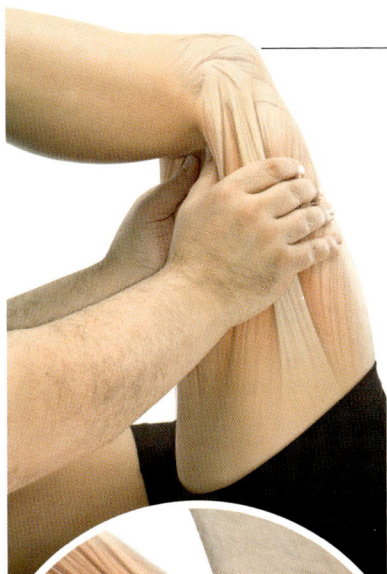

Popliteal area. If tension occurs in the area of the popliteal cavity during exploration, friction maneuvers can be made in traction (i.e., outwardly with both thumbs in order to relax the area).

Thigh-pelvis relation

There is a clear relationship of fascial continuity between the hamstrings and the sacrotuberous ligament. In fact, it is impossible to anatomically separate this ligament from the structures with which it is related: above are the sacral fascia and erector muscles of the spine and below with the ischiofemoral group.

The deeper part of this ligament is more boney and serves a support. It is essential for the maintenance of standing posture and pelvic integrity. Conversely the superficial fibers' function is to transmit movement, activity and tension.

The instability between the right and left ligaments leads to torsions in the sacrum, which will move with the rest of the spine.

Evaluation tactics

It is easy to feel the sacrotuberous ligament moving diagonally from the ischial tuberosity to the lower sacral angle.

Palpation by comparison to the tension present in both ligaments, right and left, allows the therapist to establish which one presents a greater degree of stiffness. The tighter ligament requires more attention during massage in order to adjust the tensors that are subject to the sacrum.

It is important to note the insertion of the gluteus maximus, which is also inserted into the outer sacral border.

In cases where the athlete has pelvic anteversion, the ligament must be moved downwards by downward pressure. If the pelvis is in retroversion, this process should be reversed.

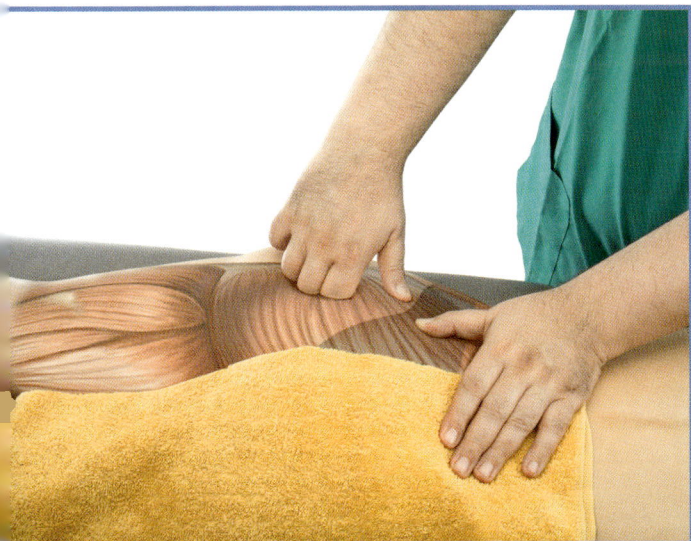

Sacrotuberous ligament zone. Use friction with pressure from the knuckles, working areas of tension in the sacrotuberous ligament, coccyx and sacrum areas. This creates space between the skin of the underlying planes and mobilizes the tissue.

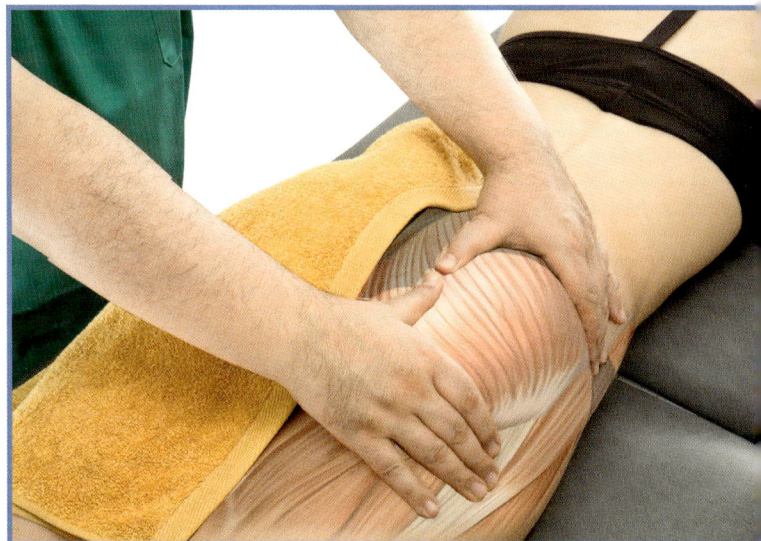

Sacral zone. Massage with the thumbs of both hands extended over the medial border of the strong gluteal muscle tissue. The sacral border must be observed.

Upper body I: lumbar column and thorax

3

General Techniques

◆ Oblique pressure
◆ Friction and drag

Devices

◆ Palm of the hand
◆ Forearm

Connection area (posteroanterior and superoinferior)

A historic pelvic or thoracic backward tilt can both lead to a greater lumbar extension. This may lead to a decrease in hip extension ability if the hip flexors are shorter than normal.

In this situation, the masseuse may believe that the extension available is normal (e.g., a step when walking, as the whole limb extends from the lumbar spine and not from the hip). This can be objectified while inspecting the later oscillation step phase, making the abdomen more prominent.

Evaluation tactics 1

If the hip flexors are at a normal length and the pelvic tilt and consequent lumbar extension persist, a step in the extension phase of the limb will be relatively greater. Conversely the previous step phase (bending) will appear to be at a lower range. The step will be longer during extension than during flexion.

Additionally the pelvis performs rotational movements during the step and other movements, which make it possible to lengthen both the anterior and posterior steps during the gait, for example. This rotation will move upwards toward the spinal column. During this rotation each spinal segment presents variations according to its capacity of movement in each plane.

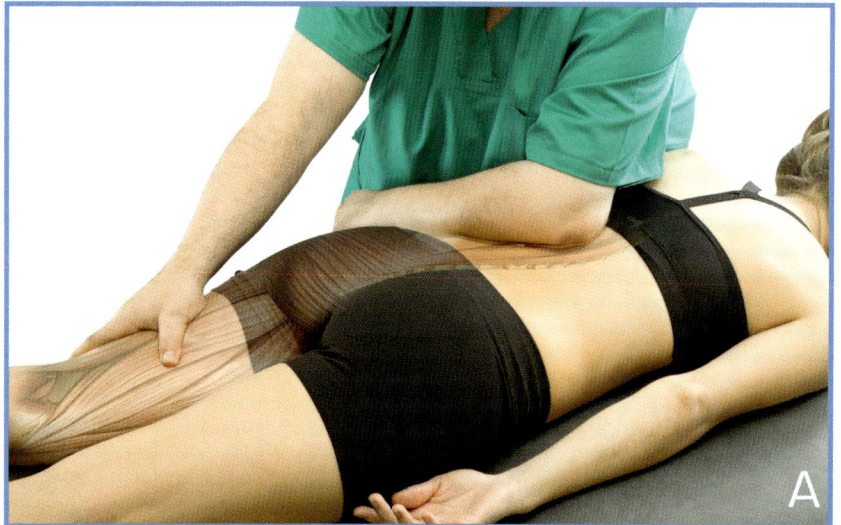

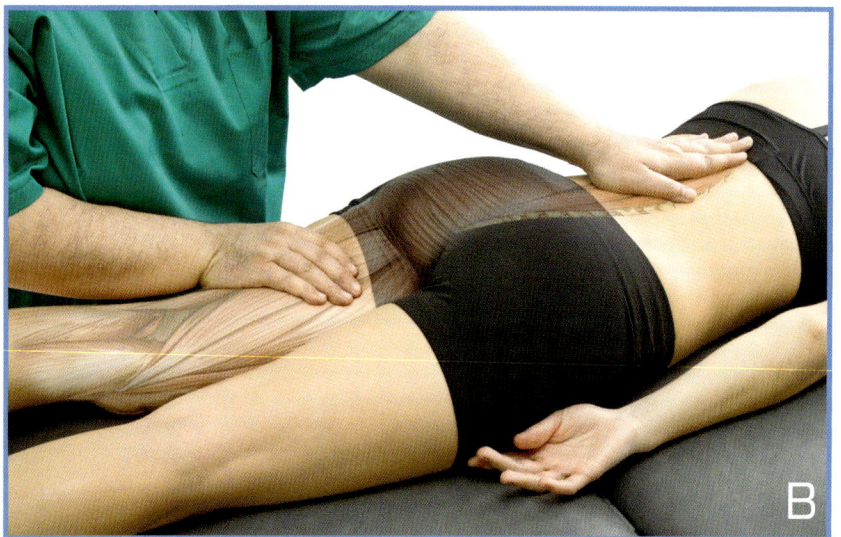

Lumbar paravertebral region. *Massage with friction and drag with the palm or forearm in the paravertebral region. The lower extremity on the same side of the body can be mobilized by applying traction toward the cranium (B) relaxing the muscles which it works, or friction is applied with the forearm in the caudal (downward) direction (A) to apply tension.*

Precautions

◆ The coccyx alignment may be displaced by trauma. It may be useful to perform massages on fibrous areas on one side or the other, always with the permission of the athlete.

Massage on the sacrum brings deeply profound relaxing effects, since the fascial tissue covers the region and maintains it in close relation and influence with the iliac and lumbar spine.

Precautions

◆ Since the lumbar square is connected to the diaphragm, it may also contribute to respiratory dysfunctions, particularly during exhalation. Excessive tension in the lumbar area can cause pain in the lower ribs and along the iliac crest.

Evaluation tactics 2

The athlete rests on her ischium, back straightened and feet well supported. She is asked to progressively flex her entire spine, starting from the head, so that all the vertebrae are flexing one by one. The therapist observes these movements from behind. It is common to find rigid areas where several vertebrae move in a block, as if they were inseparable from each other.

Help offered by the masseur is very important in such cases, as he can press on these rigid portions as the athlete draws his attention to them, helping her to recover flexion and movement capacity.

Work is completed by massaging the intervertebral canal, between the spinous apophysis and the transverse.

The lumbar spine is not designed to perform rotations. Its morphology allows it to primarily perform flex-extension movements. Therefore, pelvic rotations should send their rotational capacity to higher areas, especially thoracic.

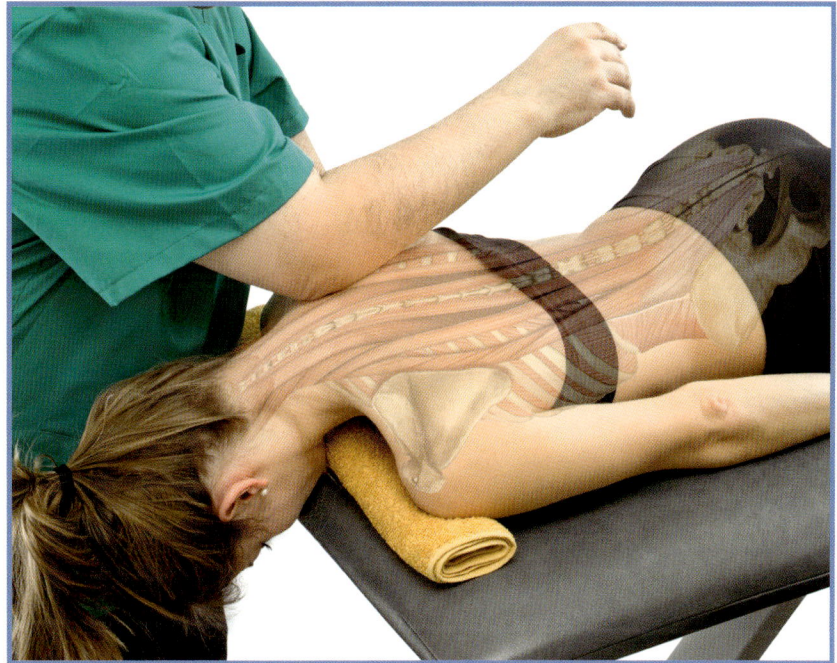

Paravertebral thoracic region.
In order to perform a massage along the paravertebrals, positioning of different segments of the spinal column must be taken into account. In contrast to kyphotic areas (posterior incurvature), the musculature is displaced outward from the midline. Application of pressure will result in the tissue moving toward and centering on the spinous apophysis. The vertebrae protrude due tension in the tissue, and this is why the massage will cause them to retract.

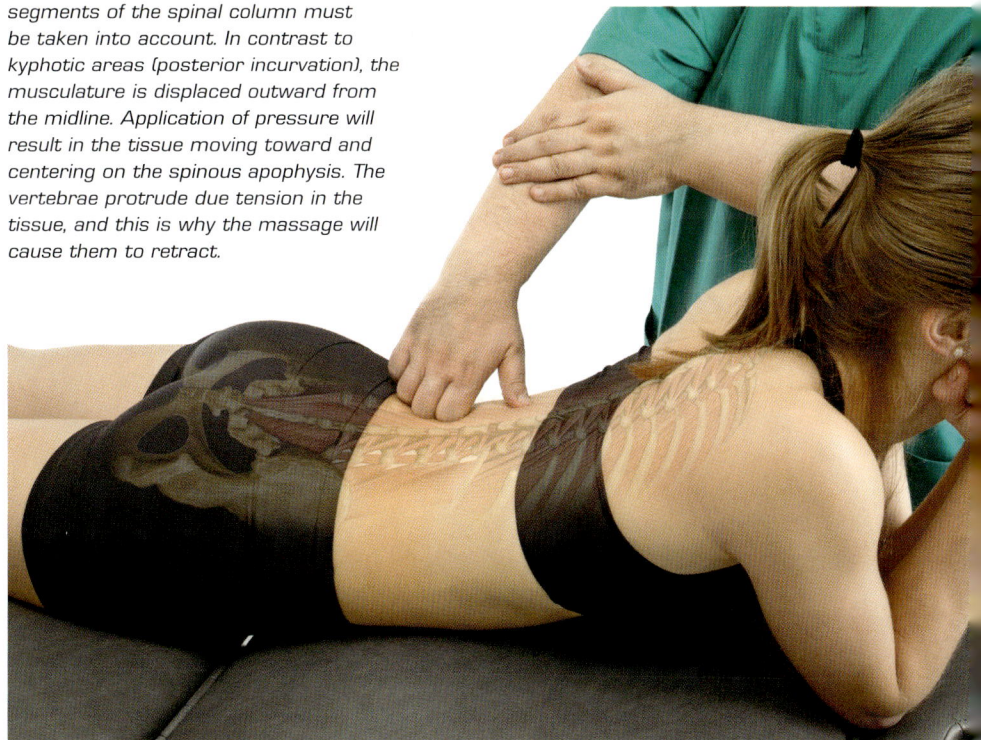

Paravertebral dorsum-lumbar region. *If instead vertebrae are too deep (anterior curvature), this maneuver is performed from the center to the periphery, and from superficial to deep. The aim is to adjust tissue tension and make room for the vertebra to readjust (posterior).*

Upper body II: head and neck

General Techniques

- ◆ Friction and drag
- ◆ Friction
- ◆ Maintained static pressure

Devices

- ◆ Thumb and fingertip

The oculomotor link

Head movements in relation to the body and environment influence tone throughout the entire musculature and provide dynamic and facilitating energy for movement. In addition to acting on the muscles of the trunk and the upper extremities, cervical tonic reflexes also affect the eyes.

Various curvatures of the spine (kyphosis, lordosis, scoliosis) aim to place the head and neck in their normal posture, in order to keep eyesight horizontal.

The body's response to stress manifests itself physically by creating defensive body patterns. Response patterns (shoulder elevation, head extension, knee flexion, etc.) are similar regardless of whether the trauma is physical or emotional.

Postural indicators

A forward posture or limitation of thoracic spine extension shows that the functionality of the posterior chain is affected. The connection between these chains dictates that we constantly evaluate them, as limitations in movement will generate overloads in their components (tension by shortening), forcing rotation chains to create internal or external compensatory rotation strategies.

There are numerous stabilizing techniques in this situation. These include the correct traction and stretching maneuvers on the neck tissues, which can be combined with the massage.

The suboccipital musculature is very rich in sensory receptors. Its connection with eye movements and its coordination with the rest of the body's dorsal muscles give it a central role in overall coordination. Massage in this deep musculature (the line below the skull) allows the masseuse to release tension in this deep neck tissue, especially in patients who present with a head forward or a habitual hyperextension of the neck.

Tactics 1

In relating the neck to the upper extremities, it is common for the muscular pairing of the rhomboid and scapula levator to show patterns of weakness, first, of shortening and second, tension. With the athlete in supine position and tilting the head and neck to one side, the levator muscle and the trapezoid on the side are being stretched and evaluated. Next push the shoulder down on one side and if a smooth and easy sensation of movement is observed, the muscle is not shortened. But if a sudden limitation of movement without a feeling of elastic return at the end of the movement is noted, the scapula erector is probably contracted.

Cranium. The cranium rests in the hands of the therapist, which must be together. This elevates and applies a soft traction in a cranial, ascending direction.

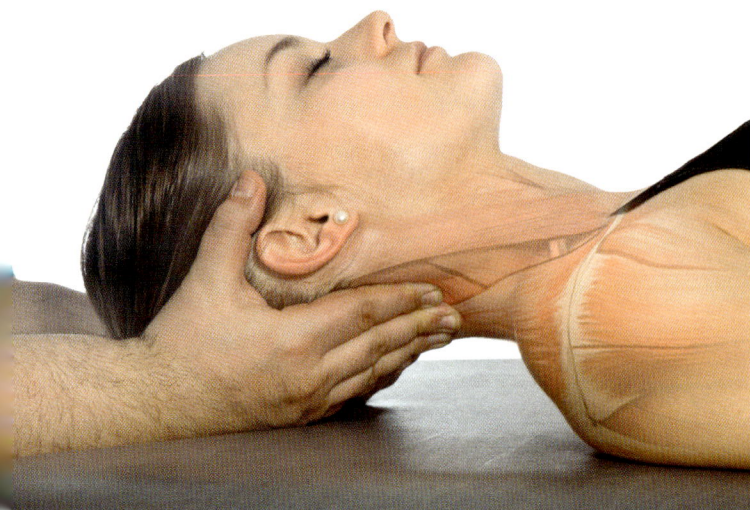

Cranium. Place the hands under the athlete's head to apply the maneuver described above, adjusting excess tension in the suboccipital area.

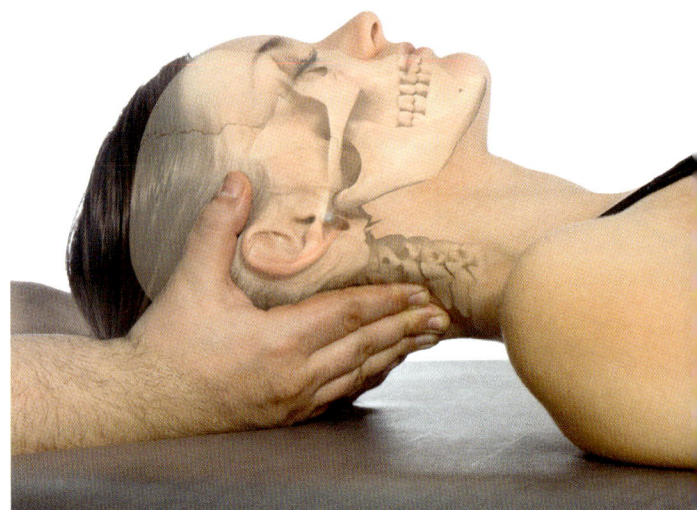

Tactics rating 2

Athlete stretched and prone, in a position similar to that of performing a push-up. Scapular movement is assessed at maximum elevation while the chest is approaching the ground.

If under observation the scapula moves laterally and upwards (toward the head) and assumes a "winged" form (detachment from the inner border of the scapula of the thorax), stabilizing muscles (rhomboids and trapezius medial) may be weak and the upper neck muscles (upper trapezius and scapula lift) shortened.

Precautions

◆ The following maneuvers should not be forgotten during neck massage, as they allow a deep-plane adjustment and release of tension.

Cranium. Massage into the suboccipital deep musculature (semispinosus and splenius of the head), located below the line of the skull. Individuals with anterior head protrusion or chronic neck hyperextension will benefit greatly from this very relaxing technique.

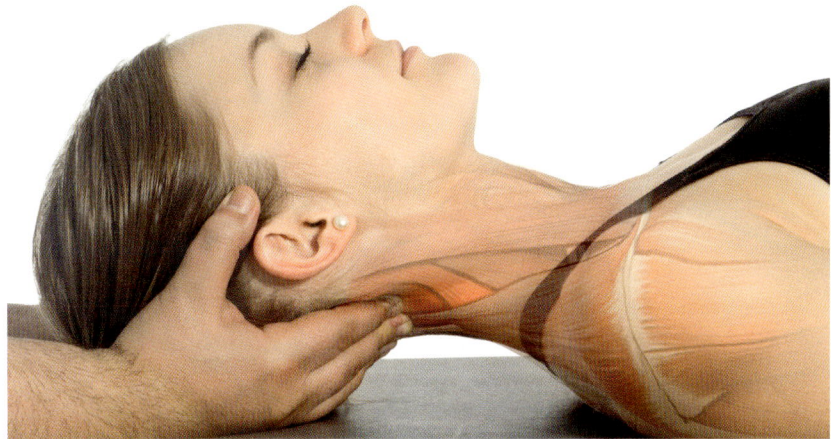

Cranium. Perform friction on the posterior region of the ear (mastoid) and the bony border of the occipital (occipital crest) as well as on the scalp. Search for and ask the patient about painful areas and apply a static and maintained pressure to eliminate them, being careful not to provoke a painful flight response.

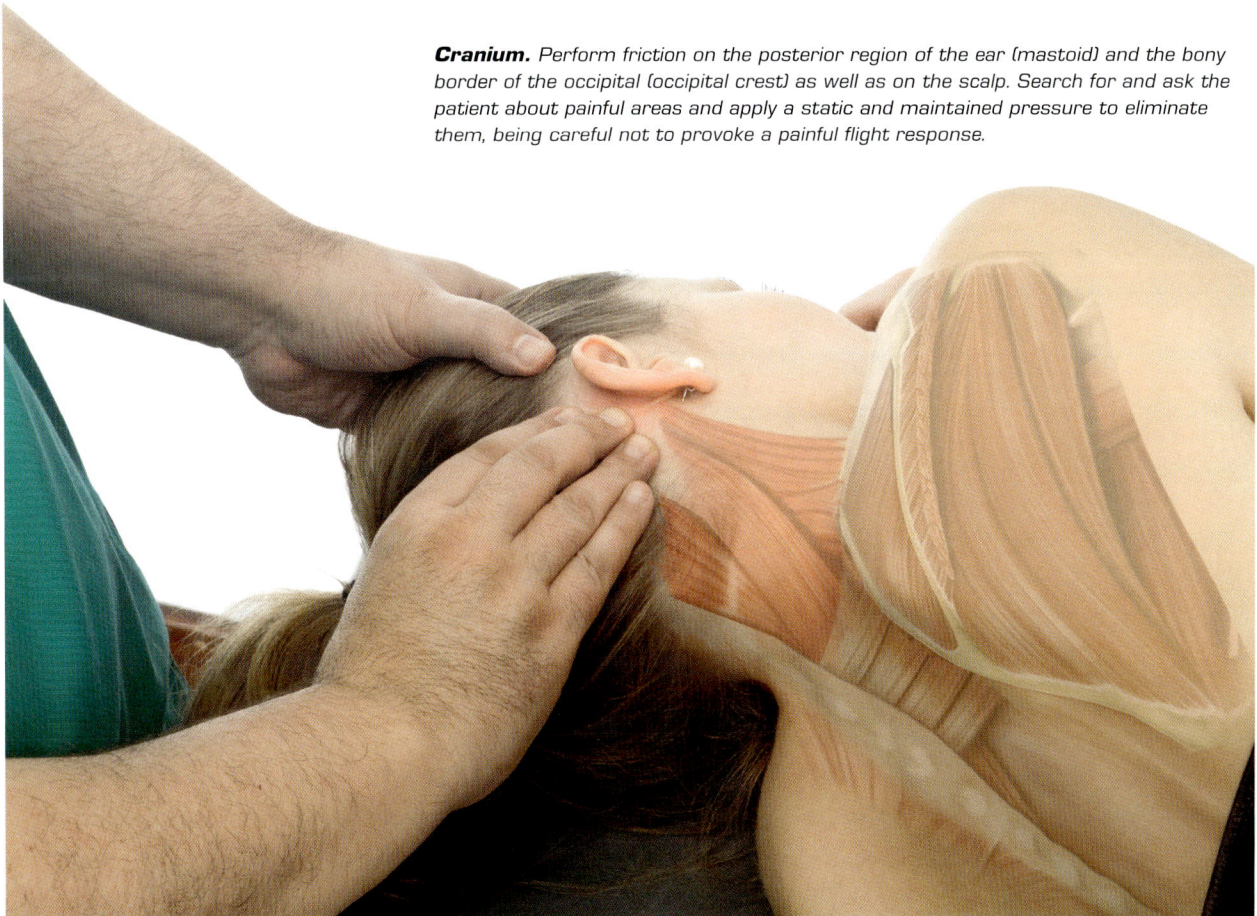

MYOFASCIAL CHAINS

3

Upper body III: neck, shoulders and forearm

General Techniques
◆ Deep friction
◆ Friction

Devices
◆ Fingers (first and second phalanges)
◆ Knuckles

This type of injury occurs after sporting activities, as suggested by the names "tennis elbow," "golfer's elbow," "swimmer's shoulder," etc. But they may also appear after performing everyday activities at work: overuse due to repeated actions in the upper extremities, such as polishing, scrubbing, pruning, using a screwdriver, etc.

Related structures
The anatomical and functional unit between the neck, shoulder and upper extremities should be kept in mind whenever any of these areas are evaluated. The fascia of the neck continues along the fascia of the arms and is often overloaded. Most of the structures that resist the upper extremities against gravity and the loads to which they are subjected depend on the vertebral levels C5-C6 and C7. Correct shoulder stability also depends on good balance in the trunk and head.

Evaluation tactics
Overall mechanics of the myofascial system should be evaluated even if we are faced with localized alterations such as tennis elbow, shoulder tendinitis, carpal tunnel affectation, etc. These injuries are frequently associated with trauma in nearby areas (whether direct, indirect or through overuse), which trigger other processes through compensatory mechanisms.

Precautions

◆ Athletes with chronic scapula girdle and shoulder problems rarely receive treatment below the elbow. However it should be remembered that all fascias are interconnected and run throughout the extremities. Most of us have suffered falls on our hands and wrists. This should be considered whenever we evaluate any of these areas.

Take automobile accidents resulting in "whiplash" as an example. At the moment of collision the driver tightly grips the steering wheel, which creates a vector of traumatic force that passes through the injured hands toward the shoulders and neck. This is often forgotten during treatment, which focuses only on the neck and shoulder areas.

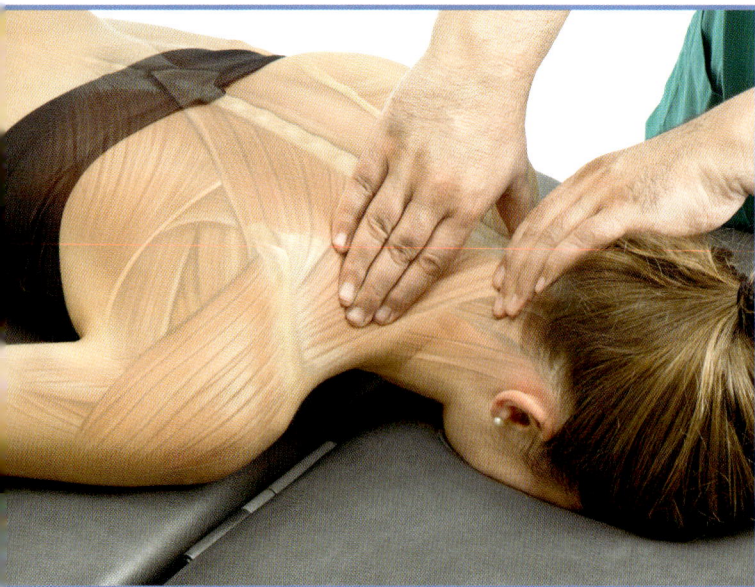

Neck — cervical spine. *Friction of the myofascial deep plane from the posterior base of the neck, directed toward the scapula.*

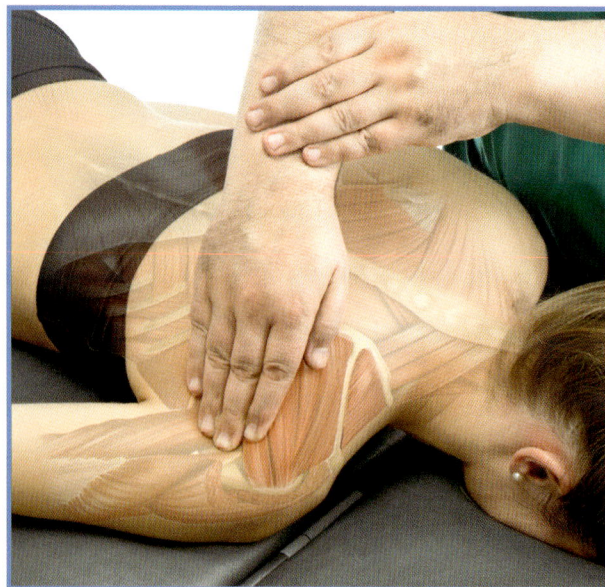

Deltoids. *Friction of the myofascial deep plane in the posterior and lateral deltoids.*

Overview

When treating the following case scenarios, disregard treating of the entire limb, including the wrists and hands as there is a possibility that this could cause treatment-related sequelae, secondary injuries apparently unrelated to the original injury.

Treatment strategy

Massaging the interosseous membrane of the forearm is an excellent starting point in the general treatment of the arm, the scapula girdle, neck, jaw and the trunk. Begin by extending the arm on the couch, forearm in pronation and the hand with the palm open. The masseur exerts pressure, sliding through the knuckles or the elbow, along the forearm between the radius and the ulna.

Applied force follows two vectors: one between the bones (in contact with them) and another looking to laterally expand the tissue from the midline.

The degree of force is applied obliquely by friction and drag in order to avoid using forces that crush vascular or nerve structures.

Combine movements

Maneuvers are applied gradually and toward the wrist, ending at the carpus bones, metacarpus and fingers.

During the massage we can ask the athlete to perform flexion and extension movements of the elbow. The combination of massage (passive) and movement (active) is very effective in treating the interosseous membrane and carpal tunnel disorders.

The skin of the posterior aspect of the upper extremity is thicker and more hairy than skin of the anterior face. This is because it is more exposed to friction, impacts and contact with the environment. Once again, the shape of its tissues responds to its function.

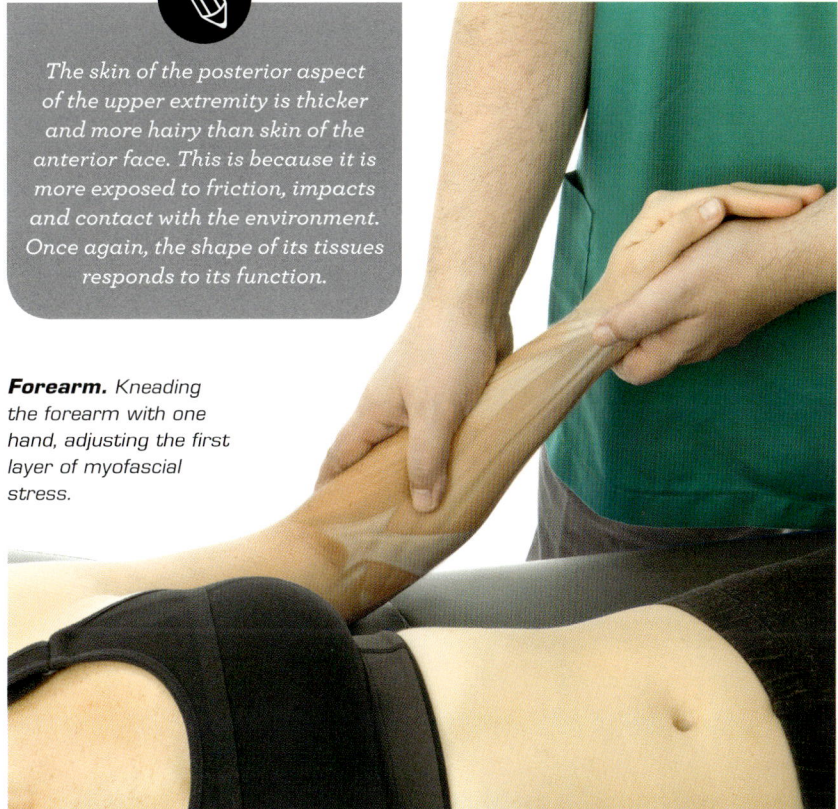

Forearm. Kneading the forearm with one hand, adjusting the first layer of myofascial stress.

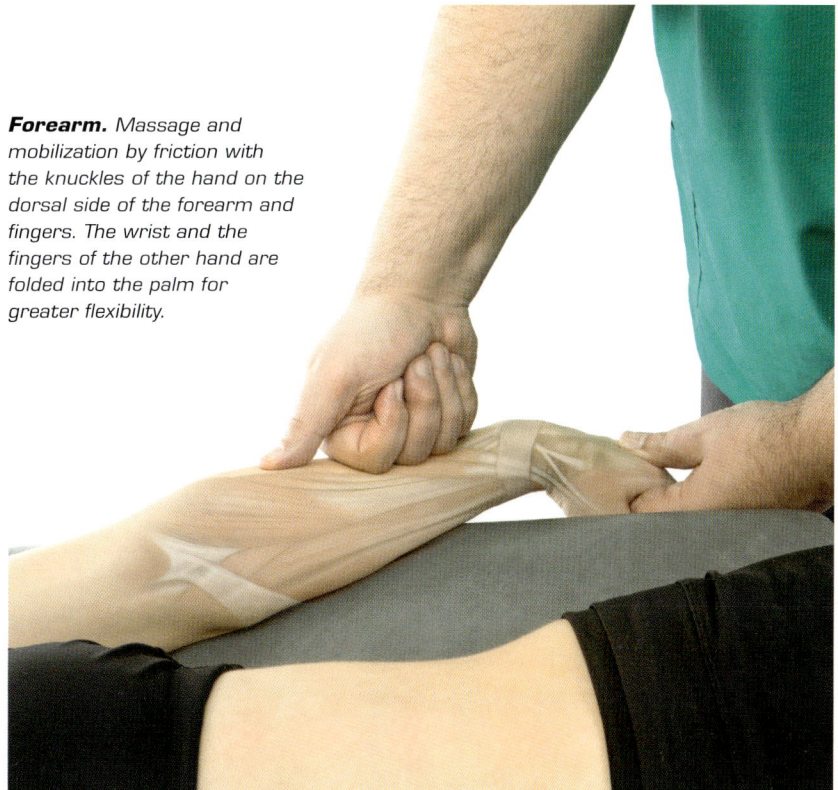

Forearm. Massage and mobilization by friction with the knuckles of the hand on the dorsal side of the forearm and fingers. The wrist and the fingers of the other hand are folded into the palm for greater flexibility.

Lateral myofascial chain

A chain for stability and lateral movement

The lateral chain runs along both sides of the body. It begins at the medial or internal border of the foot, continues along its lateral or external edge and from here begins its ascent, crossing the outside of the ankle. It reaches the pelvis, passing the lateral edge of the leg and thigh.

This chain evidences a series of diagonals that cross laterally from anterior or posterior or back, and vice versa, at different levels along the entire body. It reaches the skull at ear level.

Here we can also speak of two chains, as previously mentioned. There is a right lateral chain and a left lateral chain, sufficiently displaced between each other to understand their dynamic and stabilizing roles.

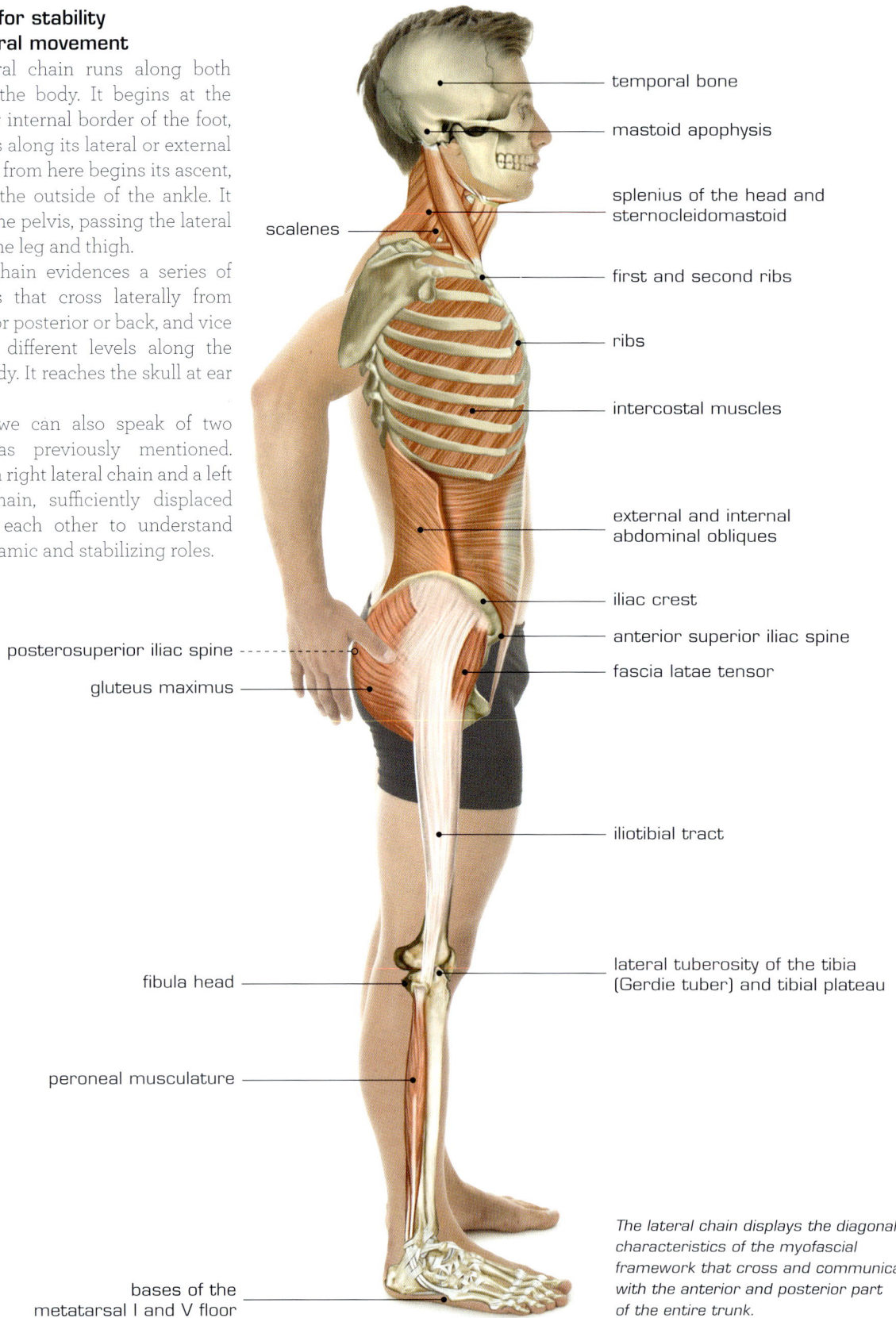

temporal bone

mastoid apophysis

splenius of the head and sternocleidomastoid

scalenes

first and second ribs

ribs

intercostal muscles

external and internal abdominal obliques

iliac crest

anterior superior iliac spine

posterosuperior iliac spine

fascia latae tensor

gluteus maximus

iliotibial tract

lateral tuberosity of the tibia (Gerdie tuber) and tibial plateau

fibula head

peroneal musculature

The lateral chain displays the diagonal characteristics of the myofascial framework that cross and communicate with the anterior and posterior part of the entire trunk.

bases of the metatarsal I and V floor

Functions of the lateral chain

The function of the lateral chain is to control body movements in the frontal plane.

Its function is to externally stabilize all major joints: hip, knee and ankle. In turn it contributes to lateral trunk flexions (inclinations), lateral elevations of the lower extremities (abductions) and ankle movements (eversion).

Thanks to fascial tissue and peroneal muscles in the leg, it is responsible for providing support to the longitudinal lateral arch of the foot (to the external plantar cavity).

It is responsible for lateral movements (left and right lateral inclinations) and afterwards stabilizing the trunk and the head, controlling and adjusting rotational movements of the trunk.

Any lateral movement of the trunk or lateral elevation in the lower extremity results in this chain stretching on one side and contracting on the other.

The role of this chain when walking — supporting one limb while the other rises during a step — is essential, as it avoids lateral instability of the body and prevents falling to one side or the other. Furthermore both inner ears, the organs responsible for maintaining the body in a stable spatial position equilibrium, are located in line with the lateral chain.

Detecting compensations

As these chains perform lateral flexion or lateral inclination, myofascial restrictions lead to an excess of tension and a shortening in the chain on the side where the restriction is located. This may lead to postures in which this lateral flexion can be observed at any level (thorax, column, pelvis, etc.).

The increase in tension in the myofascial structures that make up the chain on one side is also responsible for applying constraints and, therefore, for limitations in inclination toward the opposite side.

The lateral chain zigzags and merges with cross chains from the hip area that we will study later and with which it has a close dynamic and stabilizing relationship.

If we examine an individual from the front as they walk with their center of gravity shifted toward the supporting leg, we will see that the hip will adduct to one side when the heel rests on the ground. This implies a load for the lateral chain on one side, with the pelvis dropping on the opposite side. This pelvic drop opens the upper part of the lateral chain on that side, from the flank upwards. Assessing the coordination of the lateral chains on both sides while the patient walks or run will allow the detection of any existing anomalies and compensations.

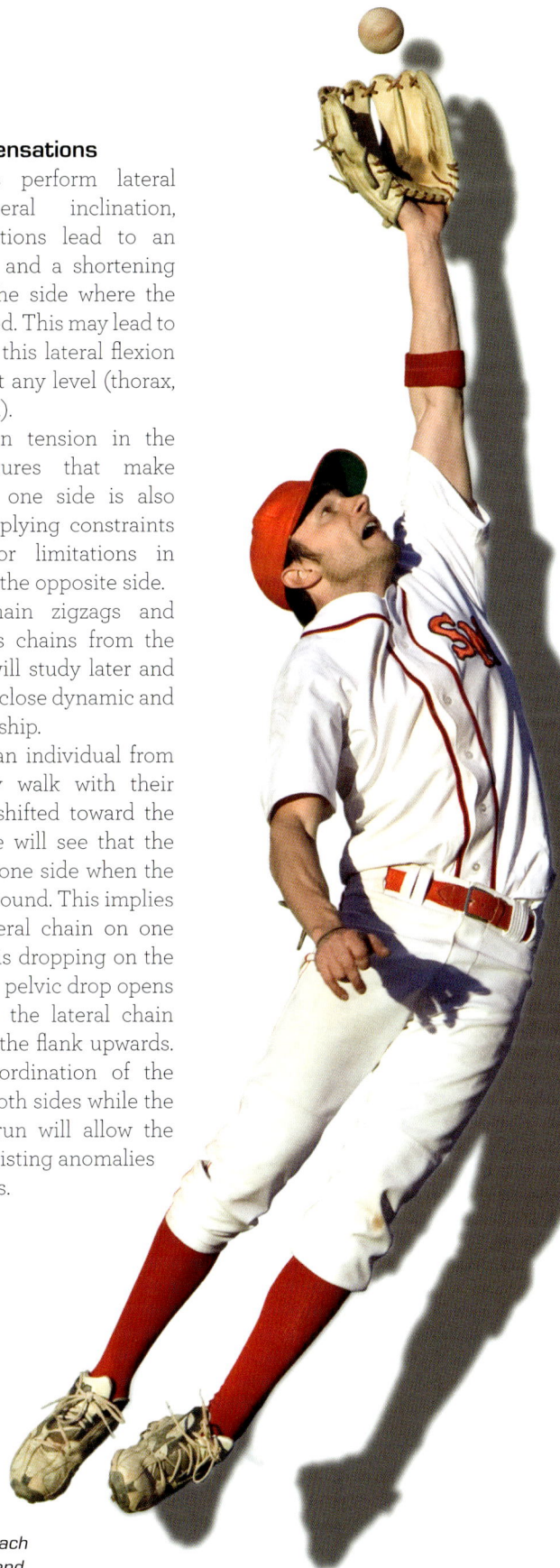

The use of the force stored within the chains allows the player to jump, laterally expand to reach the ball, and control and stabilize his movement at the same time.

Lower body I: Feet and legs

3

General techniques

- Friction and drag
- Pressure

Devices

- Knuckles
- Elbow
- Fingertips

First assess response to mobility and stability

Adaptation to the standing position and the development of all activities related to it, whether sporting or not (jumping, running, kicking, etc.), require a detailed evaluation of the foot. Despite this, the foot is a structure that is often not part of sports training plans and therefore undervalued in general. Insufficiencies or failure in adjustment of the small joints that make up the foot destabilize its biomechanics, creating compensatory systems that will alter the functionality of the entire organism.

Lack of mobility in the toes, among others (claw toe, hammer toe, etc.) is

a clear example of alteration, so it is important to maintain maximum flexibility and strength in these small joints.

From a biomechanical point of view, the external part of the foot is considered more receptive of body weight than propulsive of it. This receptive ability requires flexibility and balance of stabilizing forces. Walking on uneven terrain will test these capabilities. Movement and lateral stability of the foot and ankle are conditioned by the synergic and combined action of the lateral musculature of the foot (intrinsic muscles of the foot) and of the leg (peroneal muscles).

Foot: evaluation plan

Inversion capacity is assessed with respect to the amount of travel and flexibility evidenced by comparing one foot to the other. Both feet are held by the ankles and the outer edge of the feet with the athlete in supine position, legs extended. An inversion movement is then carried out, which consists of directing the

soles of both feet inwards, so that the two soles "look at each other." If there is a strong limitation, as well as the appearance of a sensation of tightness in the external and inferior part of the leg, external part of the ankle and external edge of the foot during the inversion path, this suggests a possible shortening of the lateral chain at this level.

The peroneal tendons are also palpated on the outside of the leg, behind the lateral malleolus of the ankle and the outer edge of the foot. To evaluate them more clearly during palpation, the athlete is asked to move the foot and ankle outwards (eversion) against the resistance of the therapist's hands. Tendons appear on the outside of the leg at the ankle. They also become visible and palpable from behind and below the lateral malleolus. It is useful to compare both feet to see if there are zones of induration, stiffness, inextensibility and pain.

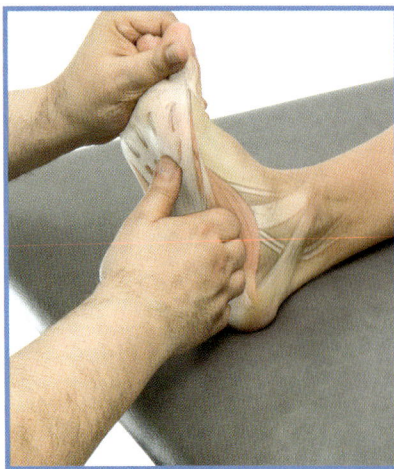

Medial area of the foot.
Friction massage and kneading from the base of the big toe and the inner edge including part of the sole. Friction maneuvers are applied with slight pressure (without triggering pain) in the most sensitive areas.

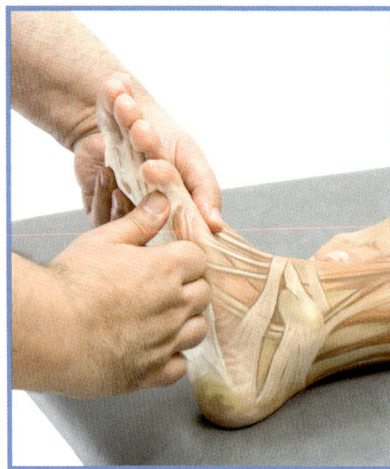

Side of the foot. *Friction and drag on the outside or lateral of the foot and ankle, surrounding the lateral malleolus of the ankle (external relief of the ankle). Treat both in front and behind.*

Sole of the foot. The prone position affords good access for massaging the sole of the foot over the external longitudinal arch and the metatarsal I and V bases. In this case, three-dimensional forces of pressure are applied with the elbow in short strokes, to flex the tissues and avoid irritation or pain.

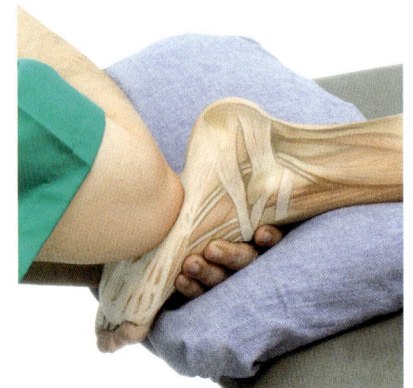

Evaluation tactics

Continued evaluation of the status of the lateral chain along the lateral compartment of the leg until reaching the knee. Compare both legs: the state of their tone and tension must be evaluated. If differences are obvious it can be supposed that the lateral chain in the hardest leg is subject to stress. This assessment can be carried out with the individual lying on the couch in the supine position (face-up).

You can also perform this test with the individual standing in his usual posture or even on tiptoe, leaning against the wall. This means that the chain is required to maintain more lateral stability. The area in which there is greater stiffness will become more evident during palpation.

The lateral or peroneal malleolus (seen here from the front) reaches lower than the medial or tibial malleolus. This explains that ankle sprains are more frequently inward (inversion) than outward (eversion). Recovery from this type of injury requires, in addition to the treatment required by the sprain itself, recovery of any stability that was lost when the sprain occurred.
Support differences in the feet can simulate differences in leg length. Thus a supine foot will give a leg a longer appearance while a pronated foot will make the leg appear to be shorter.

Anatomical precision

The third peroneal muscle, also known as peroneus tertius, is a muscle in the anterior leg group, which authors of contemporary literature consider inconsistent.

According to other authors it is a dorsal flexor, abductor and pronator of the foot. It enters into synergy (muscular activation) when the toes are lifted (extension). Recent studies have shown that it is a more common muscle than it first appears.

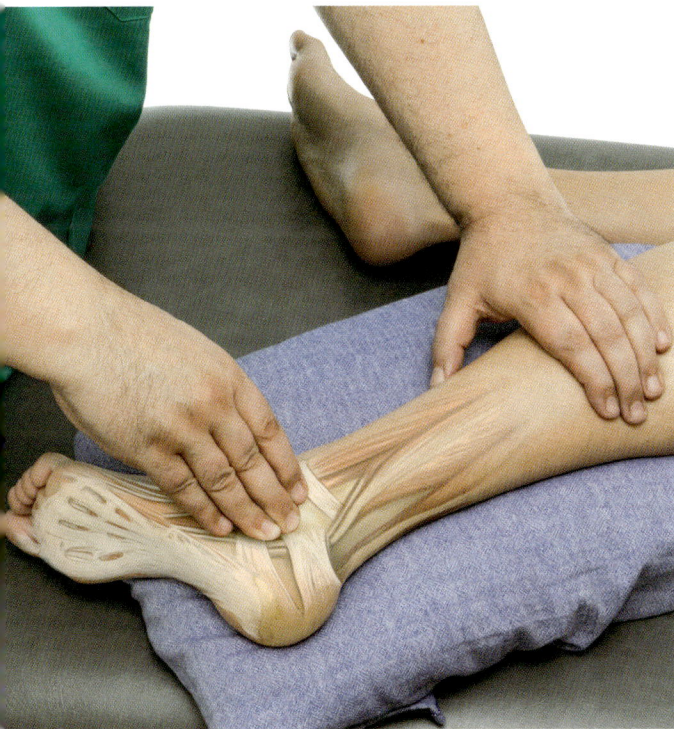

Foot and ankle. *Friction and digital dragging on the outer edge of the foot and ankle from the lateral malleolus, around it and toward the heel. The tendons of the peroneal muscles are easily perceived behind the malleolus if the athlete performs eversion of the ankle (lateral movement).*

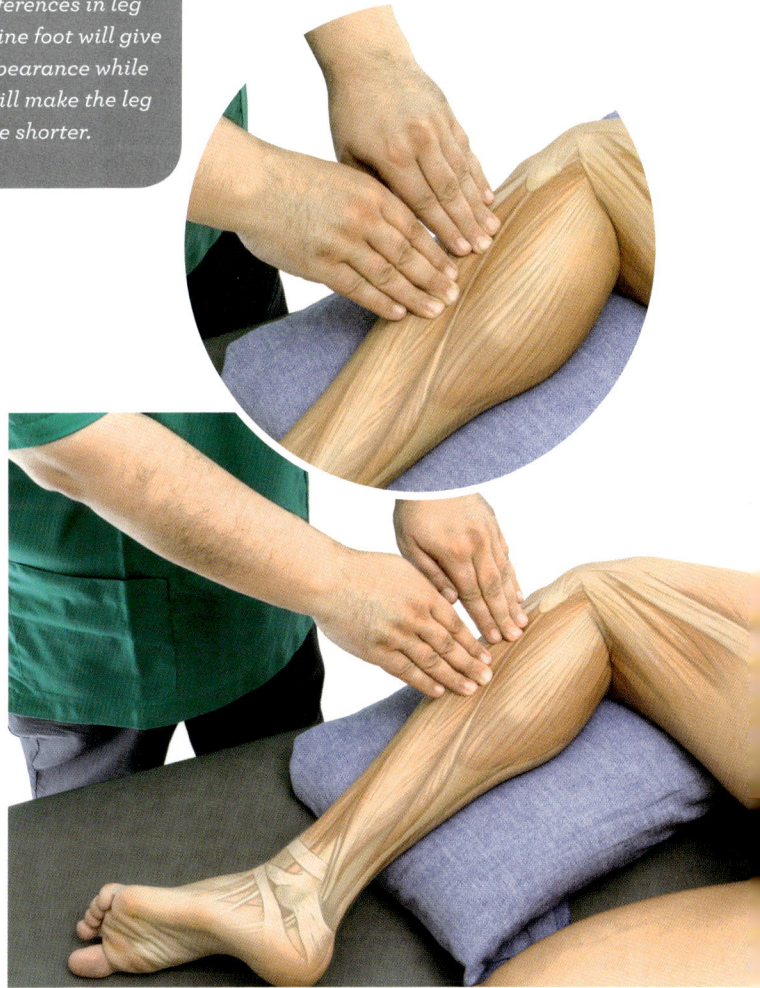

Leg. *Massage separating the anterior and posterior lateral compartments by means of transversal friction along the leg. If the athlete performs movements in flexion and extension of the ankle during the massage, better liberation is achieved in contractures and tissue shortening.*

Lower body II: thighs and pelvis

General Techniques

◆ Friction and drag
◆ Transversal or longitudinal traverse friction
◆ Pressure and drag

Devices

◆ Elbows
◆ Knuckles
◆ Elbow

Areas of relationship and translation

One of the strategies that the body uses to economize energy expenditure when walking is to make a lateral inclination of the body toward the load-bearing foot. This involves a slight fall in the pelvis toward the free extremity (the one that is not in support). Therefore a relative adduction is created in the hip of the load-bearing limb.

Pushing the greater trochanter (external protrusion of the hip) toward the lateral chain causes it to stretch from the interior of the body. On the outside of the thigh is the iliotibial tract, a structure that will withstand tension caused by this relative adduction, the separation between the iliac crest and the femur.

When we move a leg forward to take a step, several things happen. There is tension in the iliotibial tract by passive adduction

(mentioned in the previous paragraph) and by the synergy that will be established with the gluteal muscles that end in and tighten this girdle (gluteus maximus and medius). The latter are responsible for slowing down, shortening the step and contraction, which in turn tightens the girdle in which they are inserted, the latter being prepared to support the translation of body weight (relative hip adduction). Thus starts the loading phase on the support leg.

Examination of the lateral thigh group

The athlete assumes the lateral position, and with the leg that rests on the stretcher folded to provide stability, we take hold of the leg to be examined. This leg is then stretched and abducted until the tract covers the greater trochanter. We hold the

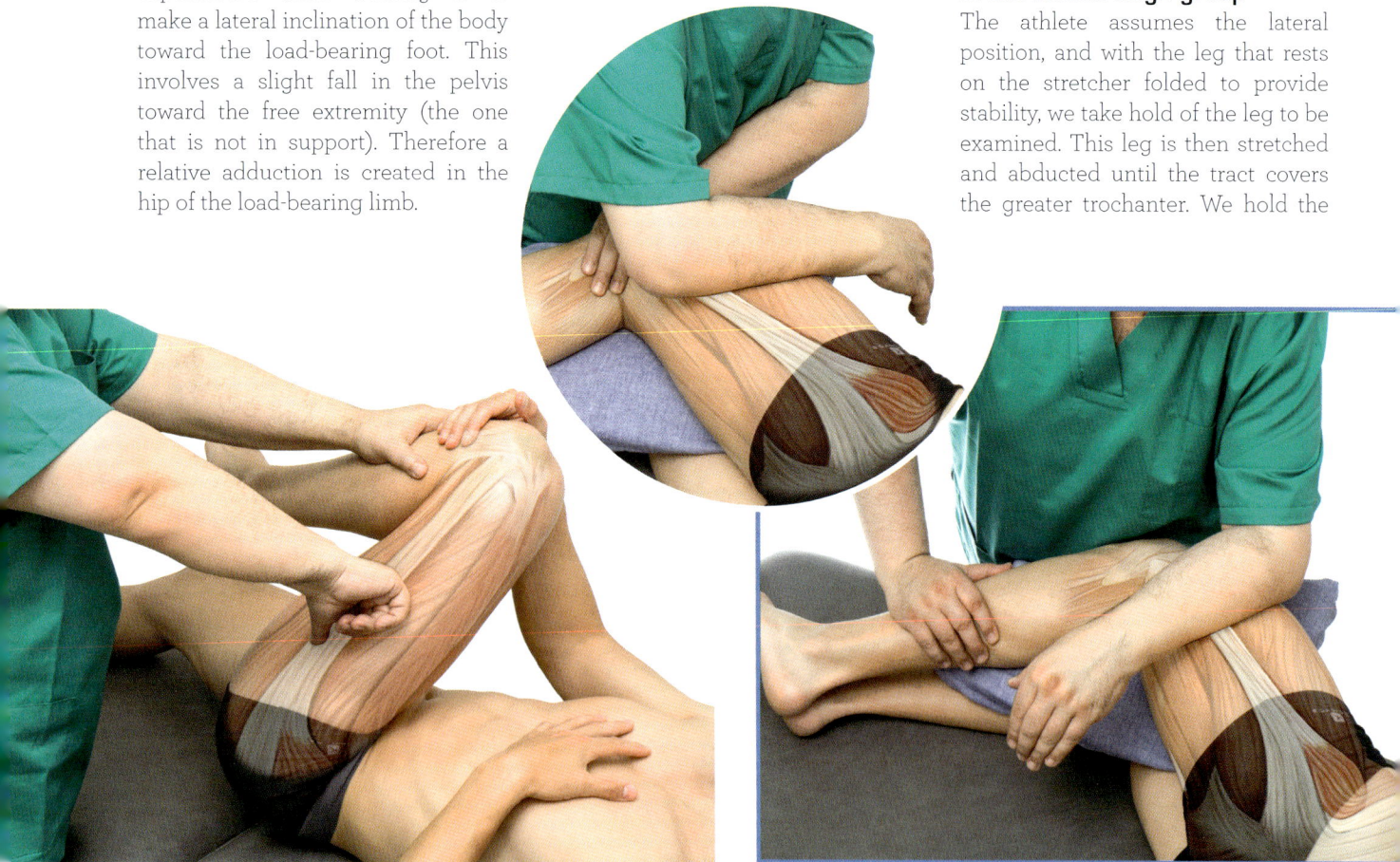

Pelvis. Massage into the fleshy mass of the gluteus maximus should be performed following its insertion along the sacral border and continued by friction, tracing the transverse in the direction of its fibers. It may also be effected longitudinally, toward the lower extremities.

Pelvis. Massage in areas of tension on the iliac crest from the anterior and posterior iliac spine with the athlete in lateral decubitus position. The forearm starts its movement at the waist, descending to contact the bony border of the iliac crest. Pressure and drag maneuvers are applied by rubbing in the caudal-inferior direction and toward the sacrum. The athlete can perform small movements of arm elevation, anteversion and pelvic retroversion in the lower extremity, assisting massage maneuvers.

leg at the ankle and knee. Keeping the limb in neutral position without extension, adduction or abduction, bend the knee slowly to 90°.

Hold the limb by the ankle and flex the knee, then remove the hand that holds the knee and let it fall back naturally. If the knee hangs in the air or barely falls, the tract is shortened. It is important that the athlete is relaxed for this test to be reliable.

The pelvis-lumbar/ spine relation

A simple test allows the assessment of mobility capacity of the pelvis with respect to the lumbar spine. This test involves a "fall" of the pelvis in relation to the upper lumbar. The athlete is placed standing in front of the therapist with feet more or less parallel and separated to the width of the hips. The therapist then asks the athlete to flex one knee while keeping their foot on the floor. This act should cause the pelvis to "fall" on the side of the knee flexion. The lumbar spine on the opposite side should show a flexible lateral tilt below the L3 level. If the lower lumbar region cannot tilt naturally, the tilt will move toward upper spine segments.

Evaluation tactics

The border of the iliac crest is at the insertion site of the abdominal muscles. Its outermost part serves as an insertion into the external oblique. The top of the crest houses the insertion of the internal oblique, with the transverse inserted deep inside. Therefore within this bony ridge lie important layers of connective tissue. Assessing the existing density on both sides of the ridges and

The base of the sacral bone may be tilted due to structural changes in the length of the legs. This will cause the spine to tilt to one side and force the tissues to shrink, giving the athlete the feeling of sitting constantly on a chair tilted sideways or walking on the side of a slope.

subsequent treatment by friction and elongations will help to stretch the lateral chain.

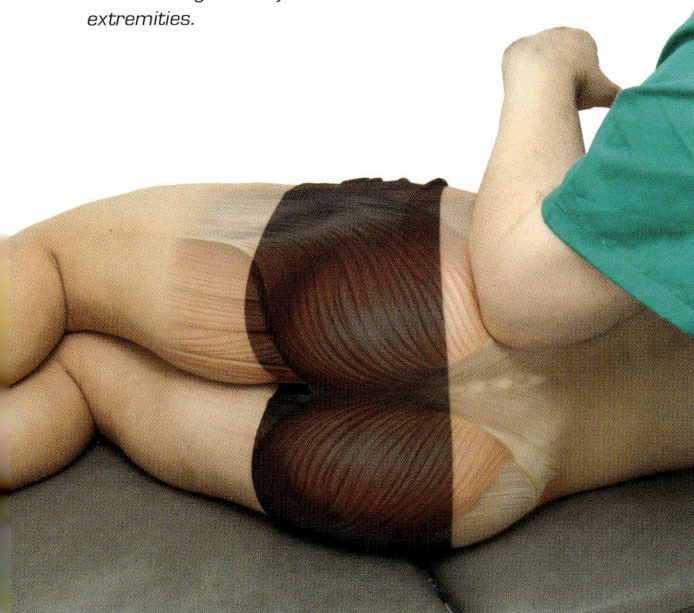

Pelvis. *Massage into the fleshy mass of the gluteus maximus should be performed following its insertion along the sacral border and continued by transverse friction traces in the direction of its fibers. It may also be effected longitudinally, toward the lower extremities.*

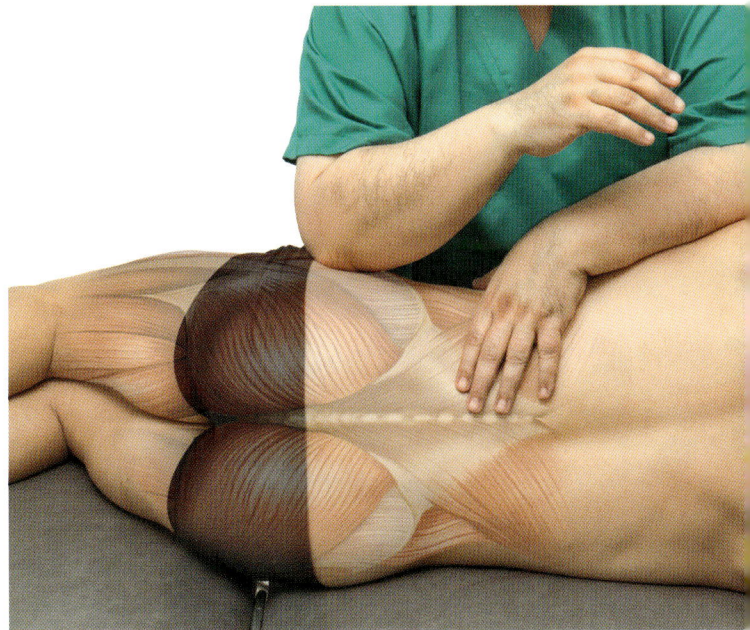

Pelvis. *Massage areas of tension on the iliac crest from the anterior and posterior iliac spine with the athlete in lateral decubitus position. The forearm initiates movements at the waist, contacting the boney border of the iliac crest as it descends. Pressure and drag maneuvers are applied by rubbing in the caudal-inferior direction and toward the sacrum. The athlete can perform small movements such as arm elevation, anteversion and pelvic retroversion (or in the lower extremity), assisting the massage maneuvers.*

Upper body I: lumbar column and thorax

General Techniques

◆ Pressure rub
◆ Friction and drag
◆ Traction

Devices

◆ Fingertips
◆ Knuckles
◆ Elbow
◆ Forearm

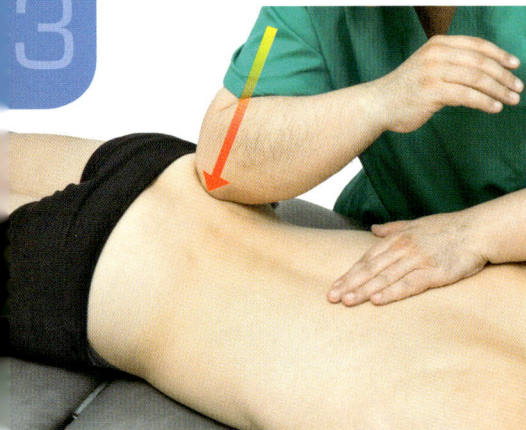

Connection area

The lateral myofascial structures affect, among others, breathing and arm movement. The diaphragm separates the abdominal cavity from the thoracic cavity and determines position and costal tension. Costal stress shapes the space occupied by the diaphragm, where ribs and diaphragms feed back into their structure and function. Both structures must be stable. Thanks to this stability, the myofascial framework that covers the thoracoabdominal region achieves a harmony necessary for fluid and sustainable movement.

The ventral or abdominal fascia tends to move sideways when it loses firmness. Many reasons may explain it: poor diet, obesity (hypomotility), overtraining, trauma, surgical scars, etc. This results in problems in respiratory dynamics, since the lack of support in the "abdominal girdle" does not allow the diaphragmatic "piston" to provide the energy necessary for the individual to pursue a healthy lifestyle.

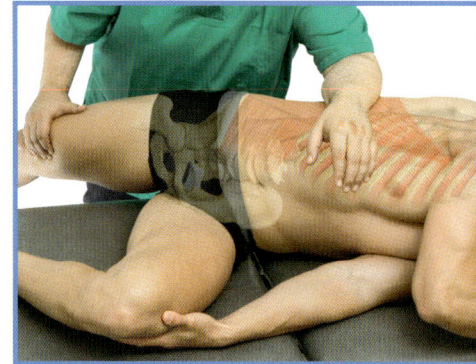

Lumbar spine. *Massage may be performed while the pelvis and shoulder girdle rotate in opposite directions. In the photo the athlete's thigh is carried backwards while the masseur applies a posteroanterior forearm drag and rub of the dorsal lumbar column.*

Lumbar spine. *With the athlete prone, pressure maneuvers are applied by elbow strokes at the medial border of the gluteus along the lateral border of the sacrum, crosswise and longitudinally in the direction of the muscular fibers.*

Lumbar spine. *The athlete lies on one side and with the upper leg forward, causing a pelvic rotation on that side. The masseur holds his arm with one hand and pulls at it, causing a backwards rotation of the trunk and the shoulder girdle. With the other hand placed at different levels the masseur performs wide rotational movements. Mobilization can be assisted with friction maneuvers and drag with the knuckles or fingers in regions where rotation is perceived to be more limited.*

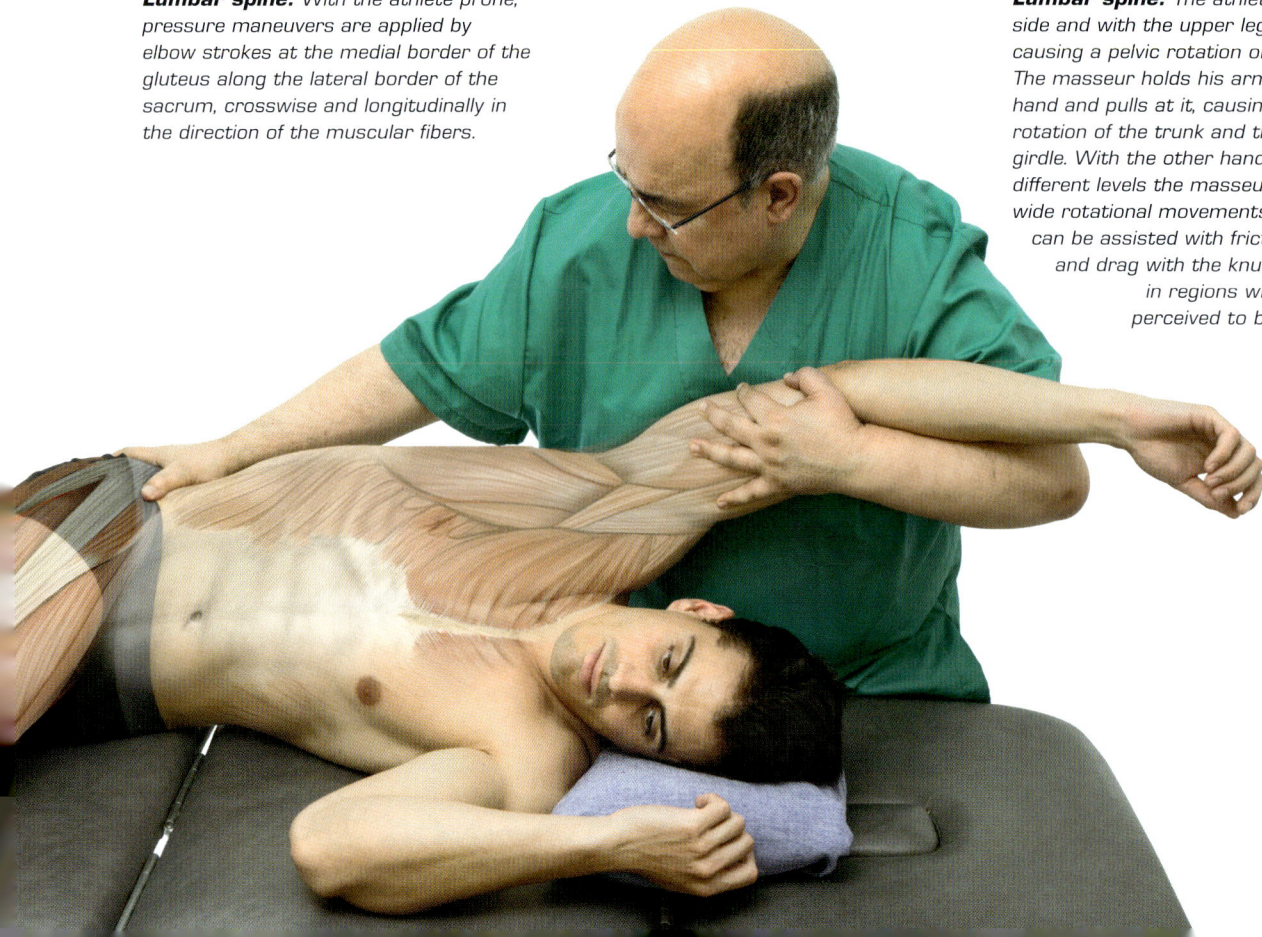

The pelvis-thorax relation

The pelvis-thorax correlation is expressed when walking. When the heel hits the ground, the pelvis rotates toward the extended leg (the leg that is about to start swinging forward). This creates a relative rotation between the pelvis and thorax. The anterior and posterior cross chains control this rotational activity. However the side chain, thanks to its "X" arrangement from the waist, contributes to the functionality of these chains by producing and controlling rotation. The obliquity of the lateral line axis on both sides of the body also provides movement and stability in the frontal plane (lateral inclinations of the pelvis and trunk), giving the body a rotational capacity and maintaining integrity as the body inclines to the right and left.

Alternating rotation between the pelvis and thorax is controlled and limited by oblique forces on each side. This coupling allows the body to remain stable in the frontal plane through different adjustments in the forces of the oblique abdominals on both sides.

Intercostal musculature of the thorax

The intercostal muscles have an angulation similar to that of the oblique abdominals.

The external and internal intercostal muscles are responsible for transmitting the rotation that originates in the pelvis by controlling the relative rotation of each intercostal segment. They activate during both walking and breathing.

As with obliques, the intercostal musculature will also collaborate with crossed chains in planes of transverse movement (rotations). However, their deep positioning in the lateral chain makes them play a more stabilizing role in tilting movement (frontal), especially of the lower part of the neck.

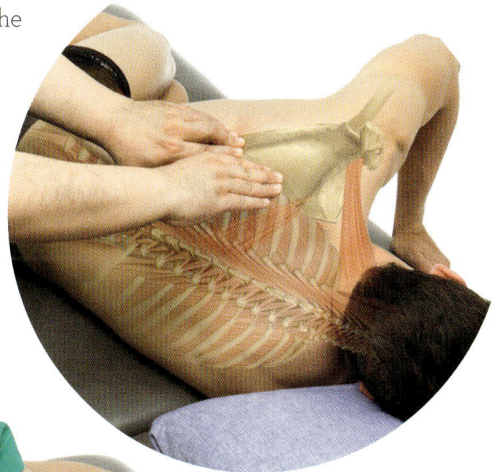

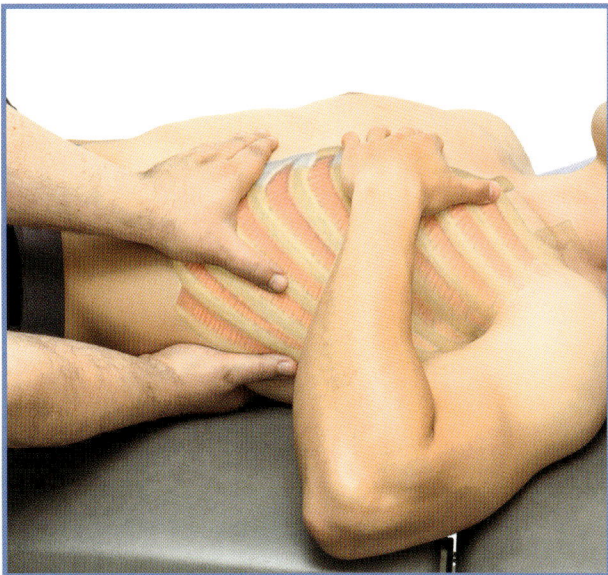

Thorax. It is important that the athlete appreciate anterior and posterior costal displacement. To do this, the masseur places his hands one in front of and the other behind the athlete's thorax. This helps the therapist to perceive the interrelation during respiratory movements between the ribs located in the posterior and anterior planes.
NOTE. This maneuver can be considered more as an exercise in awareness and relaxation than a massage technique, where it is equally beneficial.

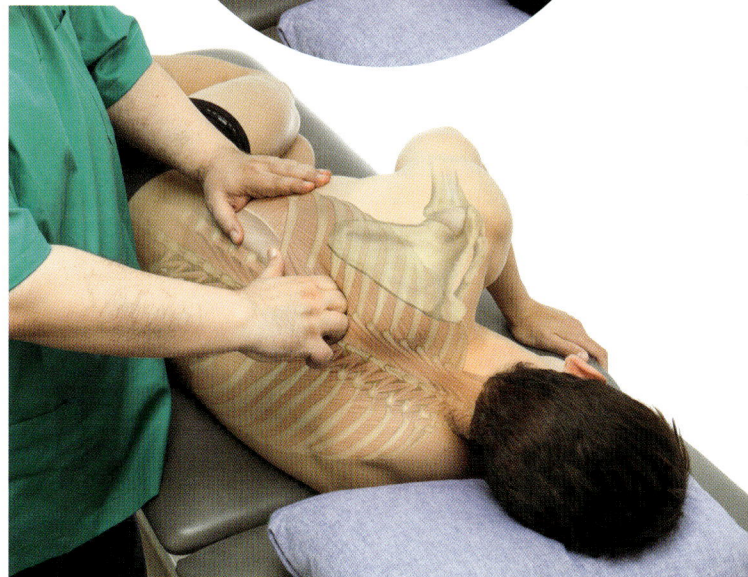

Thorax. The athlete is placed in lateral decubitus with knees slightly flexed. The masseur is positioned behind the athlete and slowly applies pressure in an anterior direction on the rib cage. During inhalation the ligaments and muscles that contact with the floating ribs are rubbed. Ascend using friction maneuvers along the ribs while the athlete tries to reach the wall in front of him and then down to the ground; this abducts the scapula, which facilitates work on the thorax.
NOTE. The athlete is asked to breathe deeply after work on each area to be better prepared for manual work.

Upper body II: head and neck

3

General Techniques
◆ Stroke friction
◆ Pressure and dragging
◆ Digital kneading

Devices
◆ Fingertips
◆ Knuckles

Chest breathing

The scalene muscles can be found in the head and neck. Due to their high costal insertion (first ribs) and their role as neck stabilizers, they may exhibit postural (tonic) behavior, although they are also considered phasic muscles. Sportspeople who hyperventilate because of anxiety, fatigue or fear may have excessively tense scalene muscles.

Evaluation tactics 1

These muscles can be assessed by placing the hands on the shoulders and the fingertips on the clavicles. The therapist's hands will let him know if when inhaling the athlete raises their shoulders, bringing them ostensibly to the ears. This indicates that the scalene muscles may be shortened.

Evaluation tactics 2

Another way to assess these muscles requires the athlete to place one hand on the abdomen, above the navel, and the other hand flat on the pectoral area.

Observe how inhalation will move the hand laid across the chest area. If this hand moves toward the chin instead of going forward, the individual is breathing with the upper chest area due to the shortening of the scalene muscles and other accessory respiratory muscles, such as the sternocleidomastoid.

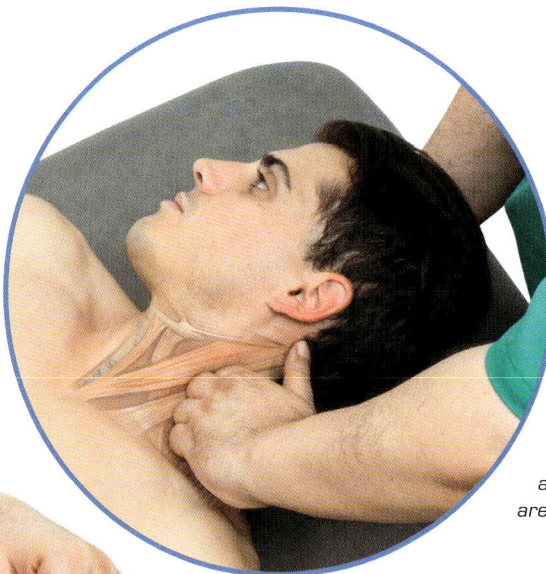

Neck. The athlete assumes the lateral position. A pillow can be placed under the head, but not if you plan to use stretching techniques. Stabilize the shoulder with one hand and gently pull, keeping it in the caudal direction. Massage the side of the neck with your fingers or knuckles while the athlete turns their head slightly away from you, elevates the chin or lifts the head from the couch.

Neck. It is very important to massage boney insertion zones, from the mastoid apophysis to the posterior occipital protuberance (as indicated by the iliac crest). This action allows drainage and a deep relaxation of all fascial structures and neck muscles including the deepest suboccipital structures. These areas thickened by liquid stasis are detectable with touch.

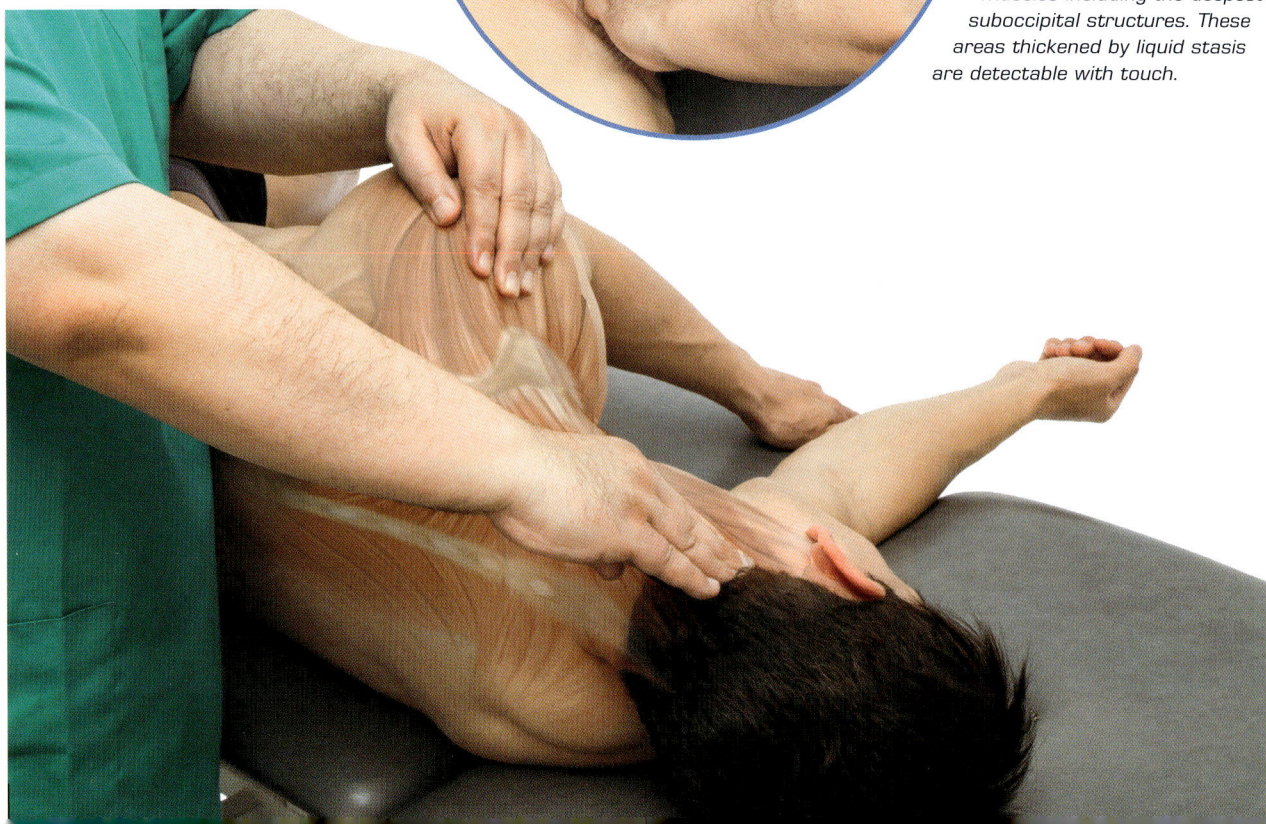

Anatomical precision

The lateral chain continues from the intercostal muscles to the cervical spine through the scalenes. The latter are located in the same deep plane as the intercostal muscles. In fact their insertion is costal, particularly on the first two ribs.

The upper cervical portion from C2 to the occiput is much more mobile than the lower portion, from C3 to C7. Insertions of the scalenes do not reach the segment of the uppermost cervical spine: C2 occiput. The head is thus freed from restraint or forces that could overstabilize, allowing the eyes and ears to orient themselves horizontally.

The splenius muscle of the head and the sternocleidomastoid, despite being part of the lateral chain by their lateral arrangement, are more superficial and more designed for movement than for stability. The sternocleidomastoid is easily palpable on the surface, both in its sternal and clavicular origin as well as its mastoid and occipital insertion.

The splenius can be felt by placing the hands on the athlete's head. With our fingers located below and slightly behind the mastoid apophysis, we ask the patient to turn their head, offering a slight resistance to movement with our thumbs. The palpation shows the contraction of the muscle on the same side as the athlete's head. The tensional state of both sides in that region can be compared.

During lateral movement there are reciprocal medullary reflexes that cause gentle undulating movements in the spine and more generally throughout the body. When the small intertransverse muscle contracts, its counterpart is stretched. When stretched, the opposing intertransverse will be forced, reflexively, to contract. This contraction now stretches the first intertransverse that contracted, which will again be forced to contract. This model of oscillating movement occurs through all segments of the spinal column. A simile could be established with the movements produced by our distant ancestors the fish when swimming. The human species maintains automatic reflexes that operate without the participation of the brain. When a baby starts to crawl, this lateral oscillation movement is clearly evident. Subsequently (when straightening up) this movement is refined thanks to the appearance of flexor extension and rotation movements typical of walking.

Head. *Place both hands together and with palms extended, supporting the fingertips in the temporomandibular area* **(1)** *or occipital* **(2)** *myofascial, that is, above and behind the ear. The small painful areas of tissue are then mobilized toward the head, making small friction and drag movements (of approximately one centimeter), allowing time for them to relax throughout the maneuver.*

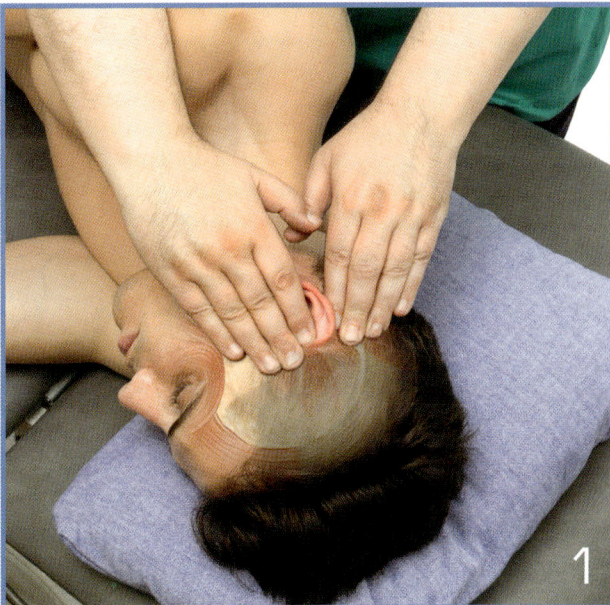
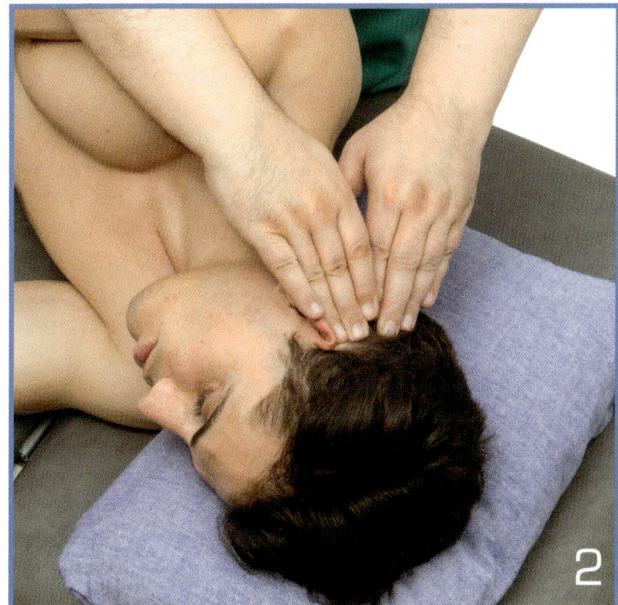

Anterior myofascial cross-chain

An integrated chain

The anterior cross-chain, as its name implies, traverses several other chains as it crosses the body. Therefore many of these structures that it is built with are part of other chains. However, in the practical section we will see other treatment techniques in addition to those explained in their corresponding chapters.

As with other chains, we can speak of two cross-chains: one on the right and one on the left. The union of the right and left anterior cross-chain forms an imaginary X, where opposite ends approach or distance themselves from each other according to a technical movement.

The chain unfolds itself at the level of the thorax, creating a path in one surface plane and another in the deep plane, allowing greater movement around the pelvis.

The chain and its axis

The anterior cross-chain is located in the frontal plane and is the product torsional motion patterns. The distribution of the tissues draws an oblique axis to the midline, and it is in this axis that torsional movements occur during their opening and closing phases. Chains respond to body movements on several planes simultaneously; therefore when the anterior chain is stretched, the posterior chain contracts, and vice versa.

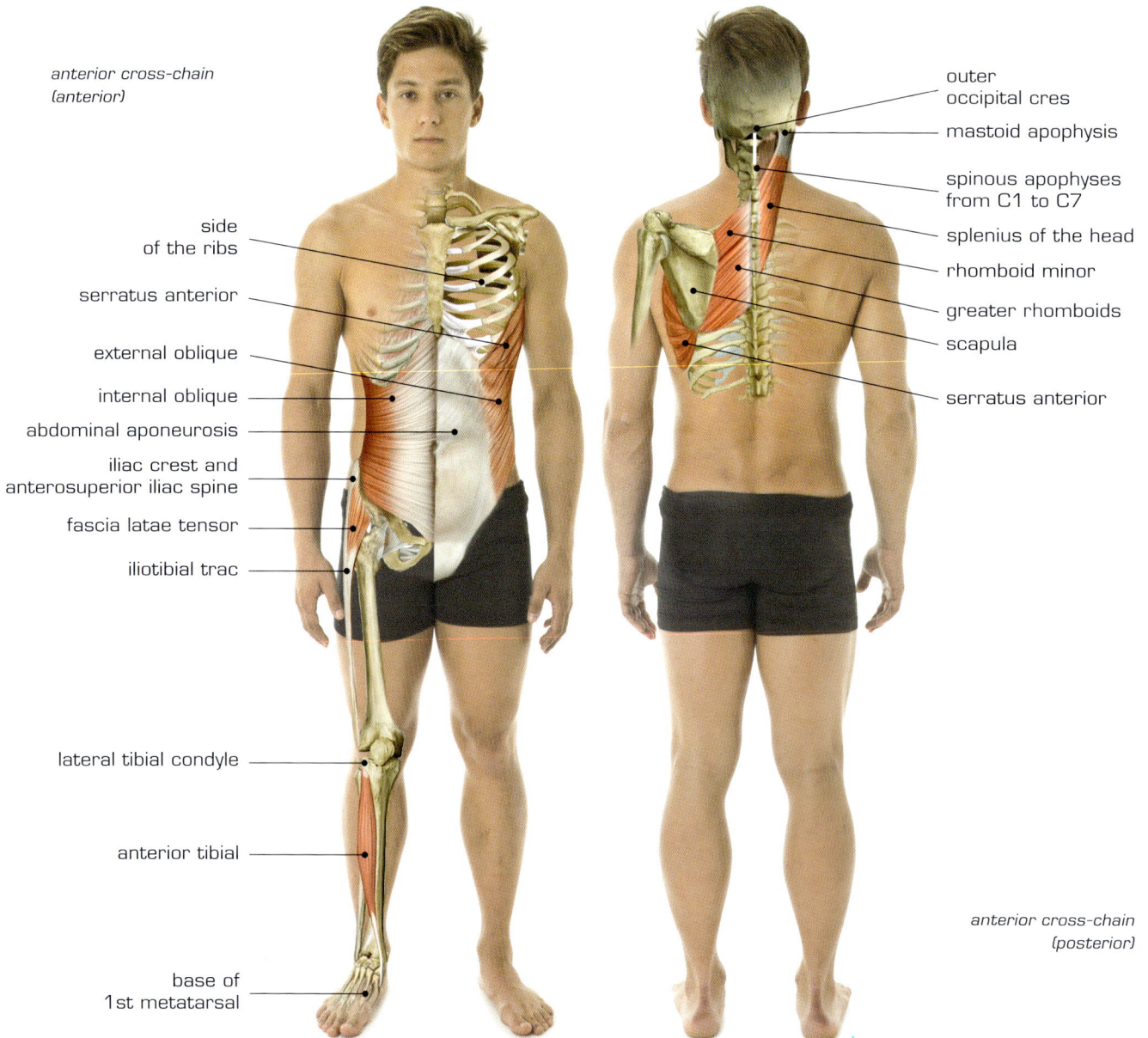

anterior cross-chain (anterior)

side of the ribs
serratus anterior
external oblique
internal oblique
abdominal aponeurosis
iliac crest and anterosuperior iliac spine
fascia latae tensor
iliotibial trac
lateral tibial condyle
anterior tibial
base of 1st metatarsal

outer occipital cres
mastoid apophysis
spinous apophyses from C1 to C7
splenius of the head
rhomboid minor
greater rhomboids
scapula
serratus anterior

anterior cross-chain (posterior)

Sequence of movement in the anterior cross-chain. As movement begins, the left-side shoulder girdle and the right-side hemipelvis are separated as far as possible, retracting at the end of the movement when the right shoulder approaches the left hemipelvis. Note. To better observe movement, it may be necessary to change the arm and leg.

Functions of the anterior cross-chain

The anterior cross-chain is representative of a physical pathway formed by a continuum in the myofascial framework of bones, joints, tendons, connective tissue (fascia) and musculature. Its function is to generate power and dynamism (displacement).

In the movement that closes the chain, the hand and the arm approach the opposite foot through torsional forces generated by the chain's myofascial framework.

Its great capacity for movement/ opening-closing generates a significant amount of elastic energy in the myofascial framework, providing a burst of strength for the technical movement, such as jumping, throwing or a tennis serve. This reserve provides greater energy savings (less fatigue) and produces a greater degree of force (greater efficiency).

Generator of energy and forces

The greater the chain's ability to move, the more effective the elastic energy accumulation and power production will be. A basic example is the action of walking. Walking results in a dissociation of girdles (pelvic and scapular). This torsion movement provides an elastic energy, allowing us to walk with less effort. When do you feel more fatigued, walking slowly around the shops or walking briskly?

The various structures that make up a myofascial chain are constantly modifying their forces of tension and compression to organize movement of the body. They adapt and integrate to fulfill three objectives: stability, economy and comfort. These measures define the body's neuromotor behavior and follow the descriptive model of biotensegrity systems (discussed in the myofascial chains introductory pages).

Detecting compensations

When a cross-chain is affected or has a local excess of tension, it forces the other chains to adopt a series of compensations. Thus new movements appear in other planes, which overlap those produced by the anterior cross-chain in the transverse plane. This creates instability and disorganizes the body, generating a loss of mobility and discomfort in the joints that can affect the structures of many different organisms.

Process

The observation of the predominance of a chain and intention of movement, together with the tests to assess the degree of mobility and the quality of force applied, allows the creation of a working strategy.

For example, in detecting the direction that indicates asymmetry in the pelvis, the cross-chain under more tension is objectified. This imbalance will disturb body mechanics, affecting their functionality, and consequently corresponding motor dysfunctions will appear.

Lower body I: feet and legs

General Techniques

♦ Friction and drag

Devices

♦ Fingers, intermediate phalanges
♦ Knuckles
♦ Elbow
♦ Forearm, ulnar border

Response to mobility and stability

During the heel's support phase, the lower section of the chain undergoes a very rapid stretch as it passes through the leg due to an eversion movement of the ankle. Tension changes due to a reaction as the heel impacts against the ground (rising force) and forces of gravity (downforce) are evidenced. These changes in tension and compression force not only the anterior tibial but also the surrounding soft tissues to adjust their tone in response.

Most of the proprioceptors responsible for receiving information on variations in stress and shearing forces are found in and around the intermuscular septa. Essentially, these areas experience most of the biomechanical relationship changes, so it is only logical that they accommodate mechanoreceptors.

All this explains why a corrective rethink is necessary, taking into

Electromyographic studies show that the most active phase of the anterior tibial muscle corresponds to the phase prior to the takeoff of the foot (eccentric). The proprioceptors have "indicated" the muscle that must lift the foot as the oscillatory phase of the event or the race has begun.

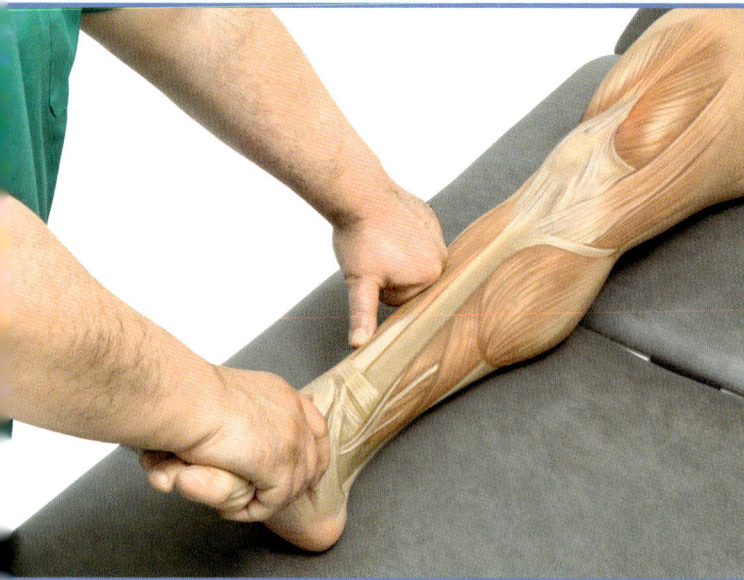

account the proprioceptive aspect of injury, sports-related or otherwise, that affect the ankle and the leg.

Evaluation tactics

The quantity and quality of the eversion path in the foot and ankle are analyzed in order to assess the elastic capacity of the anterior tibial. The degree of tension that appears on the anterior section of the ankle and leg should also be explored.

Other interesting tests are those that require the patient to assume one-leg positions, allowing load stability of the inferior region of the anterior cross-chain to be assessed. These tests allow the therapist to detect instabilities that indicate failures in the neuromuscular control mechanisms.

You can make these tests more demanding from a mechanical point of view by including jumps, turns, etc.

Foot. *Maintain ankle in eversion, which causes a stretching of the anterior tibial tendon and apply friction and drag maneuvers with the knuckles or fingers. There may be longitudinal or transverse movements in the structure.*

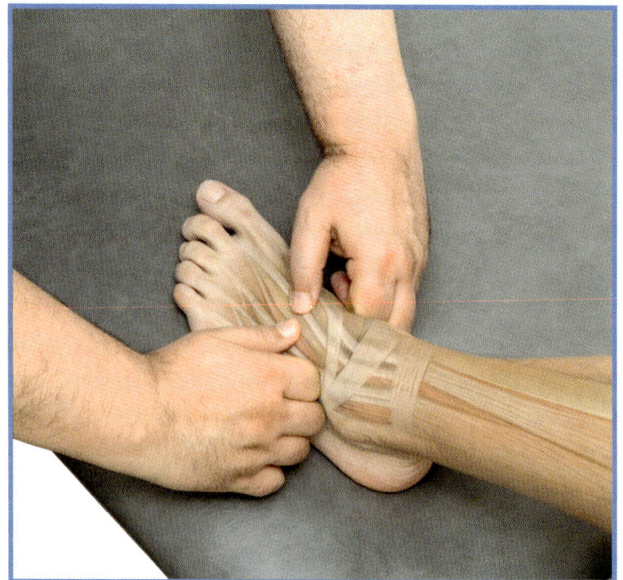

Foot. *With fingers, knuckles or even the forearm, manipulate the soft structures that are located between the malleoli. While one hand performs massage maneuvers, the other can manipulate the ankle in different movements, passively or actively, with the help of the athlete.*

While standing the soleus muscle (posterior chain) is responsible for adjusting minor forward instability, without showing electromyographic activity in the anterior tibial. In contrast, when the individual retreats (shifts the upper body backwards), the activity of the anterior tibial is manifest and clear as the muscle controls that movement (eccentrically). Also, an individual who presents pathology in this muscle may display weakness when raising the feet to walk.

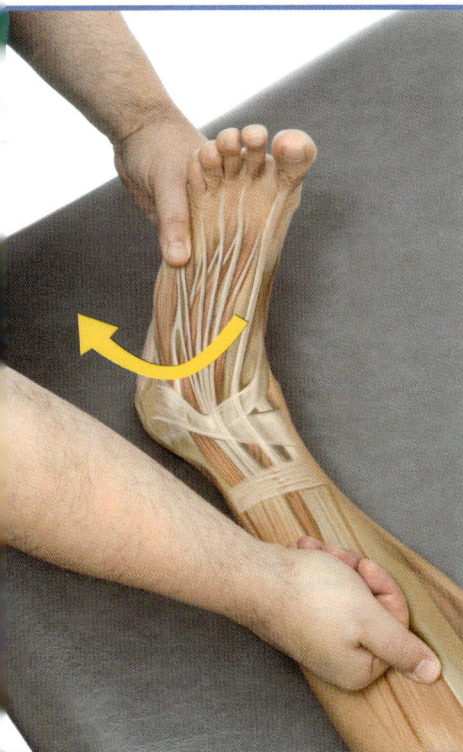

Leg. Friction and drag maneuvers performed with the knuckles, fingers or elbow from the side of the ankle toward the knee via the front of the leg. The ankle can be kept in eversion or you can ask the athlete to perform dorsal flexions (raise foot and ankle) during the maneuver.

Evaluation tactics

Both feet of the athlete are held by their inner border and at the ankles. He is then asked to perform the eversion movement, which consists of directing the soles of both feet outward; make sure that both soles "do not look at each other" but "look out." Severe limitation, as well as the appearance of a sensation of tightness in the anterior part of the leg, anterior part of the ankle and the instep during the eversion movement, suggests a possible shortening of the anterior cross-chain at that level.

Evaluation is continued along the anterior compartment of the leg to the knee. Comparing both legs, you should also note the state of their tone and their degree of tension.

In the anterior cross-chain — as in the other chains — strength is a valuable resource. For this reason it is necessary to perform these tests on the anterior tibial musculature. The athlete is asked to raise his toes and ankles in inversion and to resist any movement. The contractile force is then compared in both legs.

Therapeutic precision

Considering the involvement of the tibialis anterior in the maintenance of the plantar vault, those individuals with cavus foot (excessive plantar arch) who practice activities that impose a significant load on this structure (runners, walkers, jumpers) may present congestion in this tendon as well as tibial musculature.

Resting, relatively frequent transverse friction maneuvers on the indicated tissues, stretching and others (exercises of proprioception, orthopedic aids, etc.) complete the therapies that should be adopted. In addition, inappropriate footwear such as flip-flops, clogs or the like without anchoring at the heel should be avoided, as they force the ankle and foot to adopt claw-like positions so that they do not fall of while walking.

Leg. This maneuver allows simultaneous stretching of the ankle and leg, while the masseur applies friction and drag maneuvers in the anterior region of the leg.

Lower body II: thighs and pelvis

General Techniques
◆ Friction and drag

Devices
◆ Fingers, fingertips
◆ Elbow
◆ Forearm
◆ Knuckles

Areas of relationship and translation

The anterior cross-chain passes in front of the hip, the outside of the knee and the inner ankle, and if it is too tight there will be an internal rotation in the knee (the patella will face inwards), and at the moment the foot contacts with the floor it will produce a powerful contraction of hip extensors and abductors.

These muscle groups have undergone prestress due to flexion of the thigh and its adduction when the pelvis falls on the opposite side of the support, and increases corporal efficiency thanks to an elastic spring mechanism.

Gravity line

The pelvis moves above the support provided by the sole of the loaded foot. The line pertaining to the body's center of gravity moves from behind in the grounding phase of the heel and is in line with the foot at the stage when it is supported by the sole of the foot, finally ending up in front when the foot takes off.

When the pelvis passes in front of the foot the hip is in extension, which means the fascia latae tensor is stretched. This in turn allows the anterior part of the cross-chain at the lower extremity to perform an external rotation of the entire extremity, especially the foot, with its internal longitudinal arch rising.

Examination of the lateral thigh group

During any mobility test you must be used to perceiving whether movement develops fluidly or not. For this we will always employ a comparative assessment with the opposite member.

With the athlete lying supine with lower limbs flexed and feet resting on the couch, we let gravity separate their knees in stages. The masseur places both hands on the outside of the athlete's knees. Starting from the initial position (in which the two knees touch), the masseur progressively and symmetrically separates his two hands in short strokes and with successive movements of letting go and holding on to the knees. The ease of this movement is evaluated. This allows evaluation of the flexion, abduction and external rotation of both thighs simultaneously. We thus compare mobility on both sides to establish by contrast in what state they are in.

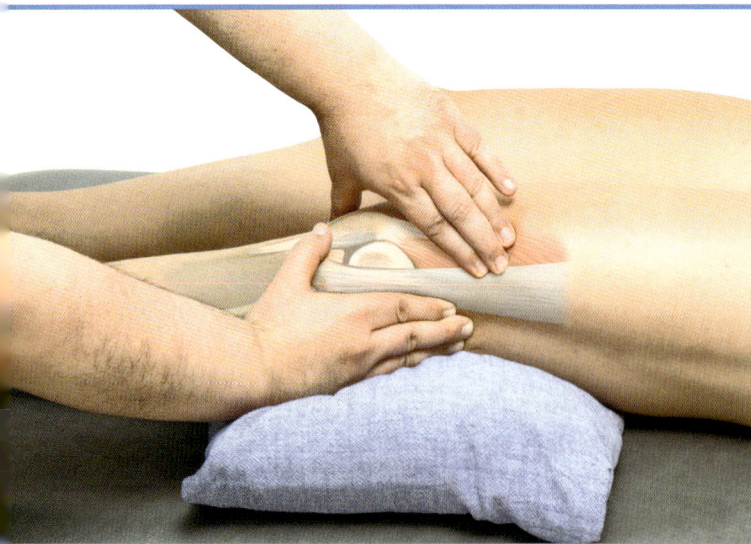

Thigh. A more specific massage at the insertion of the fascia latae into the knee (Gerdie tuber) is performed with the fingertips, to mobilize and relax the tissue.

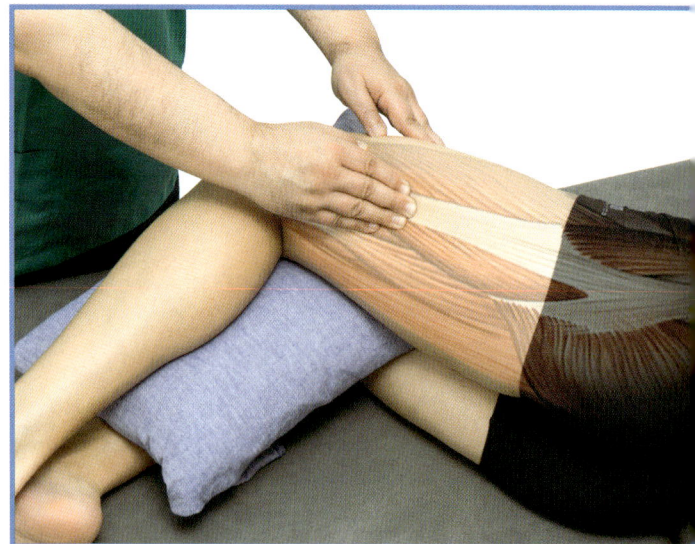

Thigh. It is important to release the fascia latae of the surrounding muscle groups from adhesions. Massage forward with the fingers or knuckles in order to release the femoral quadriceps, placing the knee in flexion (quadriceps stretching). NOTE. The same procedure is used for the hamstring group, but with the knee extended (hamstring stretching).

Thigh-pelvis relationship

The anterior cross-chain passes through the anterior superior iliac spine. This boney protrusion serves as an insertion point for many structures that are part of the cross-chains and also other chains (e.g., anterior chain and lateral chain).

Thus the abdominal muscles will move the pelvis in the direction of its traction fibers: above and below the inner oblique, above and behind the outer oblique, and the inward transverse (medially).

In relation to the thigh, other muscles that also have as insertion the anterior iliac spine will move the pelvis downwards: the sartorius below and inside, the iliac below and inside, and the gluteus medius below and behind.

Stabilizing all of these forces exerted on the pelvis requires a comprehensive evaluation that establishes which myofascial groups generate the most imbalance.

Evaluation tactics

In addition to the commonly used tests and those which we have detailed in previous chapters (shortening test of the fascia latae, pelvic drop test, etc.), ocular inspection of the region is fundamental. Understanding based on a three-dimensional exploration of the indicated zone will guide our plan of action.

Structures will be affected by unstable operation of forces present either by shortening (tonic) or weakness (phasic).

Deep-plane massage techniques must always begin on the myofascial units that present greater shortening.

Massage techniques with active movement

Activation of a muscle group inhibits the contraction of its antagonist. Therefore when applying massage in a myofascial group with rigidity and restriction, the athlete is asked to carry out a movement that will provoke the

Movement in the transverse plane is considered a new evolutionary adaptation of the human being in respect to its simian predecessors, in whom there is no counterrotation of, for example, the scapular and pelvic girdles. On the other hand, the direction of the asymmetry of the pelvis should also be taken into account. This will allow the therapist to which cross-chain has more tension.

reciprocal inhibition reflex, as this will relax the muscle group opposite to the contracting group. That is to say, if while massaging the iliotibial tract the athlete is asked for hip flexion (see photograph for the pelvic technique), the activation and contraction of the iliopsoas and gluteus maximus will relax the lateral area of the leg. This simple technique will allow the athlete to significantly increase ability of movement.

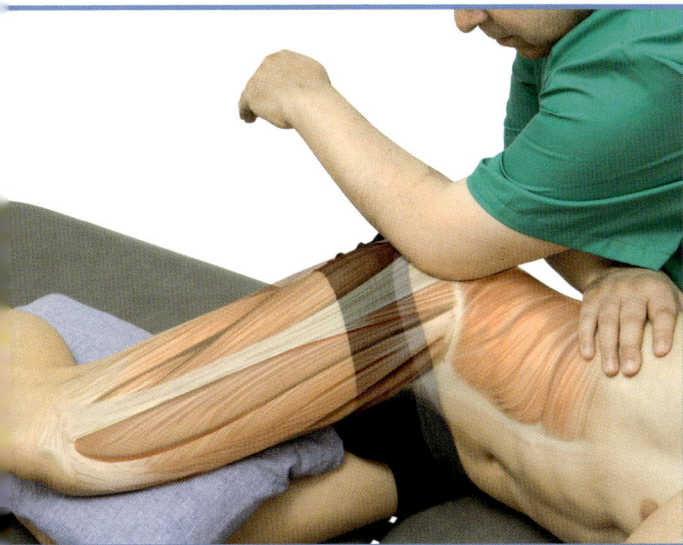

Pelvis. *Deep-tissue massage over the region of the iliac crest and anterior iliac spine while the athlete performs small hip flexion and extension movements.*
NOTE. This massage should not be performed if there is acute pain in the lumbar region.

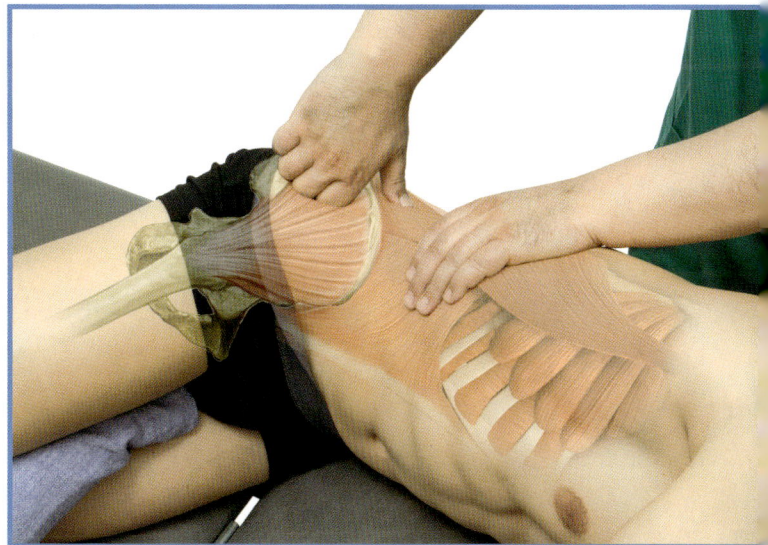

Pelvis. *A deep massage is performed at the origins of muscles in this region. It is interesting that the maneuvers differentiate the tensor of the fascia lata, the tendon of the rectus femoris of the quadriceps and the gluteal group.*

MYOFASCIAL CHAINS

3

Upper body I: lumbar column, abdomen and thorax

General Techniques

◆ Friction and drag
◆ Apply more friction and drag

Devices

◆ Fingers, fingertips
◆ Thumbs or index fingers
◆ Forearm

Connection area

When an athlete performs an abdominal movement by exercising the obliques, raising the upper part of the trunk to touch one elbow to the opposite knee, for example one of the two cross-chains is put into operation (remember that there are two chains, one on each side). This means that one is shortened and the other is stretched.

Lumbar-abdominal-pelvic relation with lower extremity

The relationship between the entire limb, pelvic and abdominal area can be perceived during the following test: standing with feet apart at hip width, performing anteversion (anterior tipping) and retroversion (posterior tipping) movements of the pelvis. During anteversion the knees are hyperextended and there is an internal rotation of the hips.

At the same time the plantar arch descends (subsidence of the plantar cavity). During a pelvic retroversion movement the opposite occurs: knees are semi-flexed, the hips rotate externally and the plantar arch increases.

When palpating the muscles in the waist (oblique) while the athlete makes a foot-launching gesture, it is possible to establish the degree of thoracic rotation available and the differences in quality between both sides.

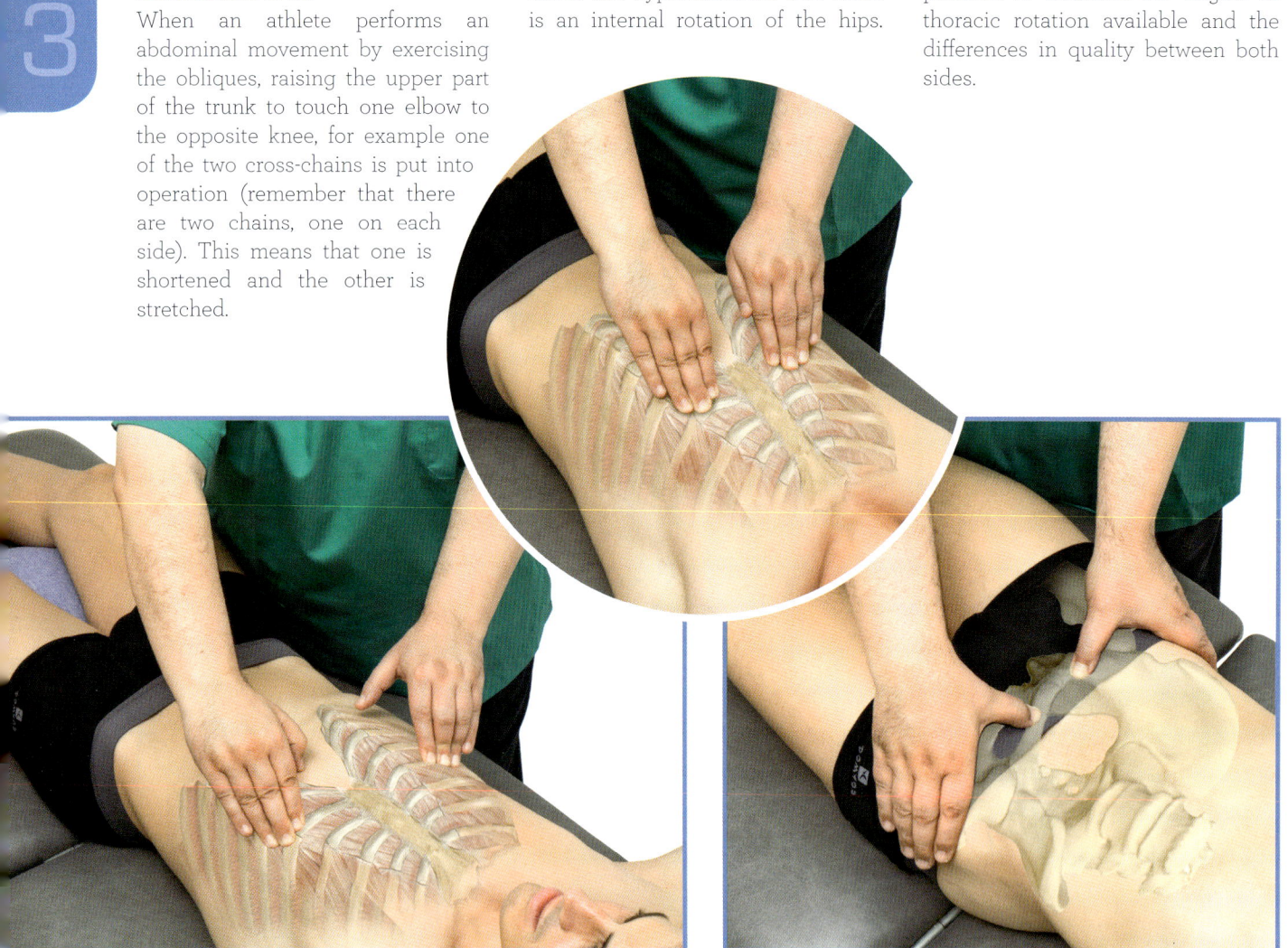

Lumbar-abdomen column. With the athlete in supine position, legs slightly bent and arms resting on shoulders, a massage is performed on both sides of the costal arch. This can also be done first on one side and then the other. Remember that the abdominal fascia ends at the height of the fifth rib (nipple height), so the massage should cease at that level. Massage is carried out from the sides toward the center until reaching the midline.

Lumbar-abdomen column. Use both thumbs or index fingers to apply slight pressure on the pubic symphysis. Increase pressure progressively without it becoming excessive. Look for sensitive points if the symphysis is sensitive. If there are no uncomfortably sensitive points, move sideways little by little, at most 3-4cm on both sides.
NOTE. The athlete may be asked to lift the head a little or alternatively flex the knees to tighten the fascial insertion, exposing any obvious problems.

Spiral arrangement of the anterior cross-chain

If we start from the internal oblique on the right side of the pelvis, we will note that the chain changes sides. Abdominal aponeurosis is densified in the linea alba, this being the functional link with the external oblique of the left side. The chain has already changed sides, producing a cross. The external oblique, now on the left side, links to the anterior serratus in the anterior and lateral side of the thorax on the same side.

Finally the chain departs from the anterior part of the thorax to join up with the group of rhomboids at the back of the body. The chain has now reached the posterior plane. From here it will cross the body again, reaching the neck and head on the posterior side — in this case the right side.

Postural considerations in the pelvic-thorax relationship

If an individual observes that the ribs are closer to their opposing hip, the anterior cross-string must be stretched with massage.

In order to assess this relationship, hands and fingers should be placed very lightly on the superficial layers of the abdominal fascia. A diagonal movement of the hands and fingers allows postural rebalancing work to begin.

Massage techniques with active movement

In addition to the benefits described on previous pages (reflex of reciprocal inhibition through massage assisted by the movement of the athlete) this technique has another advantage: It allows the masseur to maintain a free hand, which can be used to improve grip or to massage the region being treated. This is especially useful on the back and thorax, and the result provided by this technique surpasses that of any passive manual treatment.

Thorax. With the athlete in lateral decubitus, the scapula moves and stretches laterally and in a cranial direction to detach the musculature of the anterior serratus and indirectly that of the obliques, increasing its mobility.

Thorax. With the athlete in a position of lateral decubitus, massage the arch costal with the forearm while the other hand assists stretching. Holding the hip of the athlete corroborates the massage by moving the elbow toward the cranial.

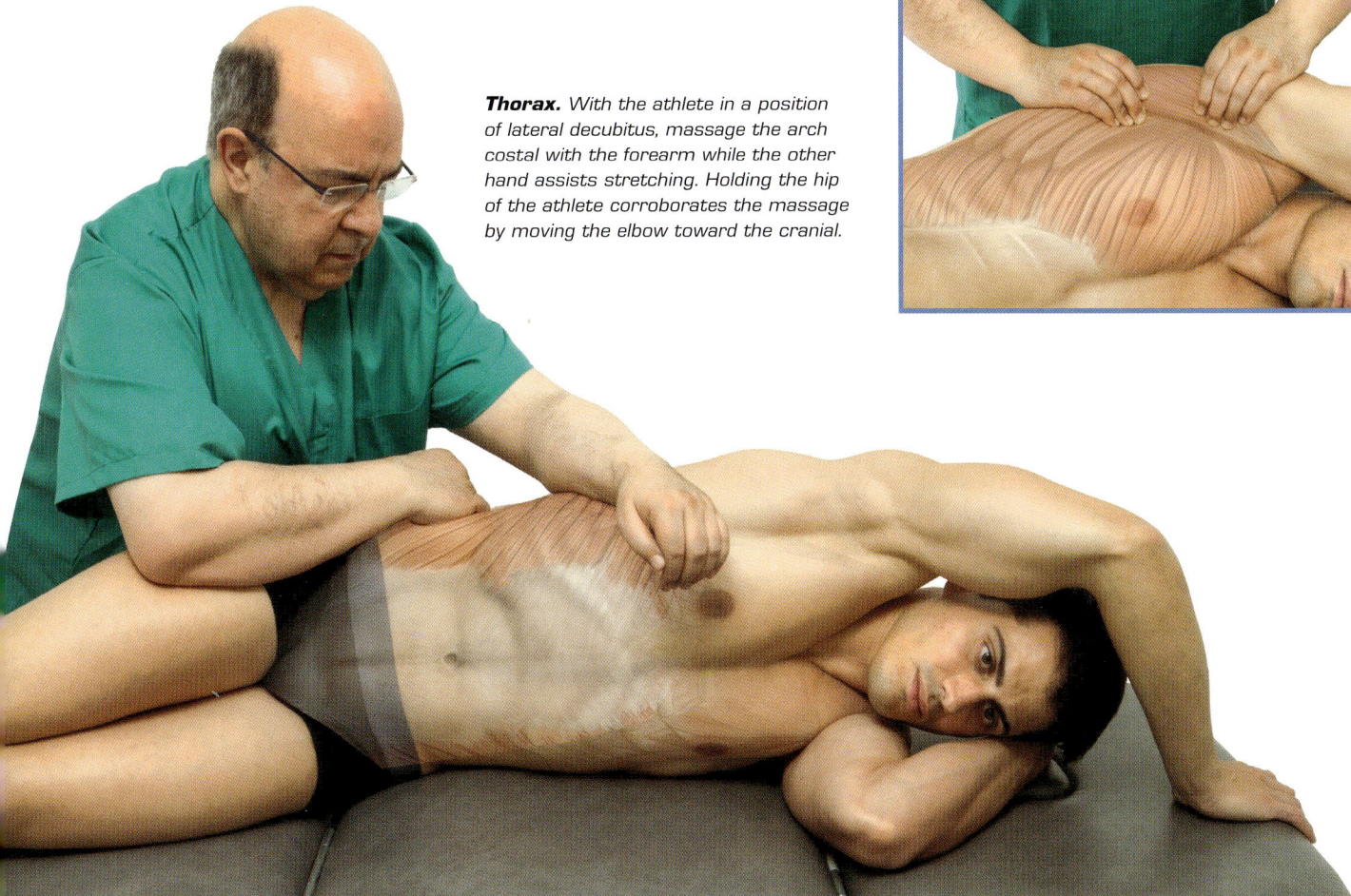

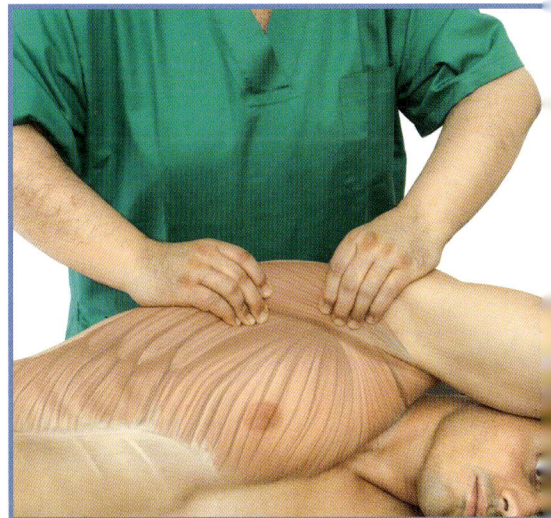

Upper body II: head and neck

General Techniques

◆ Pressure strokes

Devices

◆ Elbow
◆ Palms of the hands
◆ Knuckles
◆ Fingertips

Three spinning sectors in motion

When walking or running, three segments of the spinal column contribute to movement in the transverse plane. The lumbar spine moves with the pelvis rotating to one side. The thoracic spine, thorax and scapular girdle counteract the opposite side by exerting tension on the obliques. The cervical region and the head remain uncoupled, allowing them to remain facing forward.

Oscillatory movements of the arms stress the rhomboid muscles through the spinous apophysis to the head splenius. This process allows the upper part of the thoracic spine and the head to rotate in the opposite direction to that of the lumbar spine and dorsolumbar region.

Evaluation Tactics

It is important to establish the rotation capacity of the head and neck. The seated athlete is asked to perform rotations of the neck and head from right to left. In the case of a limitation in one of the two rotations, we can establish which chain is tenser and limiting the path. A shortened right anterior cross-chain will limit left rotation, and vice versa.

In other words, when inspecting the posture in an individual, if the head and neck are rotating toward one of the two sides, the anterior cross-chain on that side will be shortened and tense and will require massage techniques to elongate it.

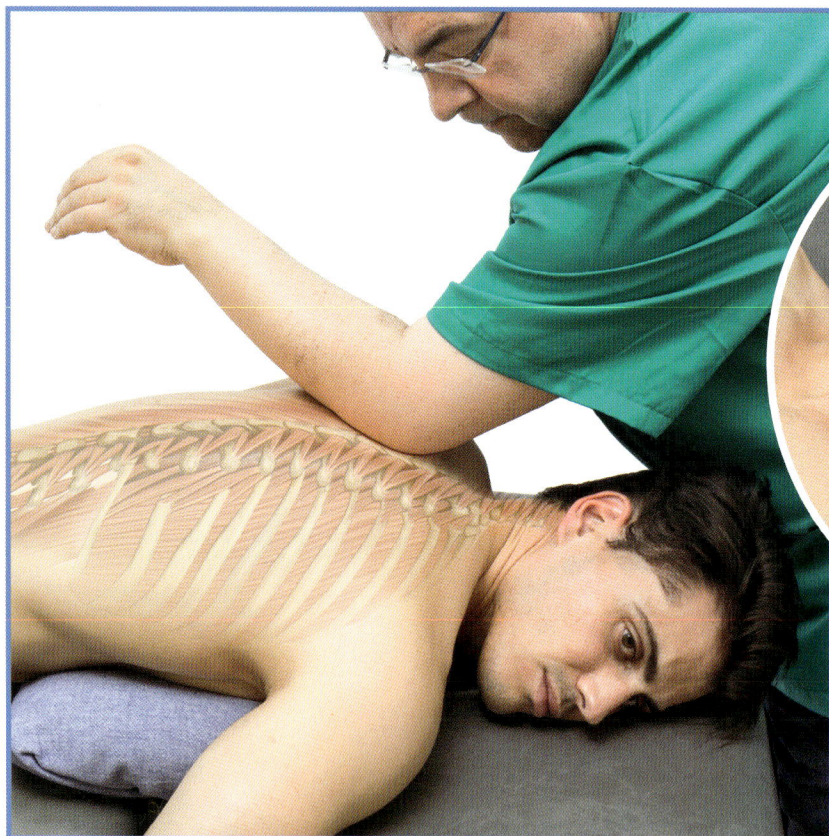

Neck. With the athlete in prone position and the head rotated away from the masseuse, slight pressure is applied on the upper ribs while the patient breathes using the contact zones. At the same time, raise the head of the couch slightly and turn it around to the opposite side from the pressure has been applied. Move laterally until reaching the medial border of the scapula. This can be combined with maneuvers on spine erectors (paravertebral) to establish connection between the neck and the head.

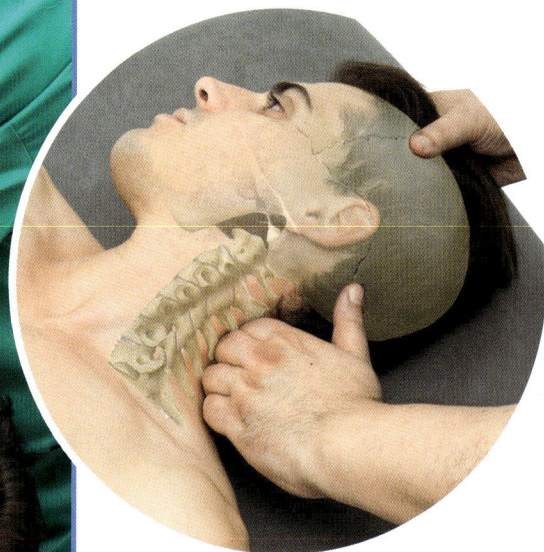

Neck. Using the knuckles with a relaxed fist, apply slight pressure on the spinous apophysis while the athlete turns the head in the opposite direction. Perform the maneuvers from the back of the neck to the lateral area.

> *The only involvement of the spine and scapula girdle and their movements, added to the myofascial tension that these would generate by action on the anterior cross-chain, could allow an individual to move forward or back without needing to use their legs and the arms, thanks to torsion movements in the transverse plane.*

Head. The same maneuvers that we examined during the massage of the lateral chain can be used for massaging the head. Use massage to relax all structures and sensitive points that appear on the back and side of the head.

Head. Make frictions on the head by moving the scalp until you locate small sensitive nodules. These are treated with gentle relaxing pressure, gently pulling hair strands or as final maneuvers in the treatment of the upper body, as they imbue relaxation and general well-being. Some athletes call this technique "zeroing" or "reset."

Head. Decompress the cranium. Place relaxed fingers and make small tractions on the base of the skull (occipital). During maneuvers the skull and head are held with the thumbs and palms of the hands. Perceive decompression by touch. To assess the degree of relaxation, note if your patient's breathing is slow and deep.

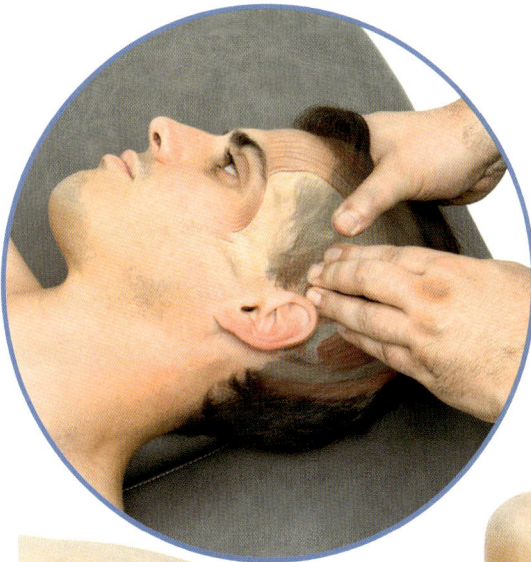

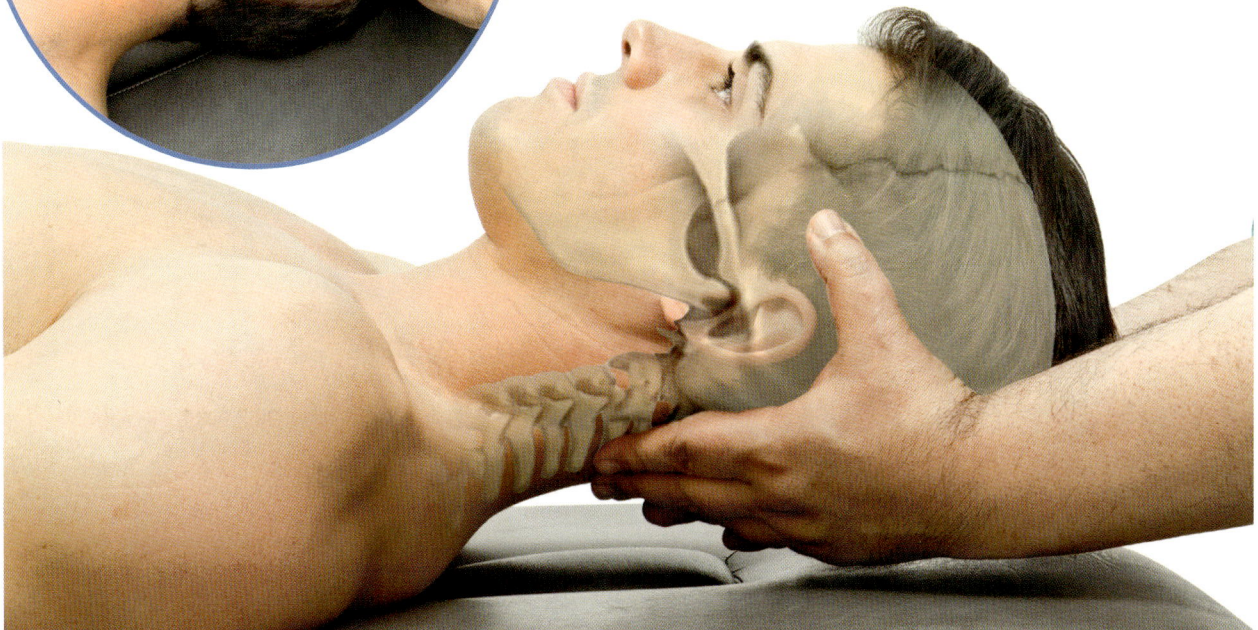

Posterior myofascial cross-chain

The same partners for the same dance

The posterior cross-chain consists of several planes of myofascial tissue in the shape of a band that runs and passes through structures of the other chains.

Thus there is one band for the right posterior cross-chain and one for the left. Both chains cross transversally and trace an "X" at the back of the body (like the anterior cross-chain), resulting in an encounter between both bands at the level of the thoracolumbar fascia. These two myofascial chains are stretched or shortened as they contract or stretch during realization of the corresponding movements.

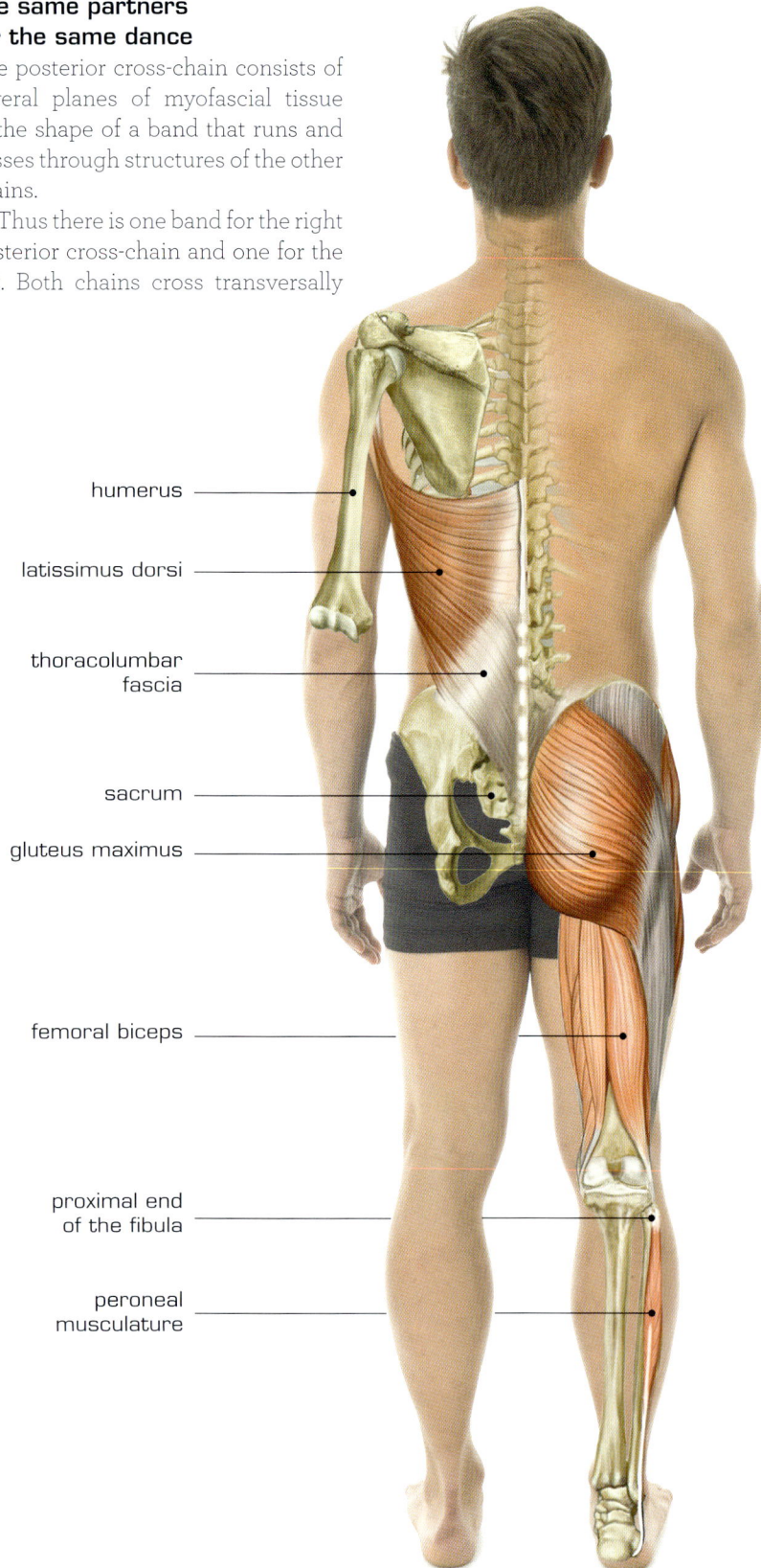

There is an obvious relationship with the previous cross-chain, since the shortening of one implies the elongation of the other.

Toward a new way of training and treating

When considering chains it is understandable to think that flexibility training (stretching, mobilization, massage, etc.) in any of the structures of a myofascial chain will increase the range of movement available in other structures related to the same chain, even if they are a considerable distance apart. Thus, the Ardha Hanumanasana yoga position, allowing extensive stretching of the gluteal region and the fascia latae, will improve mobility of the contralateral arm and vice versa.

Despite this, training protocols used to date implicate flexibility training in an isolated way (by muscular units) and should be reinterpreted and adapted to this new vision. Standardized positions still in use should be redesigned, including, for example, torsion and elongation models, which take into account the dynamics of the worked chain, as well as correctly exercised force.

humerus

latissimus dorsi

thoracolumbar fascia

sacrum

gluteus maximus

femoral biceps

proximal end of the fibula

peroneal musculature

The posterior cross-chain straddles the structures of other chains and traces an "X" in the body's posterior area, resulting in a crossing at the level of the thoracolumbar fascia region. It is responsible for torsion movements.

Cooperation to store elastic energy

In the anterior face of the body these two cross-bands are arranged to form an "X", constituting the previous cross-chain, and on the posterior side the two cross-bands form the posterior cross-chain. This cross-over formula creates a link between the right pelvic girdle and the left scapular girdle (or vice versa), and this relationship facilitates effective movement control during standing and arm-to-hip sway while walking.

Characteristics of cross-chains

The myofascial framework stores elastic energy during rotation in the opposite direction of the scapular and pelvic girdles. These spiroidal movements generate dynamism and potency, which are the main motor characteristics of both the anterior and posterior cross-chains.

A chain for flexion

The synergistic action of the two anterior cross-chains causes body flexion. In the same way, the joint action of the two blades of that "X" in the posterior plane produces extension of the body. Both anterior and posterior chains assist and cooperate with each other, fulfilling a mainly postural role.

The practice of walking with hands in pockets or carrying an object impairs the transmission of slight oscillating movements when walking. To eliminate this impaired balance, the joints and soft tissues must be ossified.

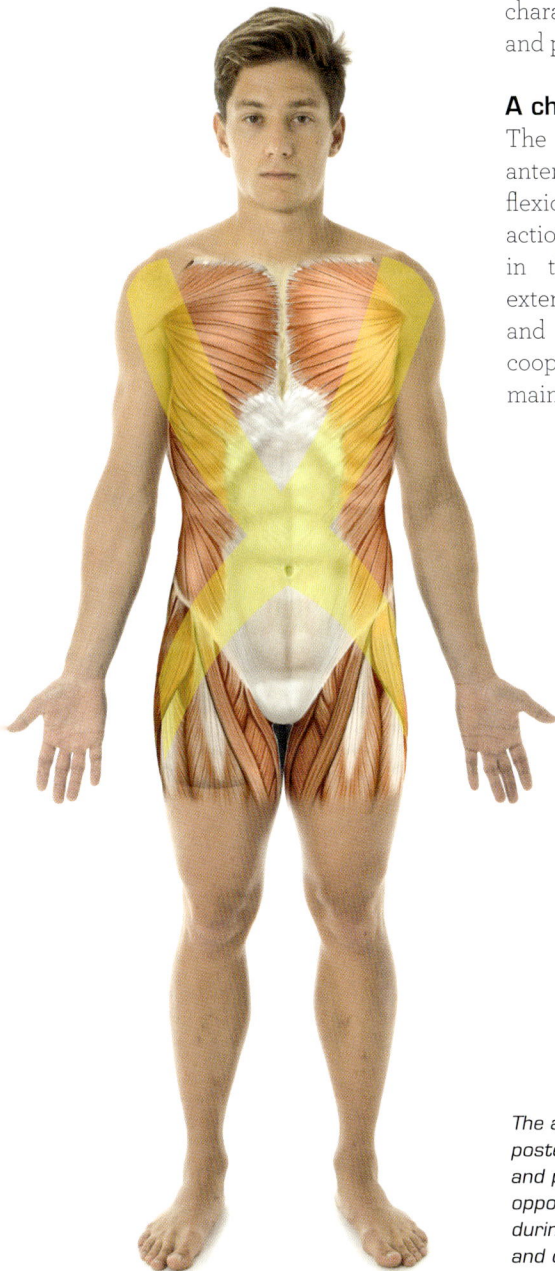

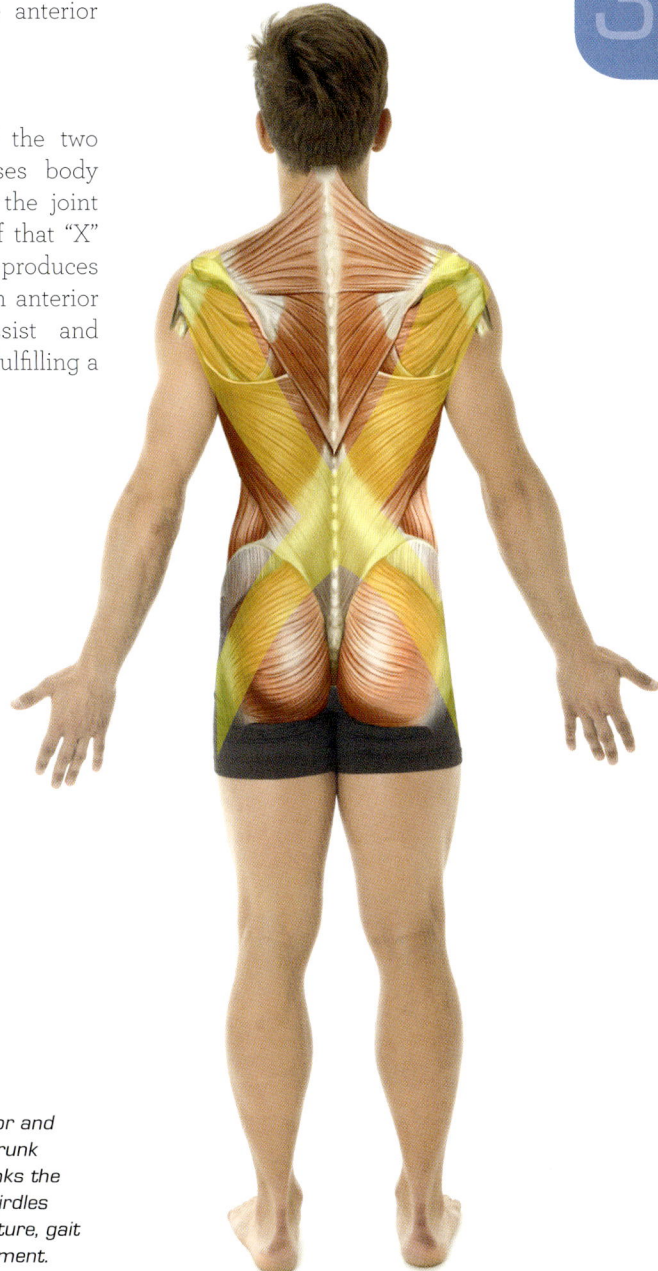

The arrangement of the anterior and posterior cross-chains in the trunk and pelvis traces an "X" that links the opposing pelvic and scapular girdles during the maintenance of posture, gait and during any technical movement.

3

Functions of the posterior cross-chain

Any action involving a dissociation of the pelvic and scapular girdles involves torsion movements. These arches provide the elastic energy necessary to perform many different actions and without which movement would lose its rhythm, harmony and grace; movements would be those of a robot, lacking the fluidity and expressiveness prevalent in living beings. Thanks to these rotations and the consequent accumulation of elastic energy, activities are carried out with less effort.

Despite its clear functional role — torsional movement of the scapular and pelvic girdles — it is associated with the corresponding upper and lower limbs, the posterior and anterior cross-chains and also plays an important role in posture, providing overall stability. They clearly collaborate with postural chains such as the anterior and posterior chains. Instability on one side (detected by proprioceptive sensory devices, among others) must be stabilized on the other side.

A torsion chain

When experiencing scapular and pelvic girdle torsion in opposing directions, the shoulder on the opposite side of the foot tends to approach that foot during the chain's closing movement. The extension of the lower right limb is associated with the extension of the upper left limb.

When the chain is opened, the opposite occurs. Rotations in the girdles are made in the opposite direction from when the chain closes. These new torsions cause the lower right limb and the upper left limb to flex.

There is thus an alternation in the closing and opening of the rear cross-chains: when the right rear cross-chain is closed, the left is opened, and vice versa.

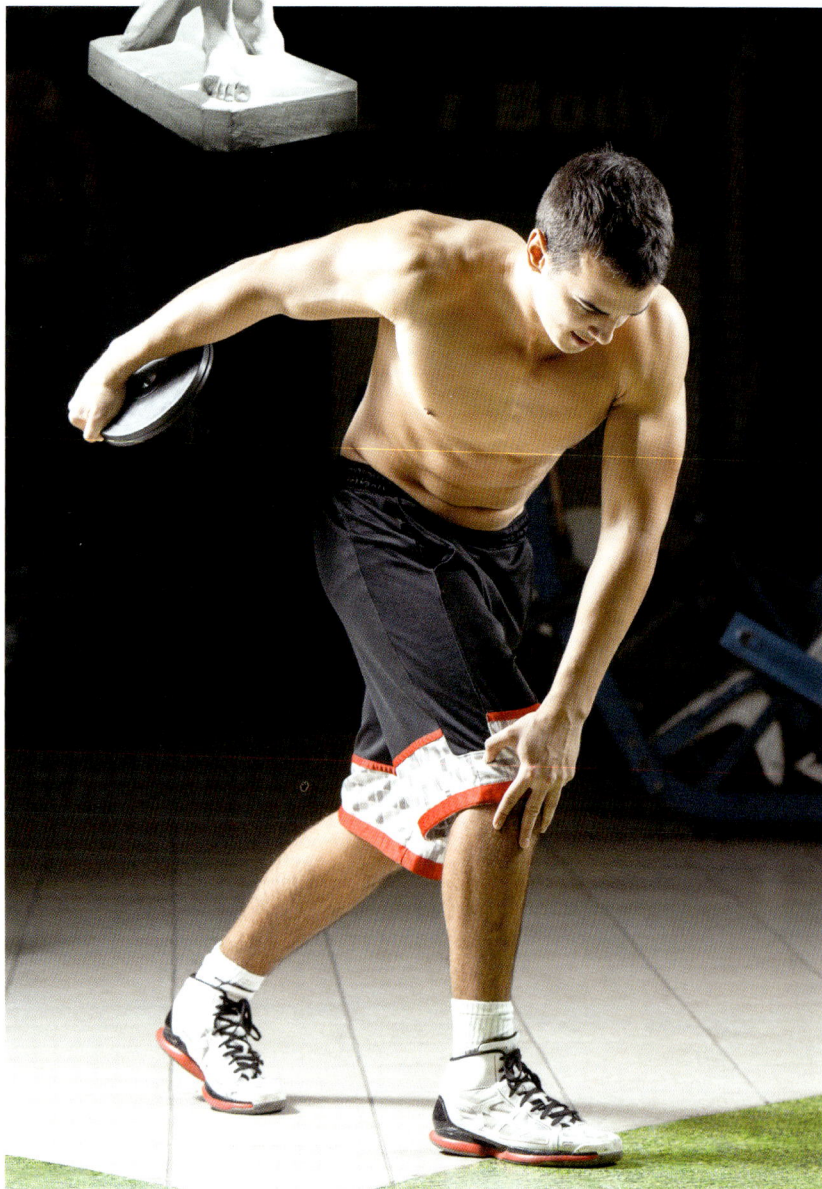

These two representations reflect tension and great dynamism in the form of balance and torsion between the pelvic and scapular waists. They express accumulated energy (in the myofascial continuum) by affecting a stable rotation, allowing the discus to be thrown with the greatest possible force and in the right direction.

Detection of compensations

The accumulation of tension causes torsion in the transverse plane that forces the body to seek compensations in order to maintain maximum functionality. However, these compensations put stress on the whole of the locomotive apparatus and other systems, which causes fatigue and, sooner or later, injury. Hence the importance of detecting such compensation as soon as possible.

Therefore, asymmetries and alterations in the athlete's balance must be sought through inspection, neuromuscular tests, periodic evaluations, etc. When the posterior cross-chain is affected by restriction one shoulder can be observed in retroversion (shoulder closest to the opposite hip). The palpation shows an excess of tension in the scapula region and in the latissimus dorsi on that side. The myofascia of the gluteus on the opposite side and the sacrum can be tense, retracted and detected at the back.

These excesses of tension somehow limit the respective movements in the shoulder and hip joints. The lumbar spine is subjected to torsion forces in the opposite direction, with the probable appearance of lumbago and other similar clinical pathologies.

All of the techniques applied to the lumbar, pelvic and thoracic regions should aim to release soft tissue (ligaments, aponeurosis, tendons, muscles). The hamstrings, gluteals, pelvic muscles, psoas, lumbar square, erectors and the pelvic region's powerful ligaments must present stability. A combined stretching and movement massage will allow rotational movements between girdles to be easy and painless.

The thoracolumbar fascia has superficial, middle and deep layers that are related to the muscles of the abdominal area: lumbar, oblique, transverse, psoas and latissimus dorsi. These structures will subject it to tension and sliding as a result of their contractions.

Precautions

◆ If preventive or therapeutic action is not performed quickly, these symptoms have a propensity to become complicated; altered joint pressures can cause joint degeneration, osteophyte growth and stiffness in the myofascial complex.

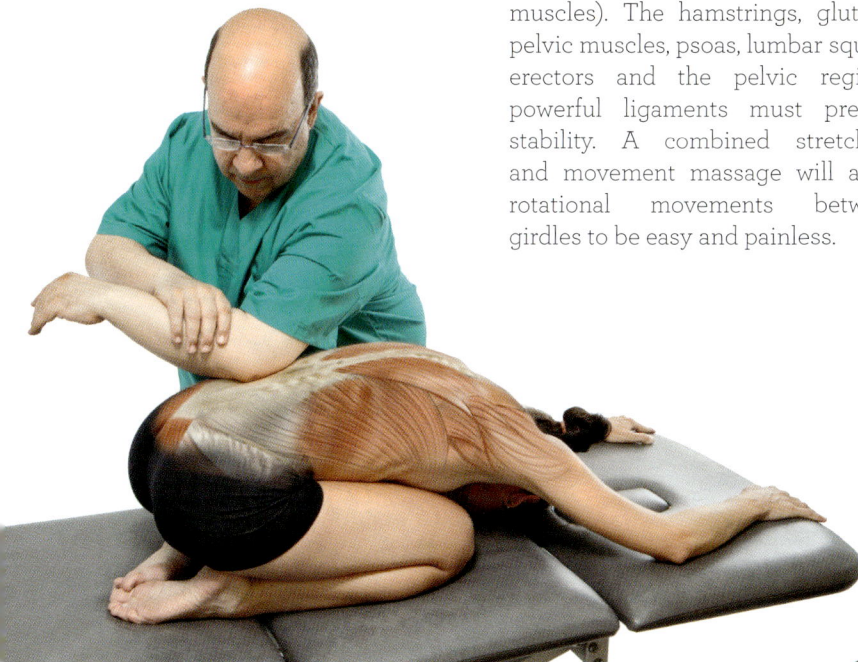

__Thoracolumbar fascia.__ One formula when working this area is to ask the athlete to adopt a variant of the genupectoral position and to work the zones with excess tension by means of friction and drag techniques with the fingertips of both hands or the elbow, according to the plane that requires treatment, while observing respiratory rhythm.

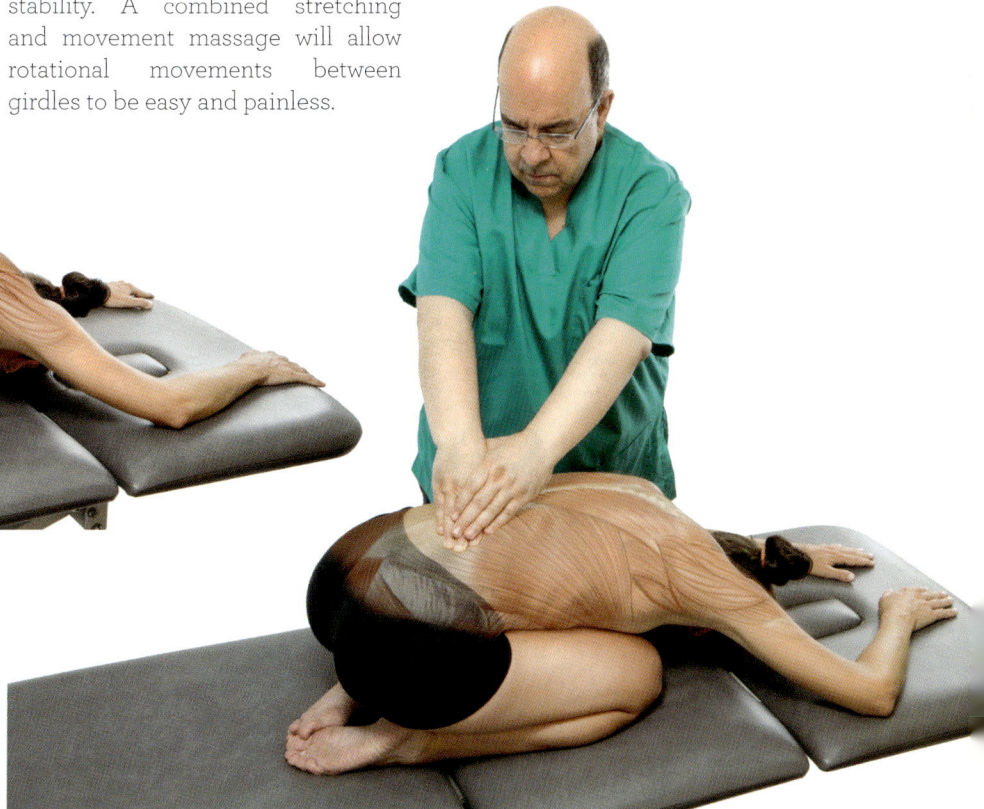

Lower body I: feet and legs

General Techniques

◆ Pressure
◆ Friction
◆ Transverse friction

Devices

◆ Fingers and fingertips
◆ Reinforced index finger
◆ Knuckles

Anatomical accuracy

The tibiotarsal or ankle joint is formed by various structures of the leg and foot (tibia, fibula and talus) and is responsible for the dorsal flexion movements in the sagittal plane and plantar flexion of the foot. One of its elements, the talus, forms the "astragaline trochlea" which acts as a pulley for the tendons that pass through it on the way from the leg to the foot. It is the only bone in the foot that lacks muscle insertions but is instead "entirely covered with ligament insertions" (Kapandji, 2012). Its articular mobilization guarantees it functions properly as a pulley in cases of stiffness.

Responding to two requirements

The foot should offer sufficient rigidity so that it can act as an impulse lever during a step and provide enough strength to support much of the body weight. It must adapt to the forces of gravity and accommodate for irregularities in terrain; it must also have the capacity to absorb impacts during the step cycle.

Orienting the plantar cavity

The tibioperoneo-astragaline complex receives help from axial rotation of the knee and behaves as a single joint with three degrees of freedom of movement, allowing the plantar cavity to orient in all directions and adapt to uneven terrain.

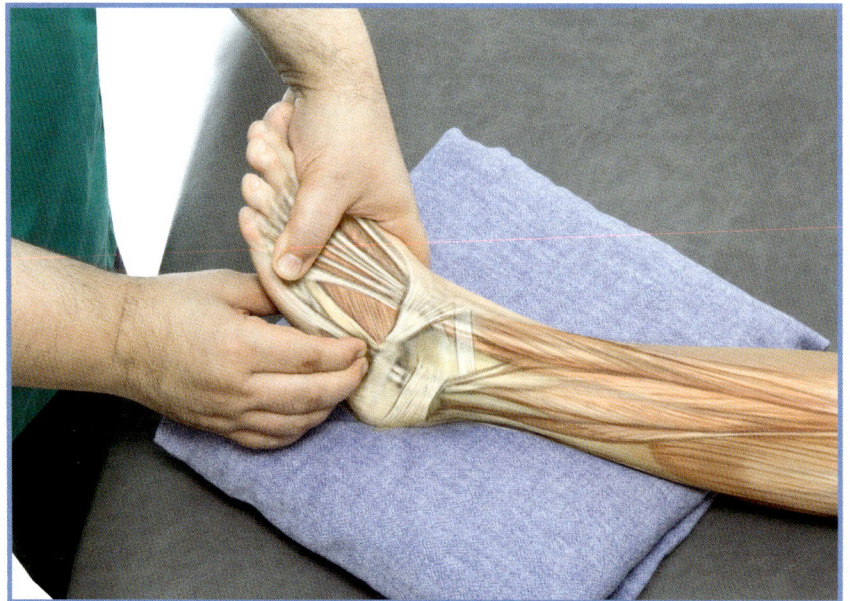

Foot. Transverse friction massage is applied over the insertion site of the short peroneus at the base of the fifth metatarsal.

Foot. Cross-friction massage over the path of peroneal tendons in the ankle region behind the lateral malleolus.

Individuals with excessive pronation may experience thickening of the epidermis or hyperkeratosis under the metatarsal heads of the second and third toe. This is due to excessive compression that these heads effect on the tissue and against the ground.

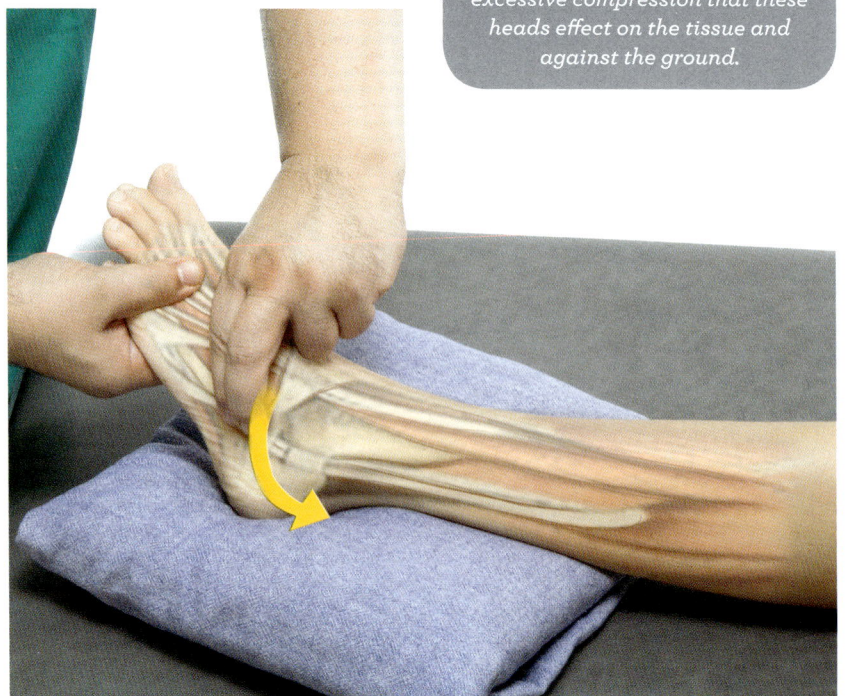

Athletes are considered to have a higher plantar arc than sedentary individuals. Conversely, activity does not influence features of the plantar footprint too a great degree.

The Helbing line or angle is an imaginary line that divides the heel into two symmetrical halves and is used to measure its degrees of deviation.

Coordinating run and walk

Long tendons that are inserted in the sole of the foot create a corrective system of the arches of the foot. The anterior tibial and the long peroneus, which generally work in opposition, cooperate to elevate the base of the first metatarsal in the internal plantar arch, since both are inserted there. The short peroneus does the same to the external arch when inserted in the fifth metatarsal.

The calcaneus leans during the heel contact phase when walking, forcing the leg to move medially. The contraction of the peroneal group participates in this eversion of the ankle and foot. The plantar arch is flattened.

Ankle sprain is a frequent incidence in athletes. This injury affects the functionality of the leg and the stability of the entire body. If an old injury of this type does not recover correctly, there may be recurrences or compensations that affect other regions near or far from the ankle, also affecting gait.

Evaluation tactics

Looking at the athlete from behind we can assess the angle that the heel forms with the rest of the leg. It is called Helbing's vertical line (Achilles tendon line), which must pass through the center of the popliteal hollow (back of the knee) and the center of the heel. If the heel is tilted inwards (valgus), it is pronated; if it is tilted outwards, it is supinated.

Other interesting valuations are the study of the athlete's tread, observation of the height of the navicular during unloading and loading, etc. All of this will allow the therapist to appreciate sporting characteristics of the use of the foot and lower extremity in each individual case.

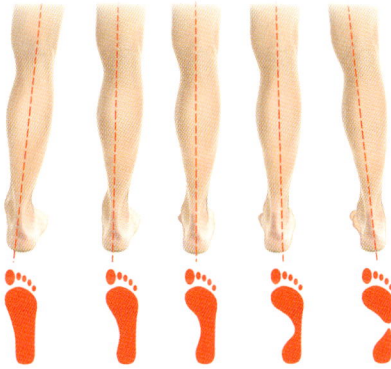

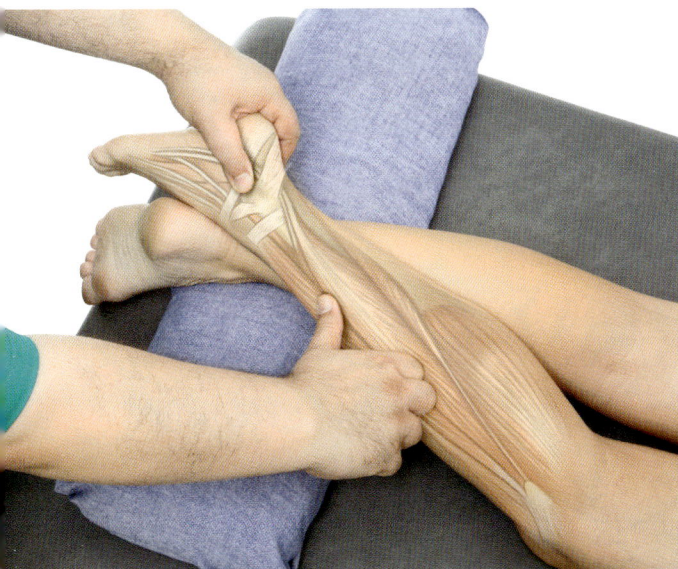

Leg. Traverse the external part of the leg (peroneal area) using pressure and friction massage techniques with the knuckles, fingers or elbow. Gently but insistently massage the areas of greatest tension.

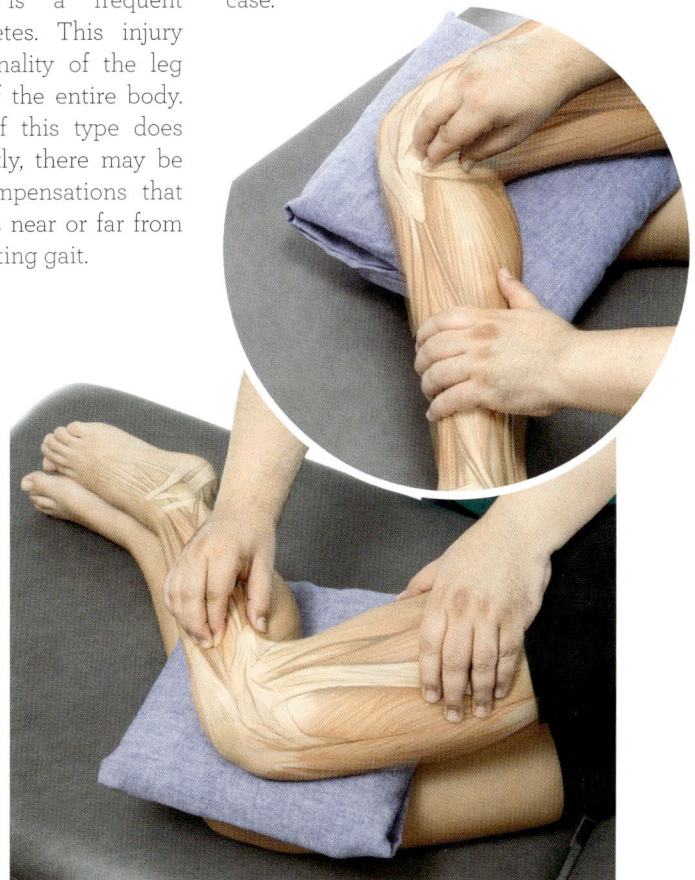

Leg. The upper area of the outside of the leg near the knee (head of the fibula) may present areas of tension and contractures. They must be located and treated with the fingers through transversal pressures maintained until the therapist perceives that they have been released.

Lower body II: thighs and pelvis

General Techniques

◆ Friction
◆ Pressure

Devices

◆ Fingers
◆ Knuckles
◆ Elbow

Relationship and translation areas

During the step cycle, an internal rotation takes place when the heel acts as a support. If you follow this rotation upwards you arrive at the lateral or external part the myofascia extensor of the hip, the "deltoid, gluteal" group. Within this group, the major trochanter acts as a "switch" that is "triggered" during hip flexion, necessary to support the heel. The gluteus maximus activates the entire external extensor area: the posterior part of the iliotibial band, biceps femoris and lateral peroneus.

A great deal of force is required to avoid falling over in the heel support phase, as the center of gravity is placed firmly behind the base of the foot on the ground. Numerous muscles and connective tissue of the aforementioned chains are put into operation; the hip extensors and biceps femora in particular realize a powerful contraction.

Evaluation tactics

Ability to lift the lower limb is assessed. The athlete is placed in supine position, with the legs extended and the knees in flexion at about 80°. It is possible to apply slight external or internal rotation in the hip to assess if mobility restriction in the posterior leg group is located in the internal compartment (semimembranous and semitendinosus) or external (biceps femoris) compartment. During this test both limbs should be compared, and it is performed at different angles to establish their state.

The vast majority of musculoskeletal problems appear to be associated with pain caused by this limitation, which can lead to misclassification by confusing, for example, pain in the posterior face of the thigh with specious sciatica.

Thigh and pelvis. Massage in the hamstring and the ischium (the boney region on which we repose when sitting), should never be overlooked when treating the posterior aspect of the thigh. A deep plane must be accessed by transverse, longitudinal or circular friction, with the index finger reinforced by the third finger; occasionally the elbow may be used. During massage the athlete can perform knee extensions to tighten the structure.

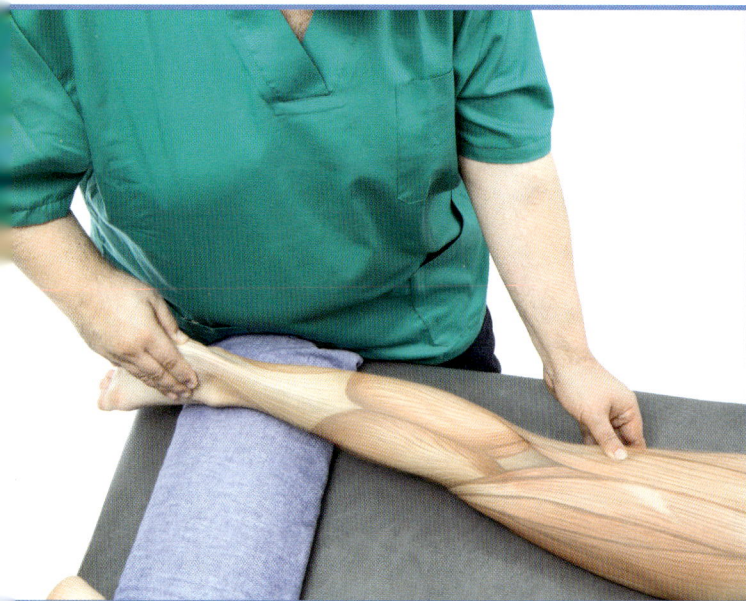

Thigh. Friction is applied with the fingers, knuckles and even the elbow on the outside and back of the thigh. So that the therapist can better appreciate the rounded section and tendon of the biceps femoris muscle, the athlete is asked to flex the knee against resistance applied by the therapist's other hand.

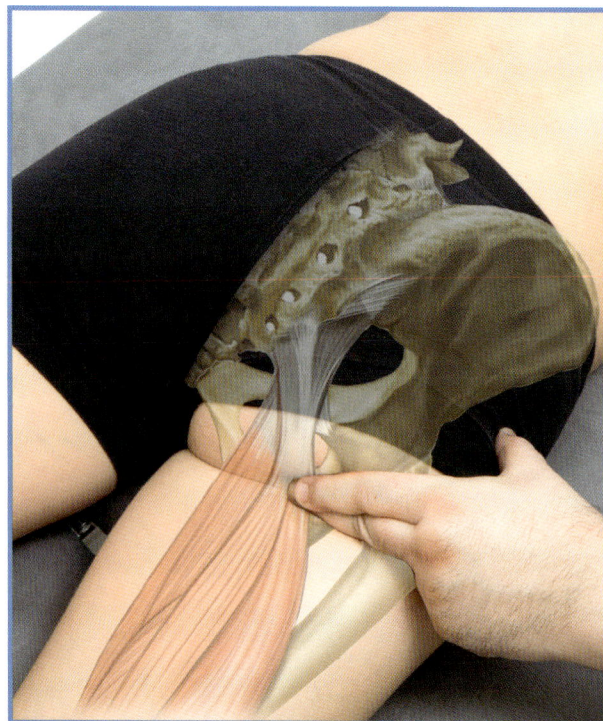

Precautions

◆ The gluteus and sacrum area requires delicate and sensitive handling in order to respect the athlete's modesty. The therapist should therefore explain to the athlete what he plans to do, rather than placing his hands on this area unexpectedly.

The thigh-pelvis relationship

When walking (during the flexion and extension that occurs in the thigh during the step cycle), the iliac muscle should tilt back and forth respectively, indicating that the posterior and anterior chains are functioning normally.

Anatomical precision

As the heel impacts against the ground the femur is pushed obliquely up and back, dragging the iliac in a posterior inclination (counterclockwise rotation). On the side that presents an inclination in the iliac bones, the sacrum assumes a position of nutation (elevation of the coccyx). In the same context, the opposite lower extremity is ready to start the take-off phase. The extension of the thigh places the axis of the femur in an oblique position. The femur is pushed forward and up, dragging the iliac on that side in anterior inclination (clockwise rotation). This return to the initial, neutral position from the nutated position is called contra-nutation (coccyx points toward the ground).

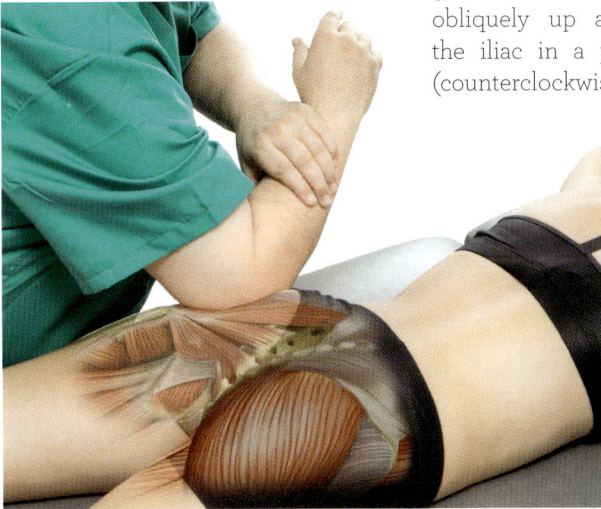

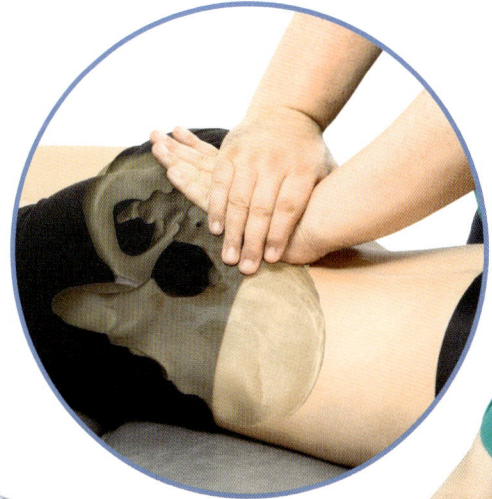

Pelvis. With the athlete prone, massage maneuvers are performed with the elbow or knuckles in the gluteus maximus. The insertion site of the gluteus maximus muscle initiates in the sacrum and in the posterior part of the iliac crest; from there we head toward the boney protrusion of the hip (major trochanter). The athlete is asked to perform both hip rotation or anteversion and retroversion movements of the pelvis.

Pelvis and sacrum. The prone position is used to perform maneuvers in the sacrum. Pressure techniques are applied to the lateral edges of the sacrum with the knuckles or elbow while the athlete performs anteversion and retroversion movements of the pelvis. One hand is placed on the sacral bone while the other acts as a support. Gentle pushing is applied diagonally and downwards, considering the rhythm of the athlete's breathing.

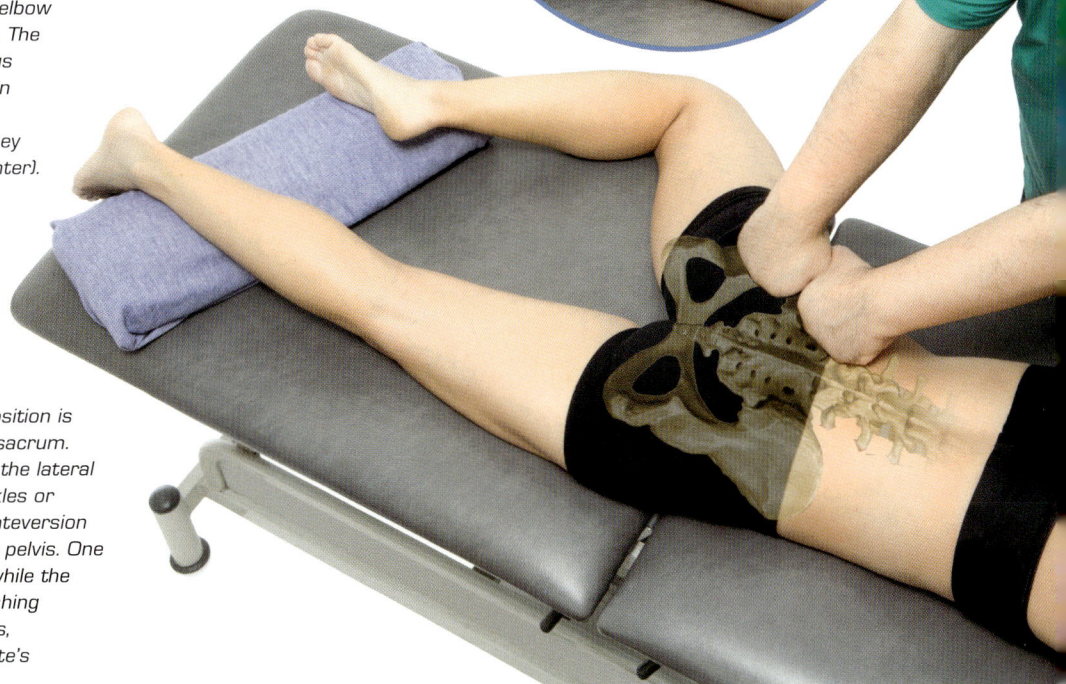

Upper body I: trunk

General Techniques

◆ Pressure
◆ Friction and drag strokes

Devices

◆ Fingers and fingertips
◆ Knuckles
◆ Palm of the hand
◆ Elbow

Modules involved in relationship

Any movement requiring rotation or stability requires a balanced musculature of the trunk. If you imagine and analyze yourself walking or running, you will notice that the arms and legs are making opposite rotational movements in a clockwise direction through the scapular and pelvic waists respectively.

This alternating balance is manifested in the extremities. When the right arm swings back, the left thigh moves forward, and vice versa. This rotational relationship allows us to harmoniously continue forward movements. For this to occur in a coordinated way there must be a more or less stable regulation zone. This regulation corresponds to the correlation between the lumbar-pelvic region and the lower part of the trunk, which controls rotational forces when necessary.

Sports that require significant rotation between the two girdles (golf, water polo, racquet sports, judo, etc.) mean that a more stable posterior cross-chain is even more essential.

Precautions

◆ Changes in gait patterns can be observed during examination, such as undulating, anserine gait, placing the feet wide apart, a marked instability in the load-bearing hip, and a side-to-side motion similar to a waddling duck. This may be due to dysfunctions in the lumbar-pelvic area, weakness of the gluteus medius or deformity of the neck of the femur or coxa vara.

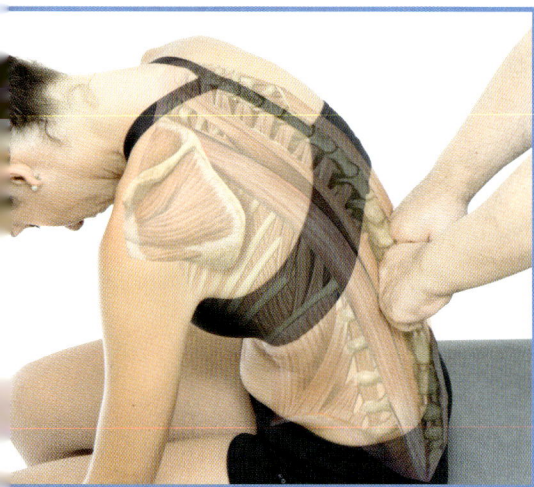

Thoracolumbar trunk fascia. The thoracolumbar fascia is treated by friction and drag with the fingertips or the elbow. The athlete adopts a variant of the genupectoral position in which he sits on his heels and stretches his arms out on the couch. This allows the therapist to treat and stretch the fascia.

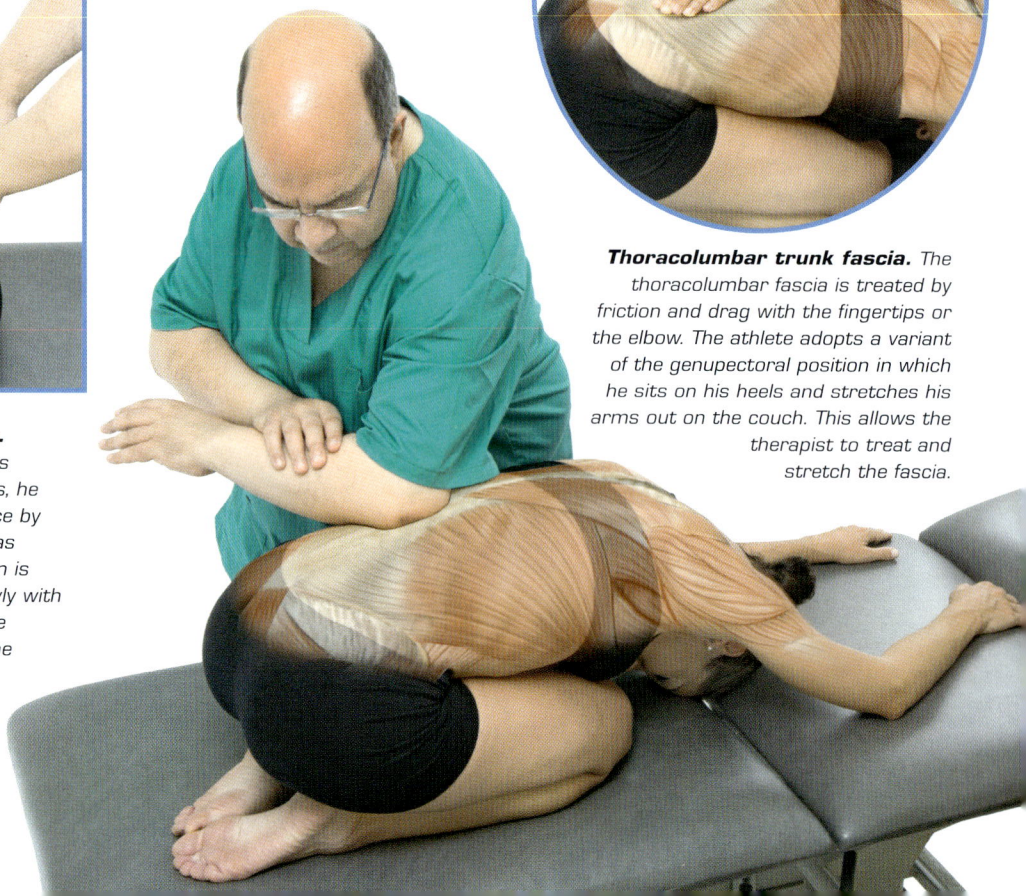

Thoracolumbar trunk fascia. With the athlete sitting and arms hanging by the sides of his knees, he is asked to exert an upward force by pushing his feet on the ground, as if he wanted to stand up. Friction is simultaneously applied very slowly with the knuckles on both sides of the spinous apophysis, directly on the spine erectors and downwards toward the sacrum or between the erectors and the lumbar square.

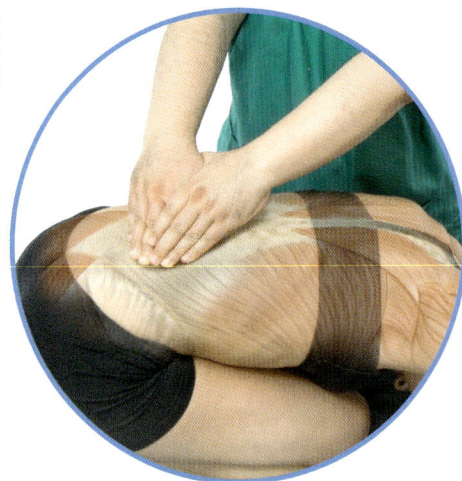

Harmonious strolling

To execute something so complex and at the same time so simple as walking in a fluid and functional movement requires a connection between both sides of the body. If one side moves more than the other, it will cause a chain discordance that will be reflected elsewhere, often in the sacroiliac joint. When trying to compensate either this is compressed to provide stability to the system or the thoracolumbar fascia is tightened. This leads to areas of excessive tension in the lower part of the back, despite repeated stretching and releasing actions.

Sensorimotor integration

The body will always achieve stability when it needs it, and the joints are no exception. If the athlete does not integrate mechanisms for the motor control system to provide the necessary information to correct a dysfunction after numerous sessions to increase and maintain range of motion through stretching, massage techniques, joint mobilizations, etc., even after having released a dysfunctional pattern, the body will enter a vicious circle. The joints, muscles, fascia and ligaments will become tense again in a fresh bid for stability.

Therapeutic strategy

The body will not know what to do when faced with a great range of motion and extensibility if it lacks the corresponding sensorimotor control of the gluteus maximus, latissimus dorsi and spine erector muscles. After obtaining an increased range of motion, the motor control must be reduced so that the sacroiliac joint closes and the lumbar spine stabilizes.

Precautions

◆ In massage techniques on the latissimus dorsi it is necessary to keep in mind this muscle's relationship with the limb and the scapula, in addition to its links (through the thoracolumbar fascia) with the contralateral gluteus major.

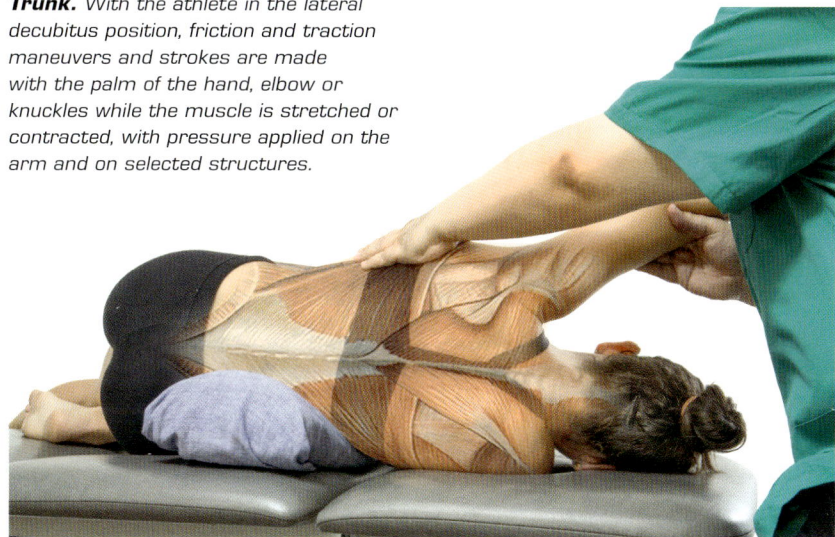

Trunk. With the athlete in the lateral decubitus position, friction and traction maneuvers and strokes are made with the palm of the hand, elbow or knuckles while the muscle is stretched or contracted, with pressure applied on the arm and on selected structures.

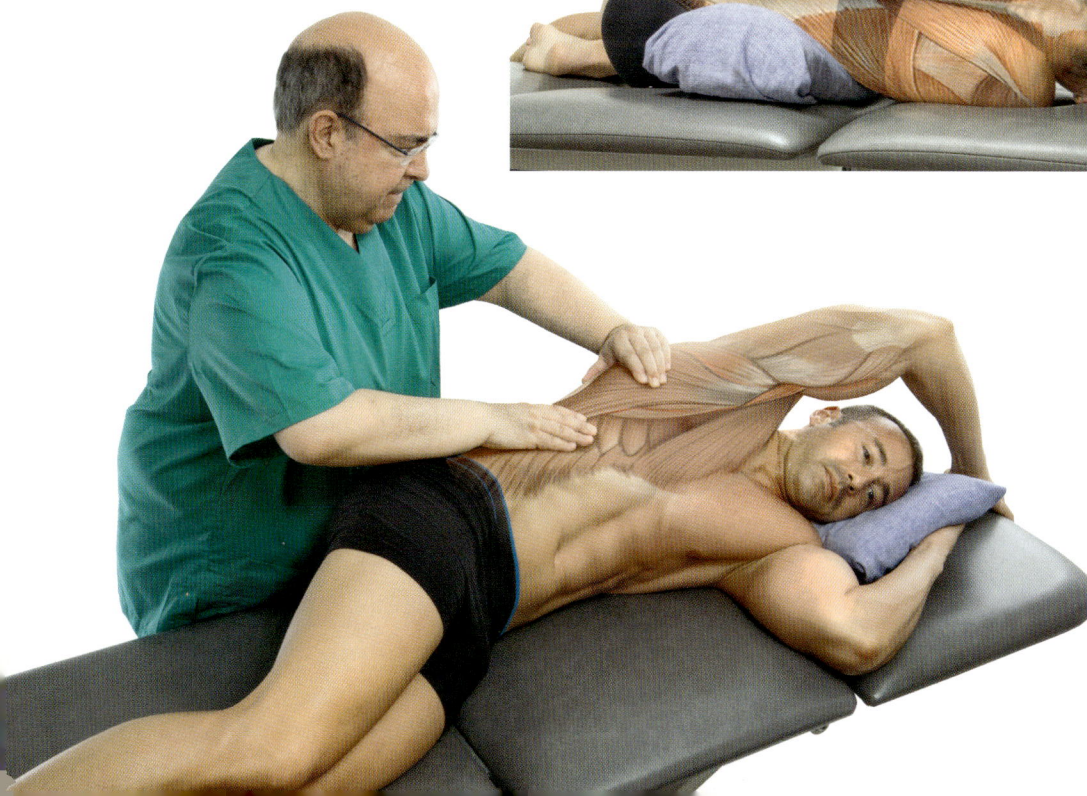

Trunk. The therapist is positioned behind the athlete who assumes the lateral position with one arm resting on the head, elbow bent and hand supported on the couch. Considering the rhythm of the athlete's breathing, perform opening maneuvers on the rib cage and myofascial framework of the treated area.

Anatomical integration

3

The integration of myofascial chains can be explained as an interconnection between all of them, forming a network of muscles and connective tissue while acting as the link between organism and musculature systems. They are integrated with all bodily organs and systems, and this is easily understood by using different approaches: anatomical, nervous, metabolic and emotional.

Anatomical continuity

Anatomical continuity of the myofascial chains is provided by the joints. Bones and muscles share links in the collagen network that forms the fascia. The joints that we find in a limb or in the trunk form a continuum in which each joint depends on the others. Therefore it is impossible to make a movement without activating that interdependence, with other parts moving and adapting to produce the movement. Collagen of the ligaments, tendons and periosteum is the "glue" that unifies the myofascial chains and the joints. Myofascial groups are organized in the form of loops, ringlets, routes, etc. They relate to other groups who have to take routes through other joints. For example the posterior chain and the cross-chain have insertions in the pelvis at the coxal and femur. The hip is the hub where these two chains are relayed and coordinated. Regulation between several myofascial chains in order to perform a correct transmission of forces is performed by periarticular bone fasteners (tendon-bone-ligament-bone-tendon).

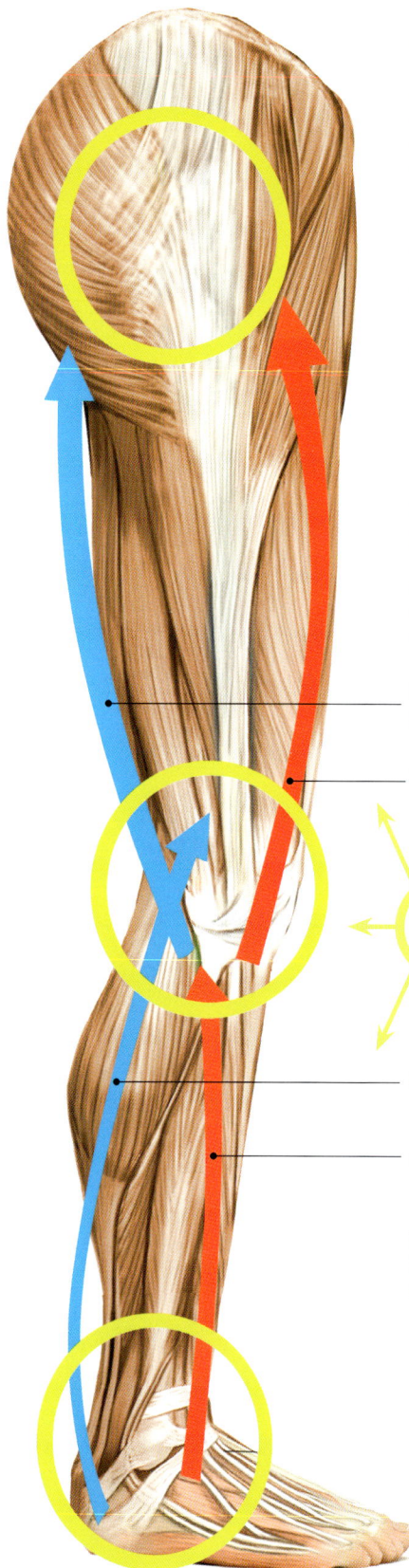

The integration of different joints in an articular chain is highlighted in yellow circles: the ankle-knee-hip joint continuum. Tendons of the anterior and posterior chain (red and blue arrows, respectively) coincide through the continuity of bones, capsule and ligaments. Integration, the "neurological order," occurs instantly and for a particular movement. For example, we can jump thanks to the cooperation between the gastrocnemius (posterior) and the femoral quadriceps (anterior). We maintain an upright posture thanks to cocontraction; muscular activations always occur in pairs (soleus-tibial anterior and quadriceps-hamstrings).

Myofascial chains (integration)

Neurological integration (muscular organization at the instance of the myofascial "network")

Stability of the pelvis and knee (cocontraction)

Triple extension (jump)

Integration of different joints into a chain formed by the continuum: "ankle-knee-hip."

Postural organization

Ankle stability

Integration by joint chains (continuous articulation by bone anchors and joint surfaces)

Transmitting energy to movement

The tension that is generated at a fulcrum is transmitted throughout the myofascial system. This tension is channeled in different directions as determined by the myofascial chains involved in the sporting movement. For this transmission of stresses (i.e., energy of motion) to be effective, the active joints must remain stable and guide the movement through the muscles and fascia. A three-dimensional network of tensions is created throughout the body, allowing us to move the indicated muscles. That is to say, the muscles apply "surge forces;" they are generators and transmitters of movement through this network because they formulate and animate movement. They are chains linked together by other chains.

The joint, a confluence zone

The joints are three-dimensional crossing points in which one chain connects with others. Muscular insertions in the boney prominences are the "exchange hubs," routing forces of movement or acting as stabilizers. When jumping, for example, the forces that go up through the soleus are transmitted toward the knee, forward to the quadriceps and behind to the hamstrings. Insertions in the knee allow a triple extension/stabilization (of the ankle, knee and hip) when coordinating an anterior chain with a posterior one. This is how several chains integrate to make a single movement. The boney anchors of the joints divide or concentrate the forces originating from a segment to the rest of the locomotor apparatus, as well as to the body as a whole.

Simulation of the relationship between the action of a push chain (left) and a pull chain (right).

Neurological, metabolic and emotional integration

Neurological integration

Nervous innervation of the muscles refines contraction and modulates the coordination between chains and the sequence of contractions necessary to make a precise movement. The nervous system harmonizes two or more myofascial chains, causing their muscles to contract at varying intensities so that the whole can accurately execute the programmed voluntary motor movement. This "program" allows the execution and learning of motor gestures as if it were a piece of computer software. There is a "wave of contractions" of different intensities and types of contraction; it is possible to activate muscles aligned in the same chain or their antagonists, and also of a chain and its opposite chain. Isometric/concentric/eccentric contractions will stabilize or move one or more adjacent joint bone segments. These complex processes, which undermine the concept of "anatomical entity," are orchestrated by the nervous system.

Metabolic integration

In order to function, the entire system requires materials and energy in the form of nutrients (digestive tract). As energy is obtained in an aerobic form during exercise or when at rest, an adequate supply of oxygen (respiratory apparatus) is also necessary. Another network of arteries, capillaries and veins directs the blood pumped from the heart to the every periphery of the organism. Blood is the means by which all energy substrates reach the muscle fibers. Conversely muscle fibers "expel" residues: carbon dioxide, lactic acid, urea, fragments of protein, etc.; a mechanism is required to evacuate them. The veins and lymphatic system are responsible for collecting these products and carrying them to the lungs, kidneys and liver, where they are eliminated or recycled. Finally, the muscle produces and receives hormones (glandular system) and other regulatory substances that serve to integrate these functions (homeostasis) and, in turn, to integrate muscular activity in the whole of the organism. Myokines are hormones secreted by the muscle itself and are involved in the regulation of metabolism, especially energy, guaranteeing a supply of "fuel." Steroids, growth hormone, somatomedins and growth factors make the muscle develop and become functional, serving the body as a whole.

Imbalance between chains generates morphological changes that alter body functions.

Movement is understood not only as a set of associated muscle contractions acting in a chain, but also as global responses or actions that fit a particular medium and state of mind.

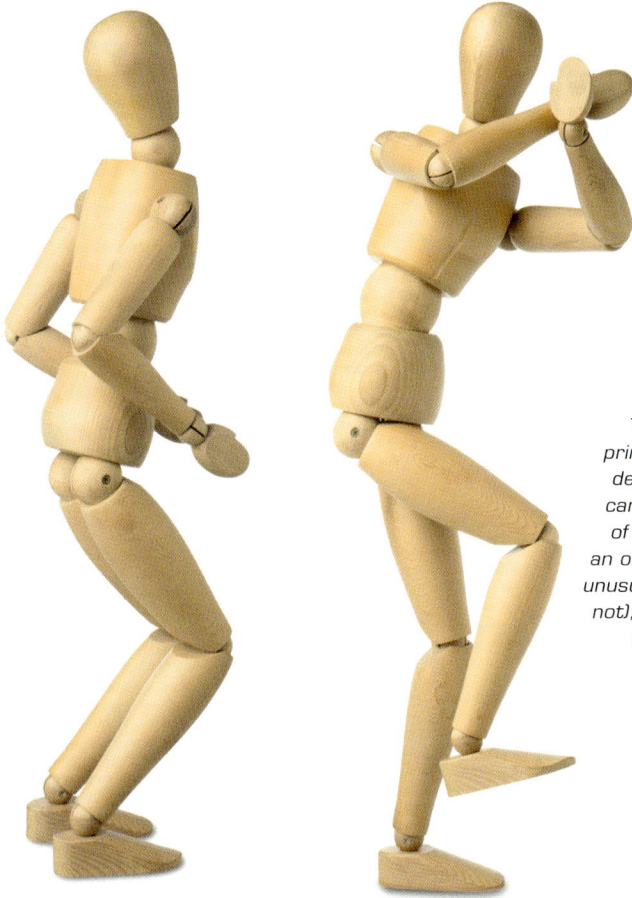

There are two types of primary motor reactions: defensive reflexes, which can involve whole or part of the body in relation to an object that provokes an unusual stimulus (painful or not); and startle reactions, consisting of a sudden movement due to a sudden stimulus.

The diaphragm and respiration

This systemic approach to manual therapy makes it essential to emphasize the importance of the main respiratory muscle, the diaphragm, and its relation with the fascia. When we submit to massage treatment, the action of respiration links the body and mind. No animal has as varied a movement capacity in different environments as the human.

Emotional integration

The goal of all motor reactions is to maintain organic integrity. For this reason we can classify motor movements as they respond to primitive or manipulative requirements. Primitive reactions have a defensive function: to move away from danger and pain. Overall, the body moves away from the source of the pain and adopts a safe, reduced or fetal posture, associated with an increase in heart rate and blood pressure. These adjustments are due to the sympathetic nervous system and our reaction, which we call fear or unhappiness. Contrarily manipulation movements are exploratory or related to the search for food, a partner or social relations. They are predominantly expansive and extensive (posterior chain) gestures in the search of well-being, and they are associated with emotions such as joy and affection.

In short, "the aim of all behavior is to preserve a certain balance (stability) between the organism and its environment (concept of homeostasis)," according to Jean Le Boulch (1992).

When a joint alignment combines the pull of gravity and an interaction with the floor coupled with the mechanisms indicated in any movement, we see how these energies are channeled into the elastic tissues (fascia and muscles). Exploiting our unstable vertical body alignment and smooth and polished joint surfaces allows movements in opposite directions.

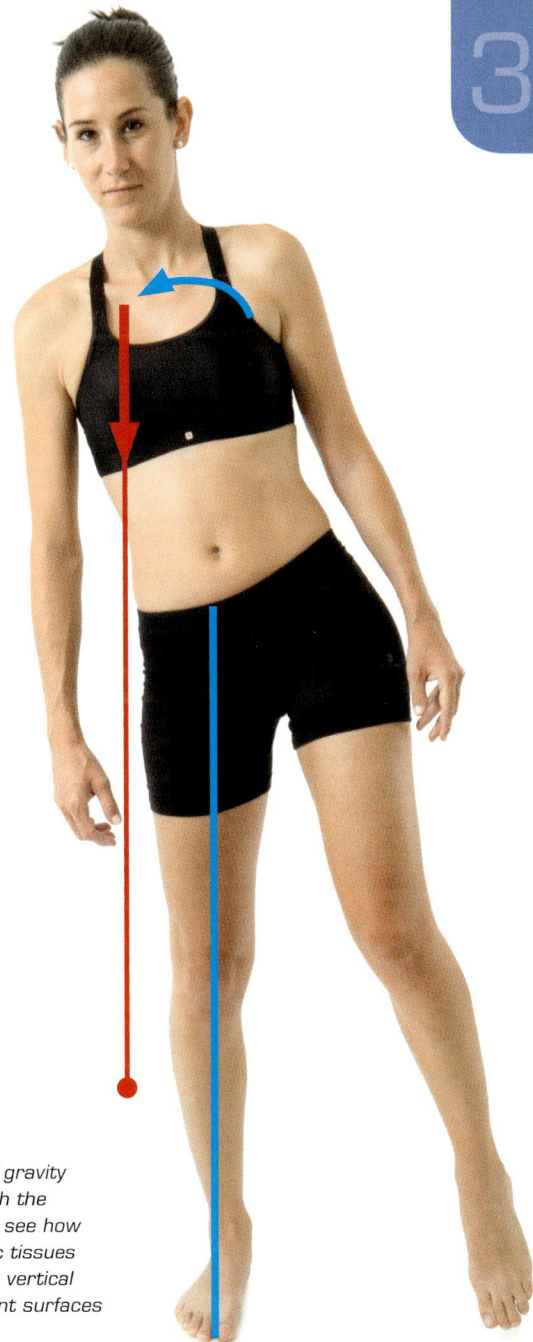

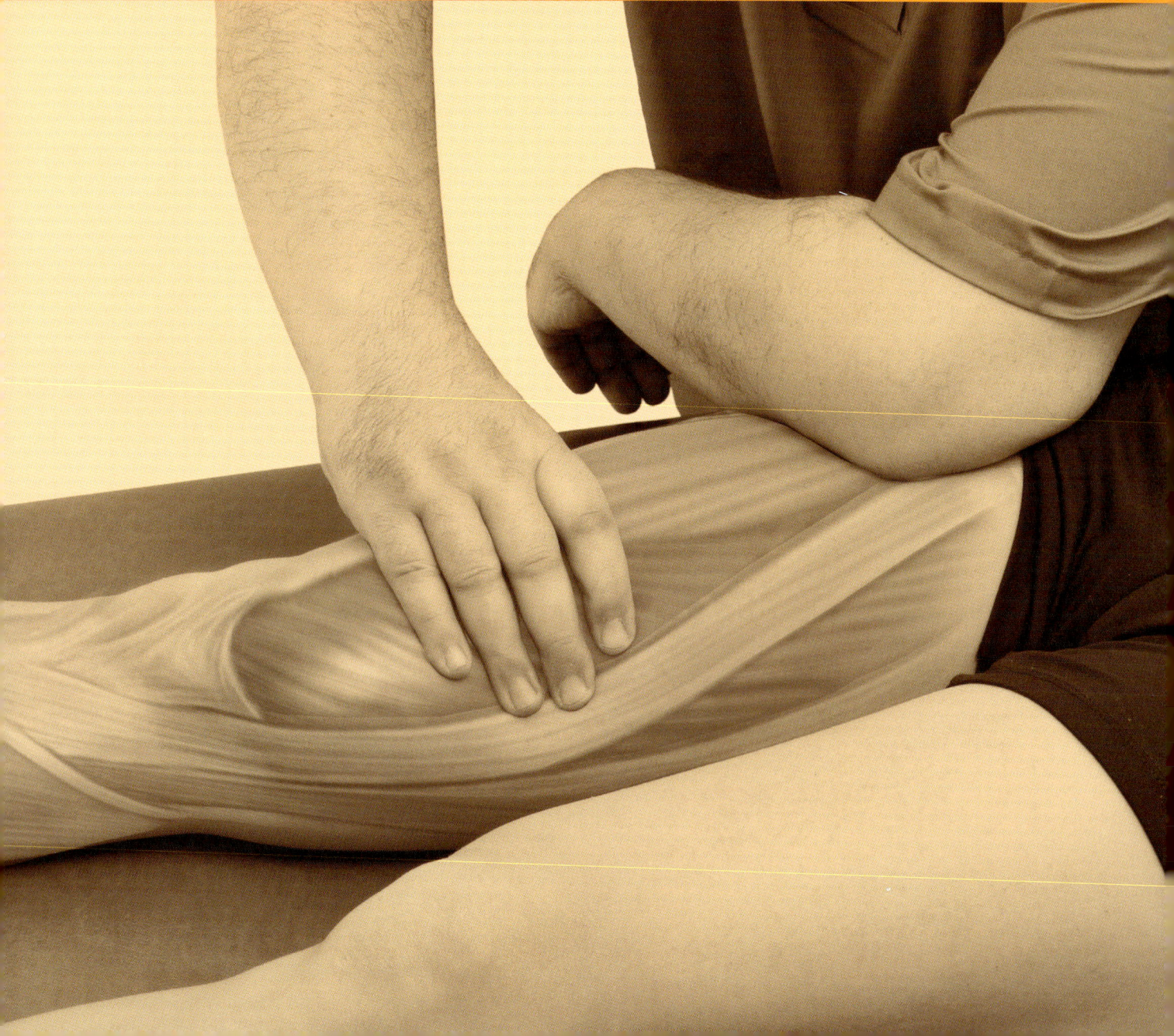

SPORTS MASSAGE: GUIDELINES

4

This chapter considers how to perform a general-purpose, sequenced sports massage employing all possible changes of position and how to work on the abdomen because of the benefits it brings to the health of the athlete.

Massage guidelines

A pattern is a set of technical movements used in each individual treatment. In fact, you can find as many patterns as massage therapists, and each therapist has his own approach and work sequences, sequences that contrast as he organizes his treatment.

A sequence for every situation

The professional therapist modifies his practice in relation to the requirements of the person he is treating. Conversely the novice therapist craves a "type" of routine that makes him feel secure in and facilitates his work, but this limits treatment options and effectiveness. This is because one cannot treat a patient presenting generalized rigidity than another with localized hyperlaxity, or a water polo athlete the same as a triathlete. A useful strategy is to write out various sequences as a test and then explore their approach.

Massage guidelines in the supine position (I)

The health questionnaire (pages 48-49) collects information about the sport practiced by the person to be treated and to differentiate which areas will be more overloaded. For example, for a cyclist the legs and thighs, psoas and buttocks and the neck extensor muscles, as well as the forearms and wrists will be overloaded. During exploration the therapist will automatically recognize which form the treatment strategy will take and devote more time and attention to these areas.

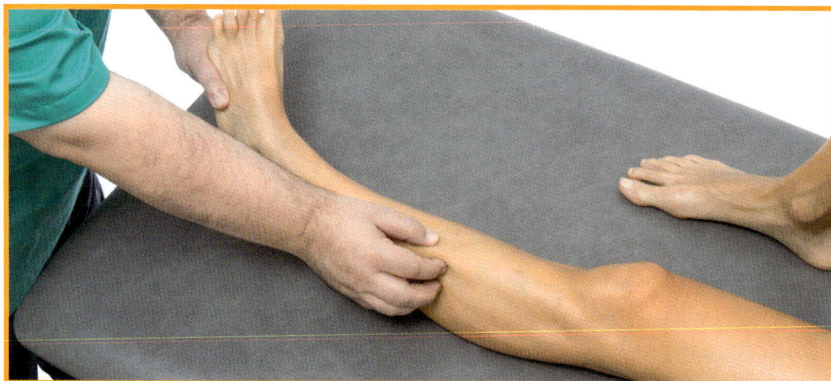

Lateral leg compartment. These tissues are loaded by traction and abrupt changes during continuous jumping or turning. A combination of digital kneading and kneading helps to relax the area.

How should the massage begin?

The athlete is first told which region is to be treated. It is normal to allocate more time to the area where the treatment begins, giving priority to the area that requires greater permeability.

Initial steps combine assessment and treatment. Long kneading passes are employed at the first tension layer; then a second "pass" is dedicated to working the already treated areas in more detail, focusing on those that are tenser or present discomfort or pain.

Techniques for general massage

During general massage employ a pattern that uses the basic maneuvers. This can start with pressure or gentle rubbing, followed by kneading combined with friction and pinching techniques, digital kneading to relax the treated area and joint mobilizations where they are needed most. The pattern is completed with more gentle rubbing or pressure, depending on the state of the athlete. This sequence allows the therapist to gauge treatment time but still reach all indicated areas while the patient's readiness and attention is maintained.

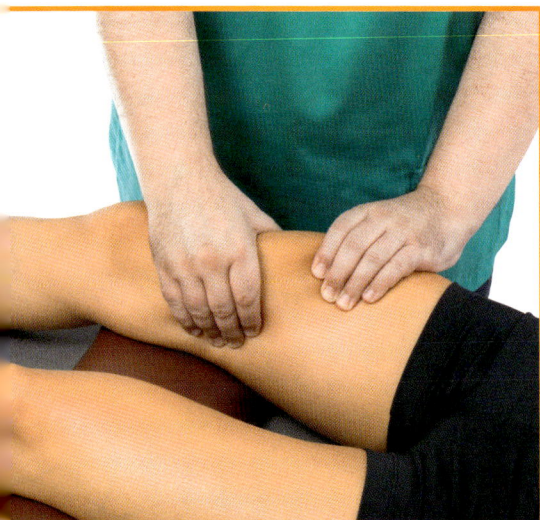

Thigh. The thigh area is gently rubbed in preparation to later treat the whole leg from distal to proximal. Initially the first layer of tension is treated and then kneading can begin in order to further increase irrigation and adjust excess tension.

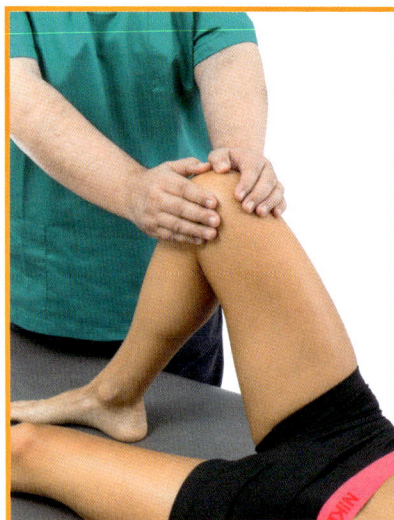

Knee. Friction and a thorough kneading with the thumb (not shown) of the knee help to discharge and regulate excess tension in tendons and ligaments.

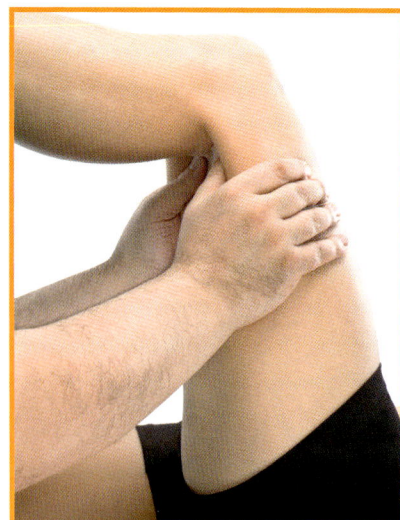

Popliteal region. The supine position allows us full access when treating the back of the knee. The athlete supports his foot on the shoulder of the masseur, so he has both hands free to apply massage in the area of the popliteal hollow.

The duration of the massage depends on the condition of the athlete and the range or extension of the area to be treated; this is limited by the time available for treatment. For a general massage to achieve satisfactory results, a minimum of 35 to 45 minutes and a maximum of 60 minutes are usually required.

Control and adjust after intense effort

Often after intense physical effort sports massage therapists make the mistake of trying to eliminate all excess tension in a single session. Some masseurs focus so intently on the tense areas that all they do is cause more discomfort and more pain. Working in this way will only create more tension, in other areas as well as the treated area. Therefore massage treatment seeks to recover myofascial tone, which may require more than one session before the excess accumulated tension is eliminated.

Massage maneuvers after competition

One of the most common maneuvers after intense competitive effort is the pressure technique. The athlete arrives at the treatment table a few hours after the competition with the tissues still warm and sore from the effort. Pressure techniques allow the therapist to dissipate excess tension without friction, and therefore without contributing more heat to the area.

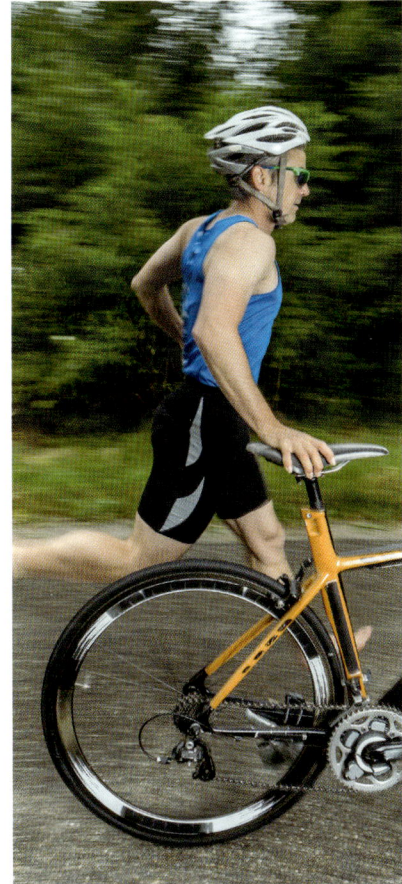

The triathlete (swimming, cycling and running) starts the last two stages with added fatigue loads. For this reason he requires a physiotherapy or osteopathy control in addition to a frequently applied general massage in which the more overloaded tissues are worked through long passes carried out at depth.

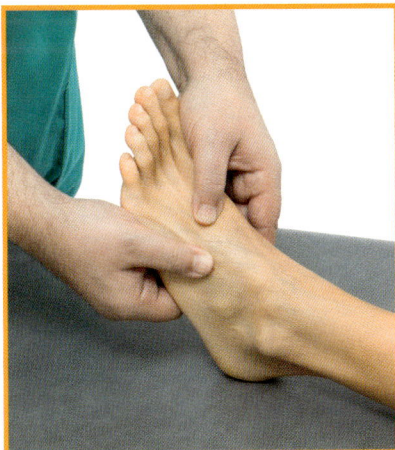

Foot. *The area of the back of the foot is often congested from exertion. Slowly applied gentle rubbing techniques provide venous renewal that facilitates decongestion and stimulates blood flow.*

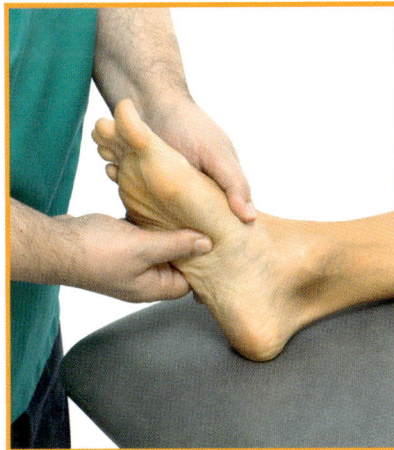

Foot. *Feet are very receptive to massage because during sporting activity they are subjected to multiple turns and impacts. Meticulous kneading normalizes and discharges the foot and the plantar fascia.*

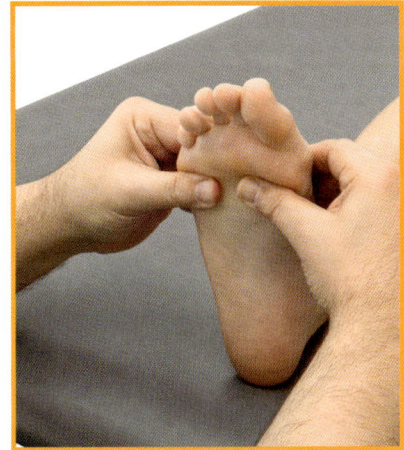

Sole and metatarsals. *The sole of the foot can also be treated in the supine position. This position also allows the therapist to perform joint micro-mobilizations.*

NOTE. Once all areas are permeabilized the entire lower limb is treated from the farthest away from to the nearest to the heart. The massage sequence then continues in the abdomen.

Massage guidelines: the abdomen

This area is often overlooked during sports massage. However it should be included in a general treatment and sports conditioning program for the benefits it brings.

Indications

Massage in the abdomen favors circulation and relaxation and allows the treatment of congestion in the organs of the lower pelvis and internal and external postoperative adhesions after a recovery period. It greatly helps to ease discomfort or lower back pain. In addition it favors the intestinal transit by promoting peristalsis through mechanically emptying the large intestine.

Before treatment

The areas to be treated must be established. A minimum of two hours should be allowed after meals before treatment and it is advisable that the athlete visit the bathroom to empty the bladder before the session begins. The masseur should rub his hands so that they are warm before placing them on the abdomen. This should be a gentle, pressing exploration; tense and swollen areas display signs of an alteration and the athlete should be referred to a doctor.

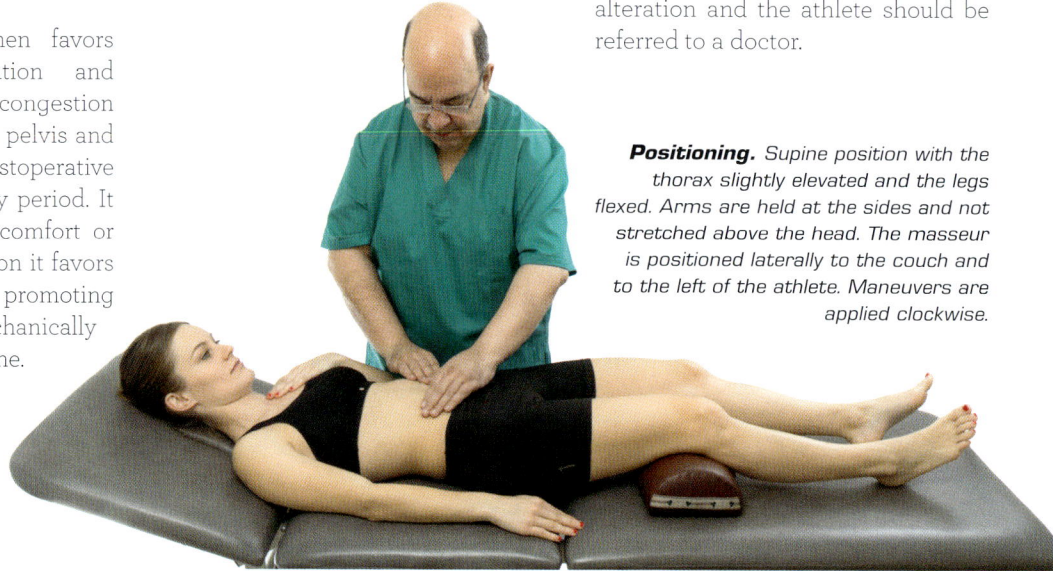

Positioning. Supine position with the thorax slightly elevated and the legs flexed. Arms are held at the sides and not stretched above the head. The masseur is positioned laterally to the couch and to the left of the athlete. Maneuvers are applied clockwise.

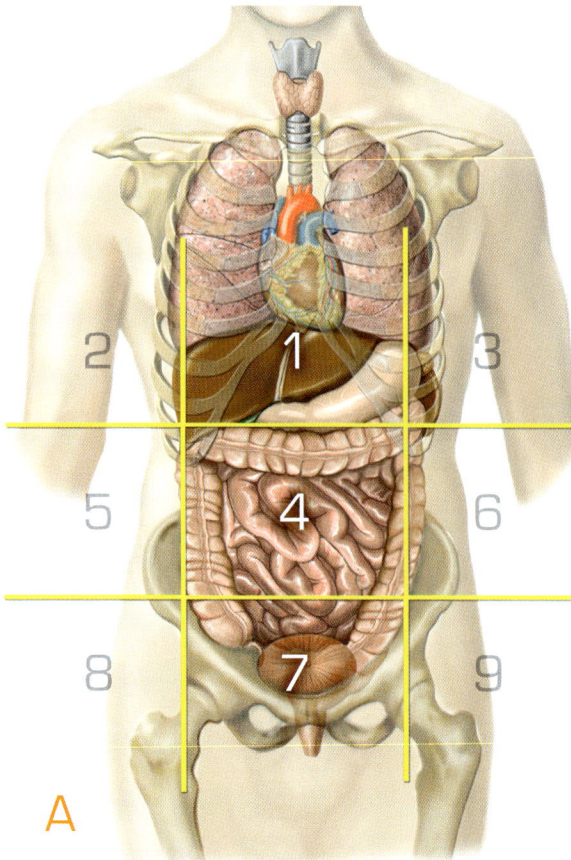

Quads. As the area of the abdomen is very extensive and contains a number of organs, anterior face A is divided into nine regions, quadrants or reference planes. The arrangement of these "plots" makes it easy to identify the different organs contained within. Image B shows the relationship of organs and viscera to the back.

The nine quadrants are:
1. Epigastrium
2. Right hypochondrium
3. Left hypochondrium
4. Mesogastrium or umbilical
5. Right flank
6. Left flank
7. Hypogastrium
8. Right iliac fossa
9. Left iliac fossa

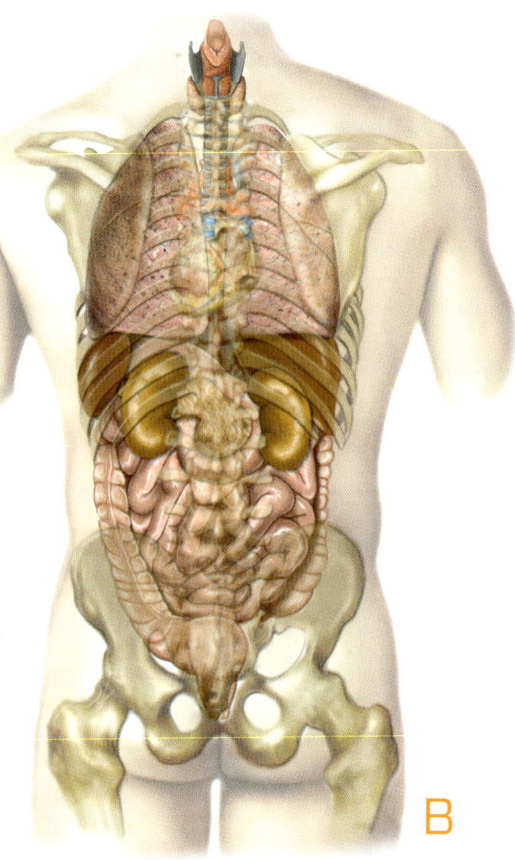

A

B

Continuous massage applied to the abdomen exerts a healthy effect on those athletes who have difficulty in evacuation because they are subjected to significant stress loads. These benefits will increase if you have previously treated the legs, buttocks and hips with massage and the appropriate joint mobilizations are applied.

Relative contraindications

Because massage increases blood flow it is avoided in the abdomen and lower extremities during the first two days of the menstrual cycle. Massage should also not be applied two days previous to the menstrual cycle in cases of very heavy and painful menstruation.

The abdomen, lower lumbar and sacral areas should not be treated during the first trimester of pregnancy and always with prior medical consultation.

Before starting

The athlete should be close to the side edge of the stretcher and the therapist will be placed to his left.

Contact is smooth as the therapist perceives respiratory rate and compass. To avoid obstructions, both hands perform smooth clockwise circular friction movements in a rhythmical, deep manner.

To increase relaxation in the zone, the lumbar area and both hips are pretreated. It is advisable to use a small amount of oil or cream, enough to allow hands to probe but not to make them slip.

Massaging the abdomen to empty the colon. *The therapist sits sideways on the stretcher and on the left side of the athlete. The pass is divided into three paths (a, b, c) that start on the left side. The first pathway comprises the iliac fossa, the flank, and the left hypochondrium (a) (that is, the entire descending colon).*

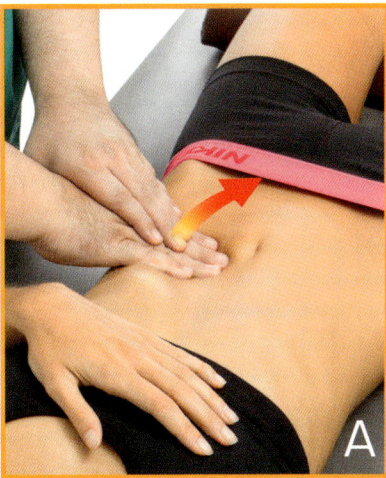

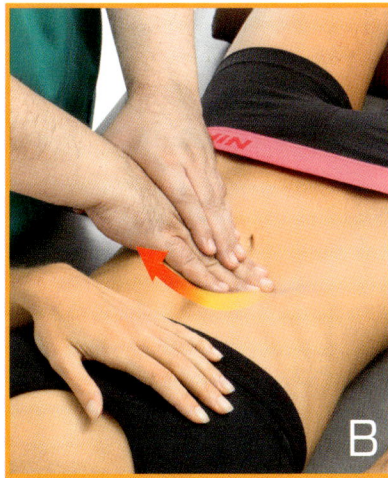

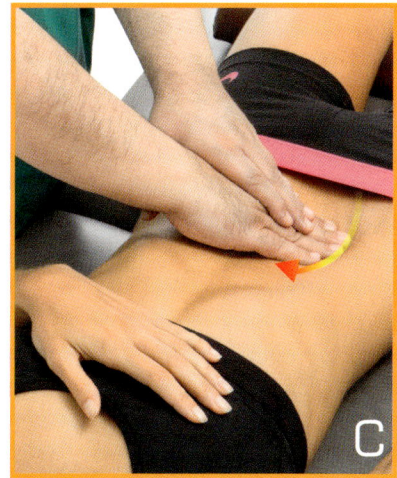

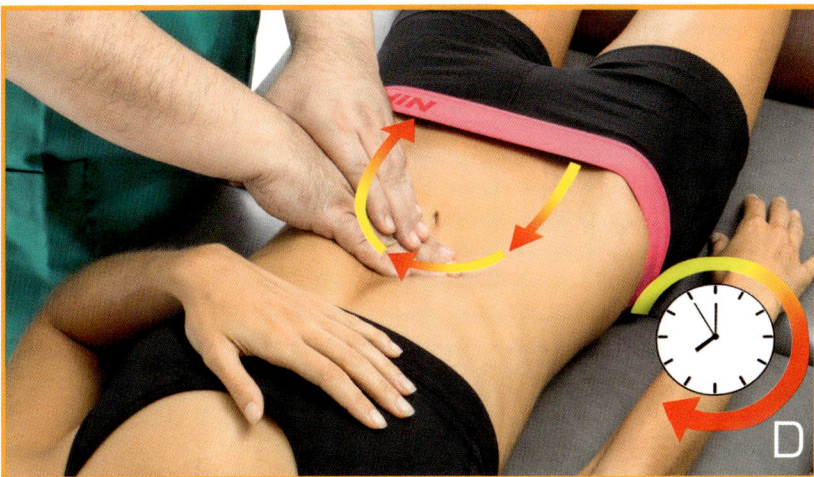

The transverse colon (b) is then treated from left to right, from the left hypochondrium to the right (i.e., the transverse colon from left to right).

The large intestine (c) is then massaged at a right angle to the ileocecal valve (i.e., the ascending colon from top to bottom).

Once the process is completed it is repeated inversely, in the clockwise direction (d): ascending colon, transverse colon and descending colon, several times.

Massage guidelines in the supine position (II)

During general massage it is important to maintain a uniform rhythm when applying maneuvers so that the athlete remains relaxed throughout the treatment, facilitating massage and increasing its benefits.

The pattern continues at a uniform pace

After the massage in the abdominal area the upper limbs are usually treated by gentle rubbing to promote venous renewal. The idea is to first treat areas that require greater permeability. To begin, pass over the anterior part of the left deltoid, the left arm and hand, then treat the entire limb from the fingers of the hand to the shoulder (distal to proximal). The same sequence is then repeated on the right side.

Changes in position

During general massage it is usual to combine work in different positions: supine, lateral (right and left) and prone, in addition to sedestation. For example the triceps brachii and lateral deltoid can be worked in prone or lateral decubitus, or in sedestation.

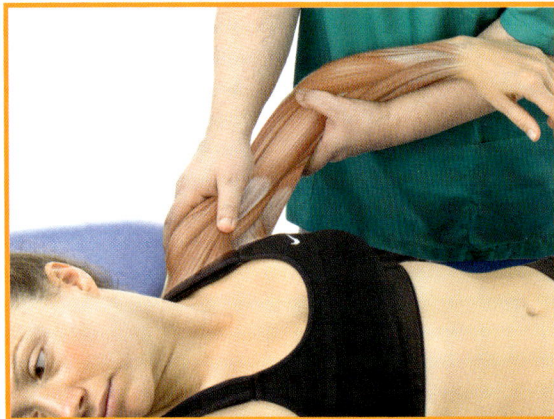

Deltoid. The kneading of the deltoid region allows excess tension to be discharged from the entire shoulder and prepares for work on the pectoral muscles.

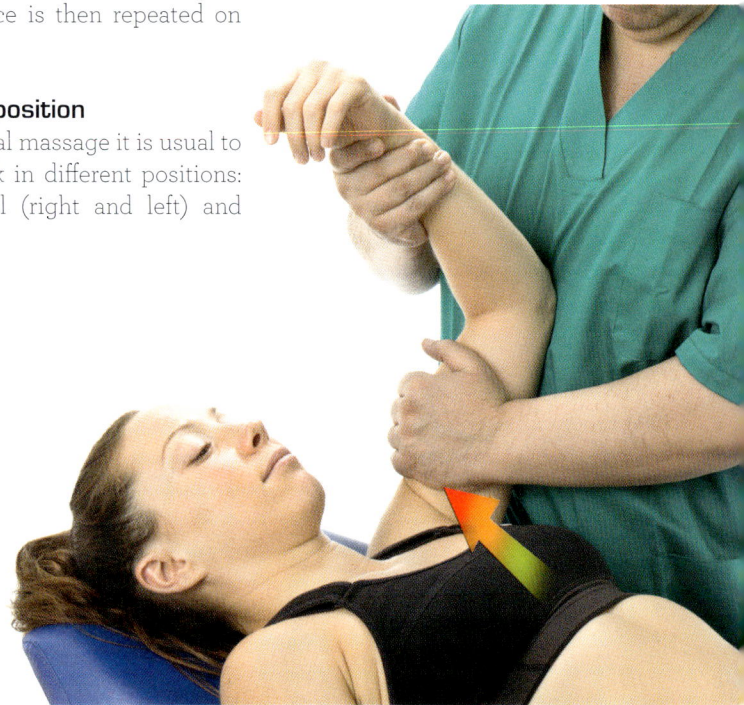

Brachial biceps. Biceps massage is important for athletes who throw or play racket sports because of the relation of the biceps to joint play in the elbow and in the case of shoulder overload.

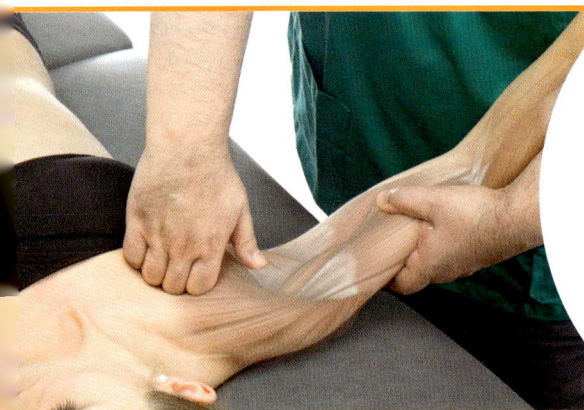

Pectoral muscles. The relationship of this musculature with the scapular girdle and shoulder makes massage very important in this area; sometimes it is not treated enough.

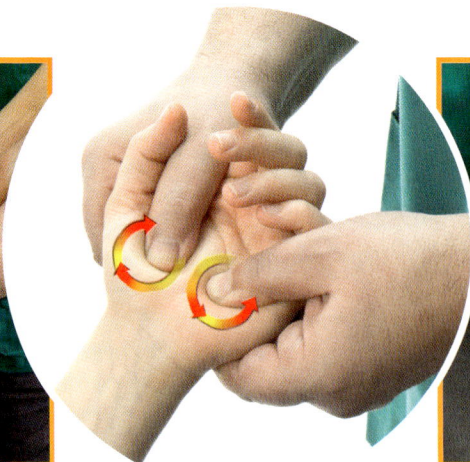

Palm of the hand. Soft rubbing and kneading allow drainage of the entire area. Once the entire upper extremity is permeabilized, treat from distal to proximal.

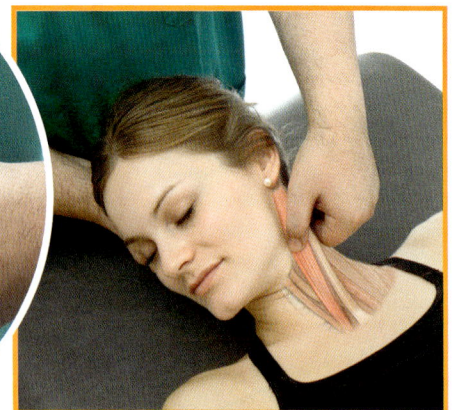

Neck. This is an area in which the massage is especially indicated when excessive cervical tension, cervicalgia, headaches and torticollis occur. Massage is applied, concentrated on the entire region of muscle insertions near the skull.

Massage guidelines in the prone position (I)

The pattern continues, treating the zones of the foot, leg and hip until it reaches the top end of the body, with the athlete lying facedown (prone position).

General and specific maneuvers

In classic massage and by extension in sport, several technical maneuvers are used that can be classified as general maneuvers: rubbing, friction and kneading. The there are the specific maneuvers: pressure, roller, transverse friction, etc.

Evaluate during treatment

During a general-purpose massage sequence one maneuver predominates all others, serving as a "bridge" between general and specific maneuvers and unifying movement during application of the massage pattern. This technique is digital kneading, which allows they therapist to evaluate while tissues are being treated. It is a gentle maneuver that relaxes an area after using a specific or "hard" technique, and also provides a pleasant sensation to complete the massage.

A general sports massage sequence consists of combining maneuvers when treating deep-plane tissues with classic massage maneuvers (joint mobilization techniques and stretching). The different techniques complement each other.

Gastrocnemius. *Kneading the gastrocnemius activates circulation throughout the body and offers a great degree of relaxation to the athlete after a high-intensity workout.*

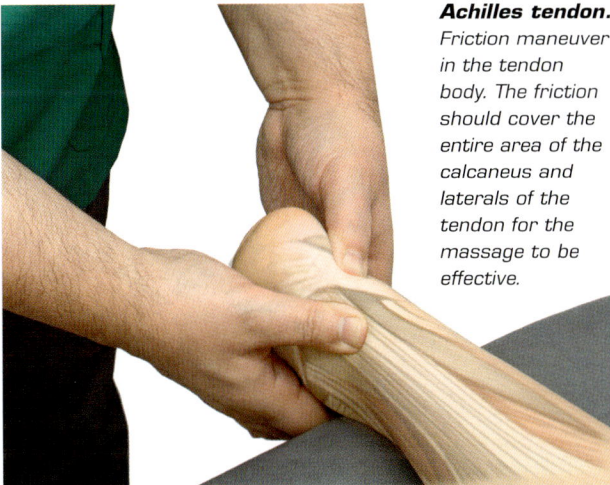

Achilles tendon. *Friction maneuver in the tendon body. The friction should cover the entire area of the calcaneus and laterals of the tendon for the massage to be effective.*

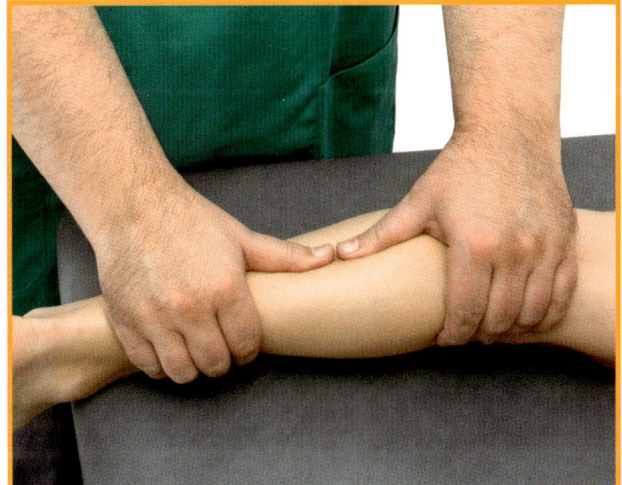

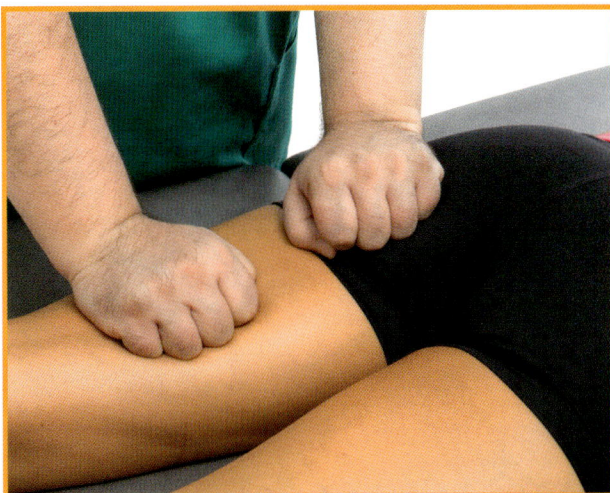

Gluteus and hamstrings. *After exertion, alternate pressure techniques on the glutes and hamstrings help dissipate tension and increase blood circulation by pumping.*

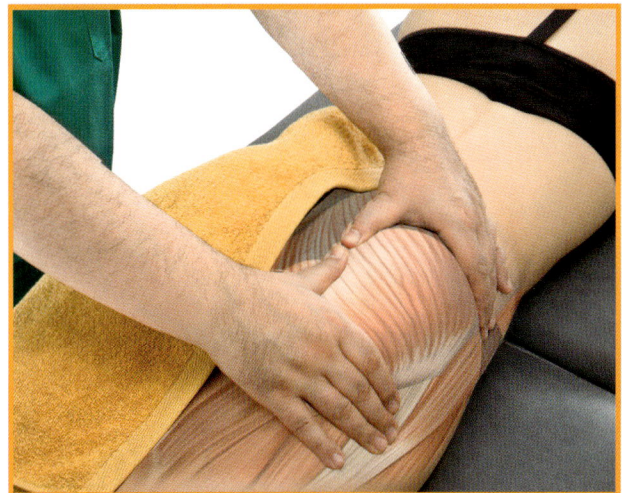

Gluteus. *Intermittent, meticulous kneading of the gluteal muscles allows the lumbar area to be relaxed and the area of hamstring insertion is discharged.*

Massage guidelines in the prone position (II)

Some sporting modalities by themselves may be a risk factor for back problems, especially if they require repeated hyperlipid positions found in some swimming styles (butterfly and breaststroke) and gymnastics. There is also a risk factor in sports where torsion movements predominate, such as basketball, volleyball, soccer and boxing, and in sports where implements are used such as golf and tennis.

Causes of back pain

These causes may be: accumulation of tension, myofascial tissue overloads, repeated microtrauma, muscle distention (to a greater extent in the lower back), problems in the intervertebral space and, in the most severe cases, protrusions, disc hernia or discarthrosis.

Factors of back pain

In sports that use one side of the body more than the other, such as racket sports, the spinal column of the predominant side is overloaded as all effort is exerted on that side, consequently triggering pain. The same thing happens if sporting activity is interrupted for a very long period (years), due to loss of muscle tone. Along with regular massage application, as prevention, exercises should be performed to compensate for the asymmetries and instability resulting from this type of activity.

Lumbar. Kneading technique that allows the evaluation and treatment of the first layer of tension and prepares the zone for maneuvers in deeper planes.

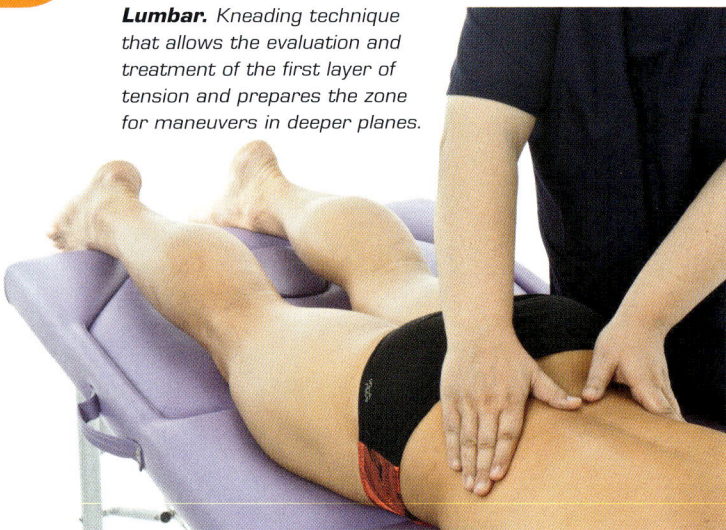

Lumbar. Technique with the forearm and opposed forces on the thigh with the hand. As it is a deeper technique it is applied more slowly than previous techniques.

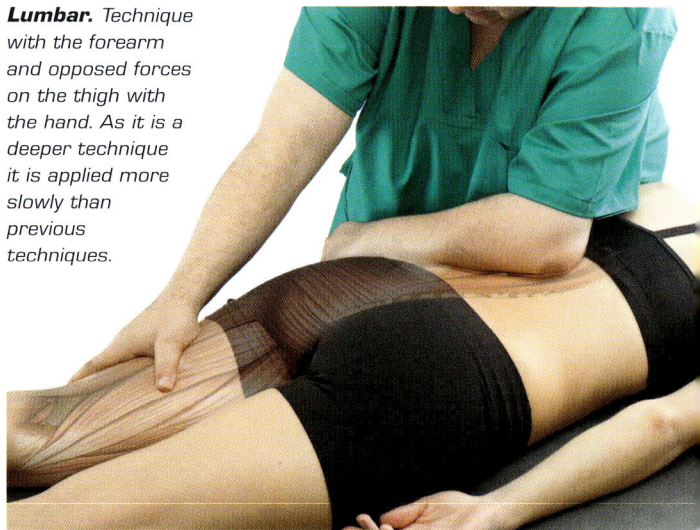

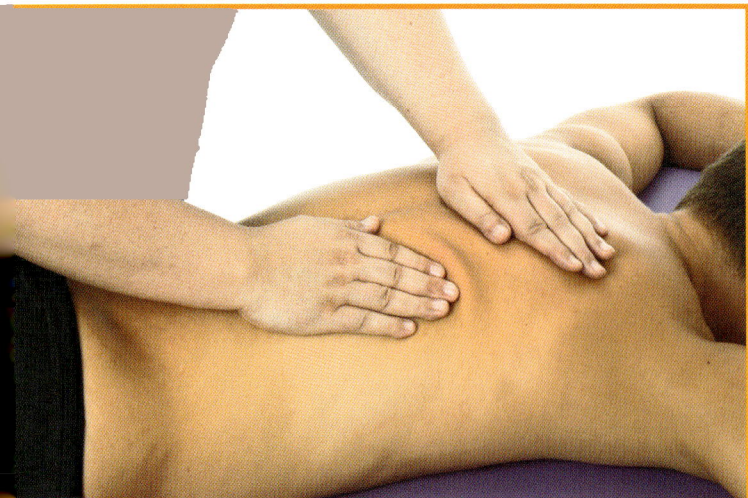

Thoracic. Friction technique with the palms of both hands used to warm the area and allow movement between tissues.

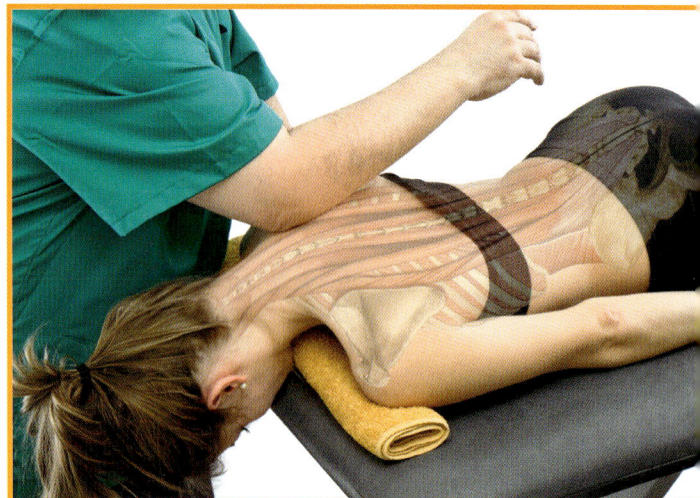

Thoracic. Pressure and drag with the elbow while the athlete drops his head. It should be applied by strokes and very slowly while asking the athlete how he perceives pressure, since this technique functions at great depth.

The head and neck region

Another area that presents problems during sports practice is the neck region. For example, when playing golf there are frequent overloads and pain in this region, but cervical injuries rarely occur. The same happens in cycling due to the position of flexion the spinal column assumes during a race and is associated with the need to maintain forward vision, which increases cervical extension, causing overloads and contractures in the neck and middle back, the interscapular region.

Continuously applied sports massage in this region allows the relief of excess tension, in addition to preventing overloads and contractures.

When playing tennis, the performance of a two-handed reverse backhand causes a violent rotation of the trunk that increases hyperextension of the spine, compromising the intervertebral discs.

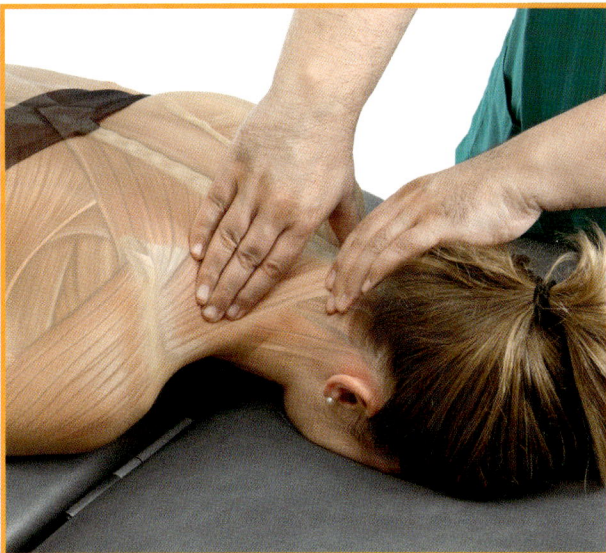

Cervical. *The positioning of the head with the face in the couch's facial hole facilitates the kneading of the neck with both hands, treating the entire posterior region of the neck and cervix.*

Cervical. *Decubito technique. Head and neck are turned to one side depending on the area that you want to treat with the kneading technique.*

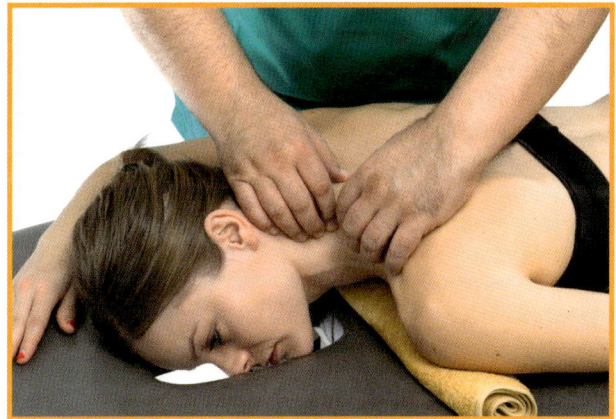
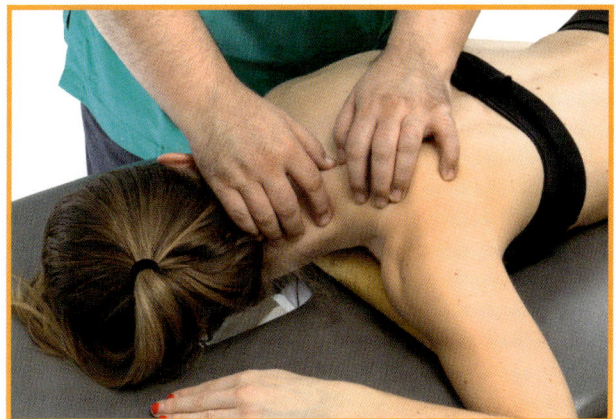

Massage guidelines in the lateral position

Changes of positions are sometimes not made because the therapist does not want to bother the person he is treating; however this is a mistake, as the benefits outweigh any possible inconvenience, and other positions also allow access to difficult-to-reach areas.

Gravitational forces and the lateral position

Besides reaching the areas that are difficult to access with the athlete lying in other postures, lateral position allows the therapist to treat different body segments aided by the force of gravity (during the application of a stretching technique, mobilization, etc.). It is a very interesting position in which to treat the lumbar square or thoracolumbar fascia, for example. It favors treatment and rest in cases where there is acute lower back pain and is also the most commonly used position in massage during pregnancy.

Iliotibial tract. Pain in the lateral area of the knee is very common in runners due to excessive tension in the iliotibial tract. As the athlete is lying on one side, it is possible to treat this band of fibrous tissue throughout its course.

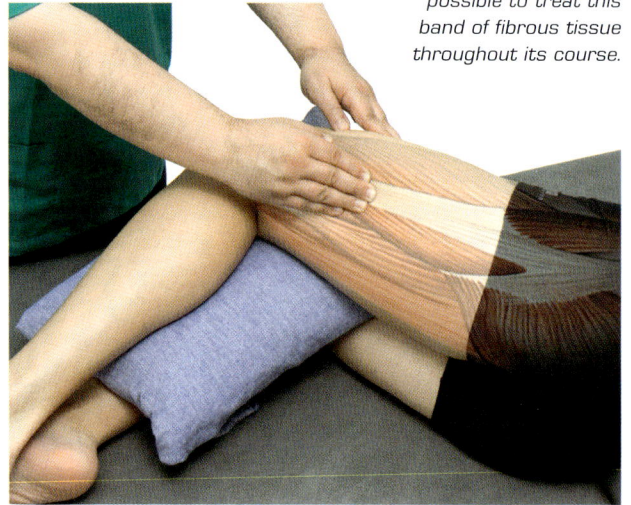

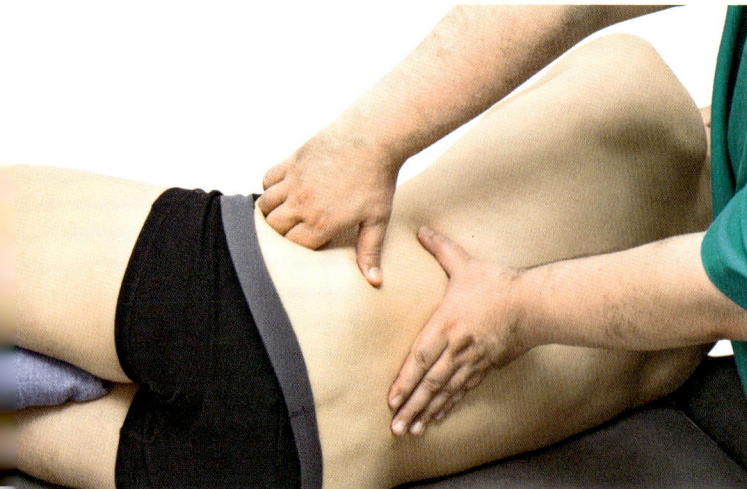

Trapezius and cervix. The lateral decubitus position facilitates detailed work on the postural musculature inserted into the skull. It also allows access to areas of the lateral regions of the trapezius and cervix.

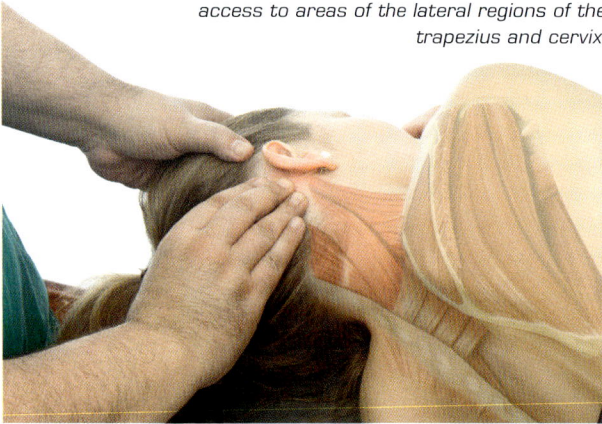

Gluteus. As the athlete is lying on one side, the gluteal region that presents the most discomfort can be treated with massage throughout its course. It also allows for treatment of thoracolumbar fascia connections with the iliac crests.

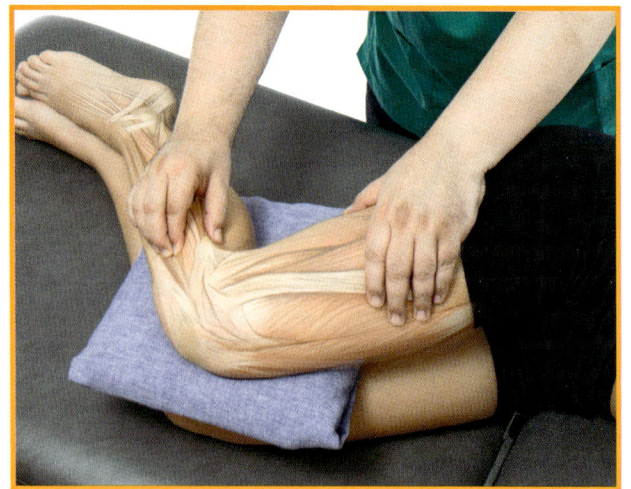

Lateral compartment of the leg. On the anterolateral surface of the proximal tibia are the tibial tubers (such as Gerdy's tubercle) and the head of the fibula. The position of lateral decubitus facilitates their manipulation, and if necessary also exposes the lame area of the leg for treatment.

4

Massage guidelines in the sitting position

To apply massage in sedestation use a chair designed for it, or the athlete can sit facing backwards on one of the courtesy chairs, supporting his head on a pillow placed on the treatment table.

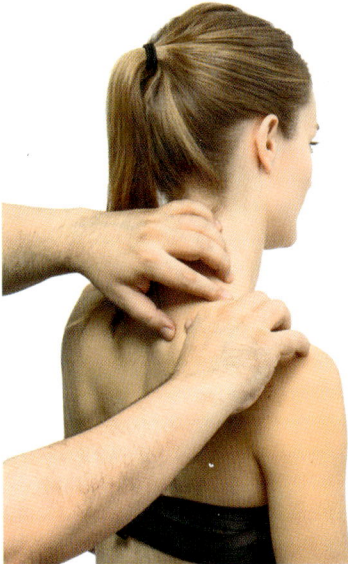

Massaging of the spinal erectors in sedestation

One way to finish the massage sequence (once the athlete has taken a few steps around the room after treatment) is to spend a few minutes applying a sedation massage on the spine erectors or smoothly kneading the trapezes.

Digital kneading. *Maintaining this position on the treatment table for some time could mean that the athlete experiences some form of residual discomfort. This is why at the end of the massage it is a good idea to softly "touch up" the trapezius and cervical areas.*

Ulnar percussion on trapezoids. *If you notice that the athlete is very sleepy at the end of the massage and needs to be more active, you can perform percussion maneuvers.*

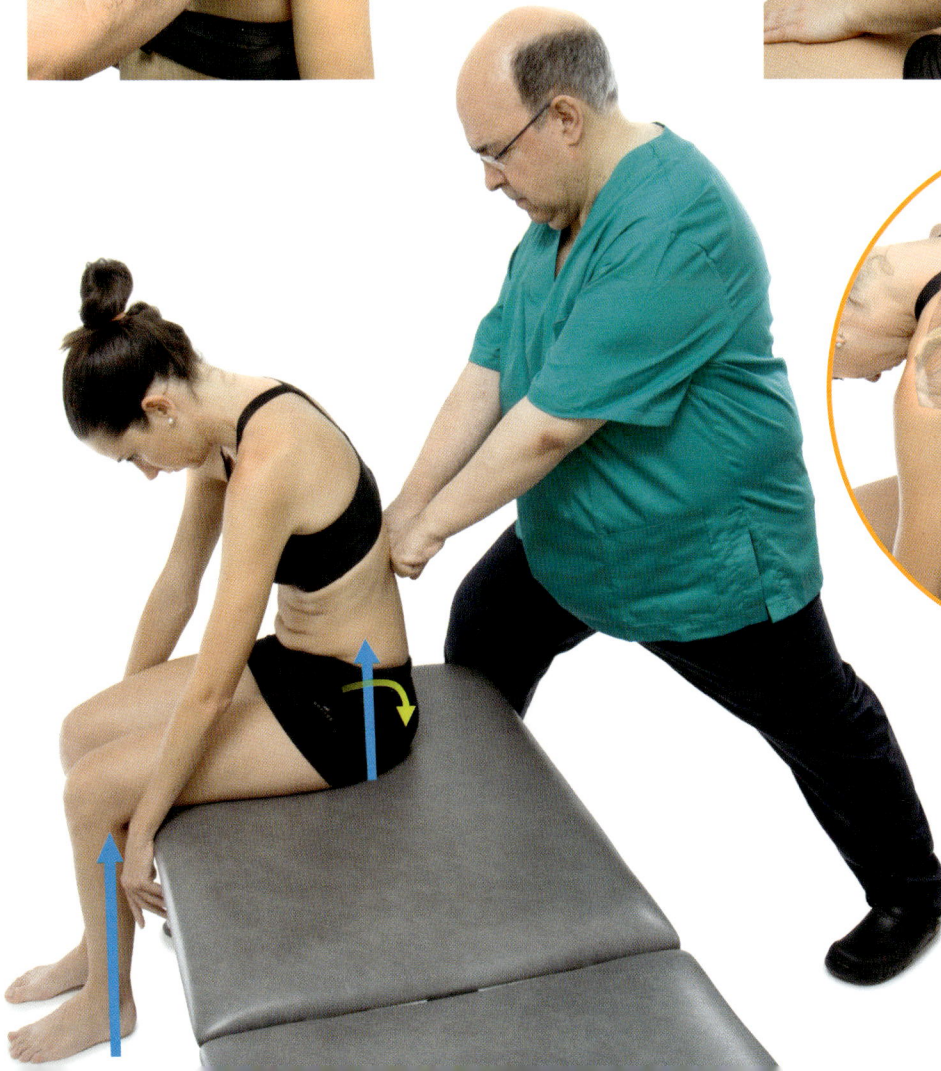

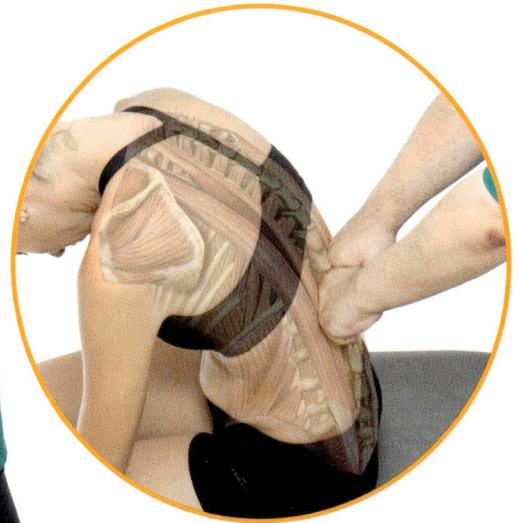

Spine erectors. *The athlete is seated with the feet well supported and arms to the sides; the trunk and the cervix remain erect. She should perform an upward force from the soles of her feet, simulating standing up, while performing hip retroversion as the masseur applies friction and drag techniques.*

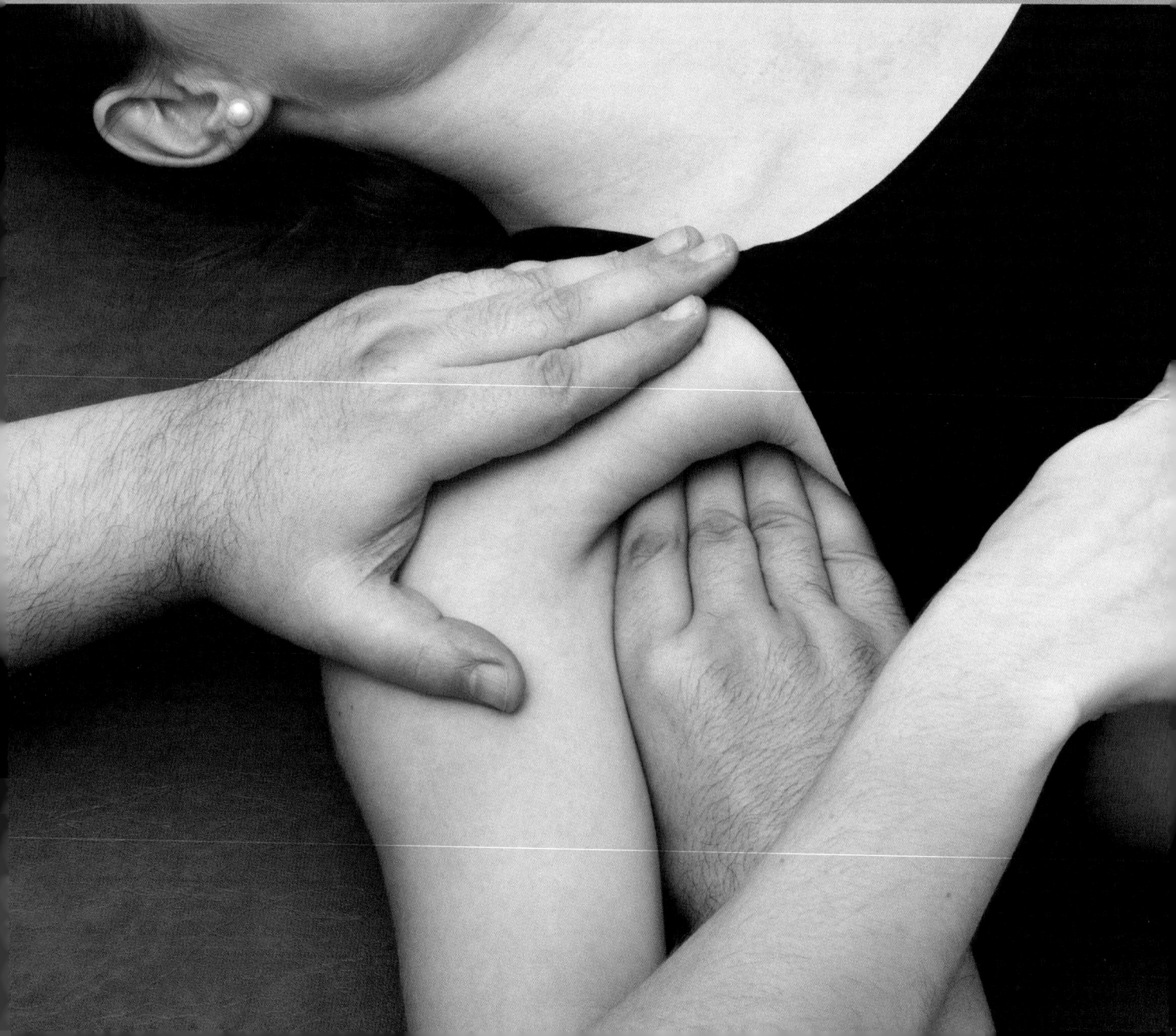

MASSAGE AND INJURY

5

This chapter describes some of the most common injuries in physical activity and sport. They have been classified according to traumatic mechanism or overload, depending on whether they present a locomotor or manipulative pattern, as well as a description of the massage techniques most commonly used for their treatment.

The massage fulfills two functions in relation to sports injury: preventive and recuperative/therapeutic (when the injury has already occurred). If we understand the massage in its broader concept of manually applied maneuvers, we find a huge therapeutic repertoire for sports injuries, especially overloads.

Some manual therapy techniques are: myofascial release, regulating and returning fluidity to connective tissue; transverse friction to reactivate the processes of rearranging ligaments, muscles and tendons; integrated neuromuscular inhibition techniques well suited to treat trigger points and repositioning techniques for the alignment of dysfunctional joints.

An overview of sports injuries

Sports injuries are generally caused by two mechanisms: trauma and overload. Trauma is caused by violent direct or indirect impact on certain body segments exceeding the elasticity limit of the indicated tissues.

Overload injury or recurring microtraumas are caused by instability in tissue strength, accumulated tension in the tissues and frequent, numerous impacts. Tendonitis, fibrosis and osteoarthritis are examples of this type of pathology.

Trauma

Injuries, whether direct or indirect, can be bone fractures, ligament or tendon ruptures, dislocations, etc., including fibril tears (internal muscle ruptures) and bruising.

Trauma represents a serious injury that must be treated "urgently," medically in the first instance. Massage can be applied and is of great importance in the recovery process and in the prevention of sequelae: scars, misalignments, rigidities, fibrosis, etc.

Chronic inflammation in overload injuries

Repeated impact on an area of the body during sports training results in "fatigue of biological materials." The cellular environment is altered by inadequate diet, age, toxins or mechanical overload, causing an imbalance that can be attributed to a nominal inflammation, which will become chronic over time if left untreated. These injuries are found at the tendon insertions or in the periosteum of the bones. Elbow tendonitis and osteopathy of the pubis are examples of these chronic inflammations.

Muscle instability

Muscles can be classified according to their metabolic characteristics as tonic (high strength and short return) and phasic (high power and wide return). Over the years overloading body weight, repetitive gestures and poor posture cause the tonic muscles to become fibrous and the phasic atrophic. Muscular instabilities appear around the joints at the scapula girdle, pelvis and shoulder and at different segments of the spine.

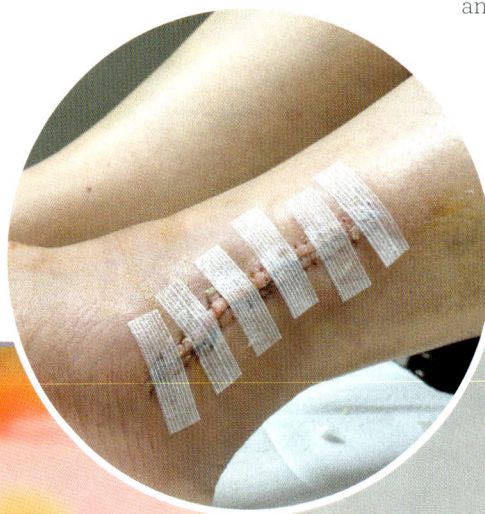

Scar

*Knee trauma
in an inline skater.*

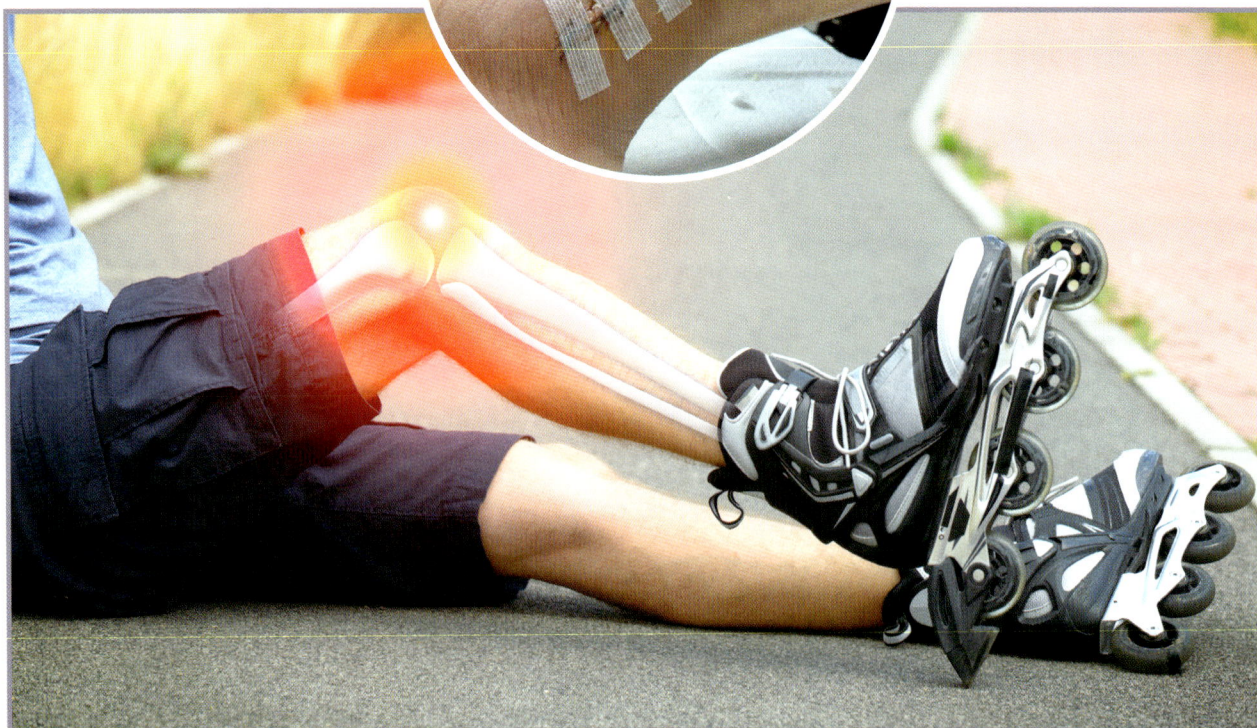

Trigger points and myofascial pain syndrome

Trigger points are muscle "knots," small spasm areas (rigid to the touch) of fibers that result in pain, which often radiate into an area completely separate from where they are located. They could be a result of overload caused by repetitive, eccentric and maximal/submaximal contractions in physical, workplace, recreational or sports activities, when muscular activity is not compensated with adequate rest. Their treatment increases irrigation and nutrient supply, reducing excess of tension and pain and increasing capacity of movement.

Scars and fibrosis of fascial tissues

Poorly treated wounds and subsequent scarring, immobility and chronic inflammation make connective tissues accumulate more collagen than necessary, reducing their elastic capacity. Restrictions due to "lumps" of collagen reduce joint mobility (both macro- and micro-movements), thereby reducing long-term functional capacity.

Entrapment syndromes

An impeded nerve or blood vessel may occur in a restricted pathway, with stress surges accumulating in several layers of tissue. The flow of material through the nerve or vessel is reduced and the part of the body segment that depends on their irrigation or innervation is compromised, reducing physical performance. Carpal tunnel syndrome and popliteal entrapment are examples of these pathologies.

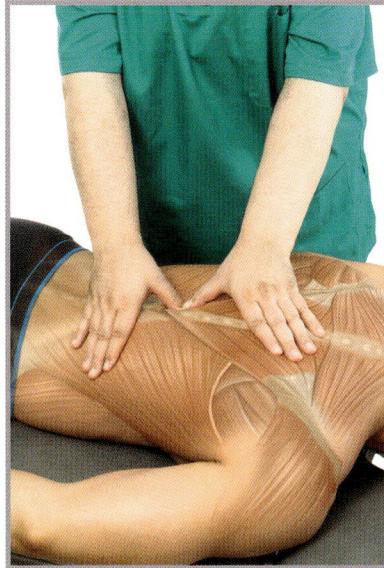

Trigger point technique.

Rheumatic joint diseases

Osteoarthritis and rheumatoid arthritis are diseases of the joints that cause inflammation, erode cartilage and elicit a reactive bone response. The space between bones is reduced, and proper alignment, cushioning capacity and joint range of motion are lost. The reason articular cartilage wears out is not precisely understood, but it is related to misalignment of loads on the joints, an immune system disorder (an inability to correctly identify the body's tissues), the presence of environmental toxins and previous infections as well as other causes.

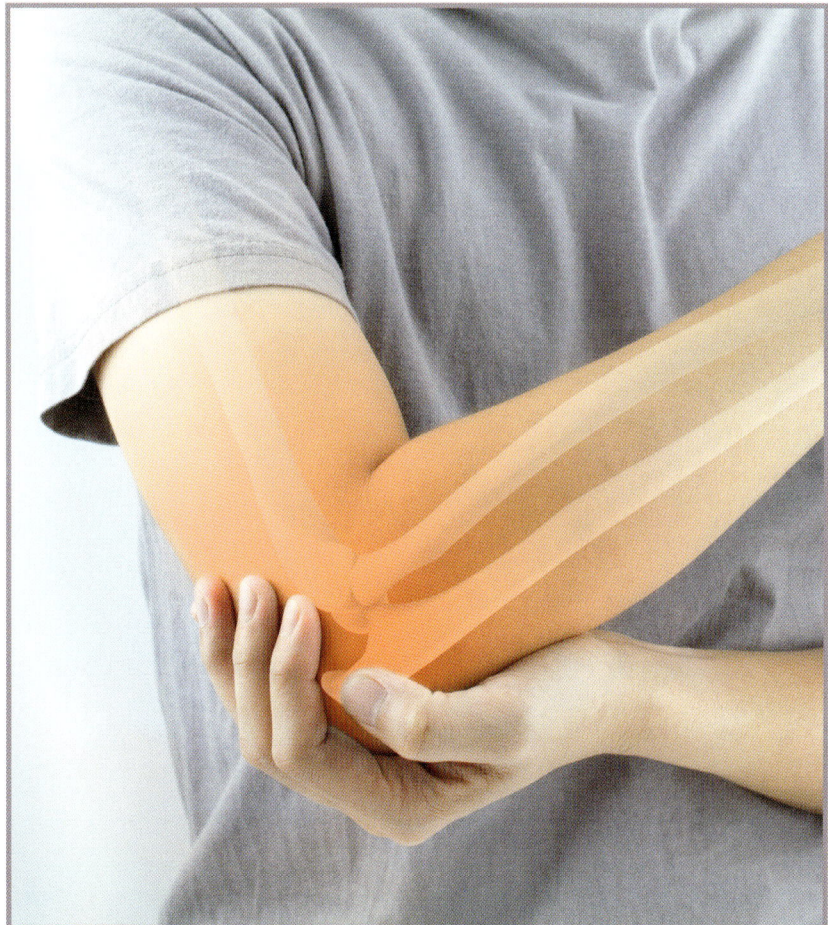

Rheumatoid arthritis in the elbow area.

Common injuries to locomotors

Plantar fasciitis
Repetitive take-off or landing of the foot when running or jumping overloads the fascia in the sole of the foot, the structure that stabilizes the plantar vault. Stresses accumulate in the calcaneus or at the base of the first toe and cause inflammation, especially in the insertion between the fascia and the periosteum of the bone. They are very persistent injuries that evolve to calcify and deform bone. They respond well to myofascial techniques and mobilizations.

Achilles tendonitis
The Achilles tendon or triceps sural transmits the force of the muscle to the calcaneus, extending the ankle and lifting the foot. The repetition

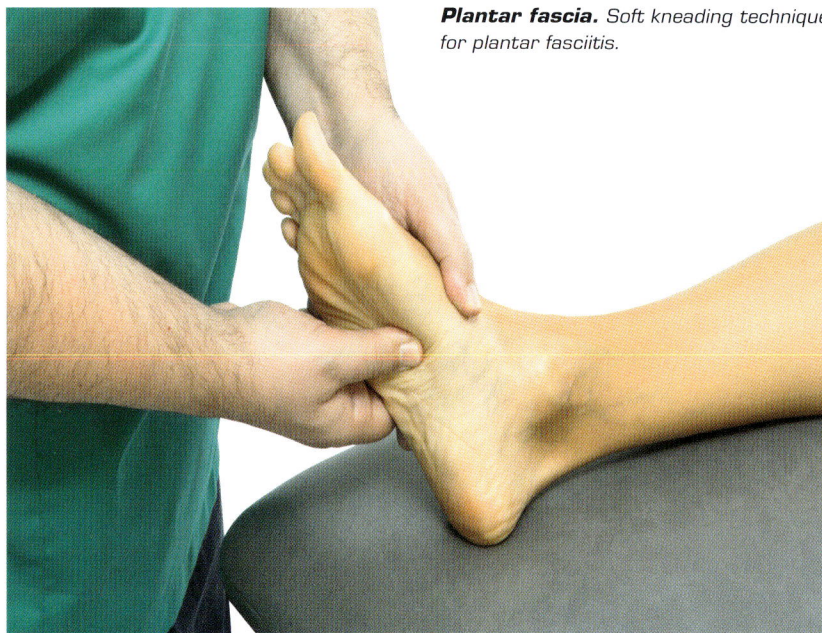

Plantar fascia. Soft kneading technique for plantar fasciitis.

of the gesture (when jumping and running) overloads the muscle-tendon-tendon-bone junction and triggers an inflammatory process.

Tendonitis of the Achilles tendon is very persistent and can easily become a chronic pathology since it is a muscle that is continuously in use. Deep friction massage and fascial pumping techniques are very effective for its treatment.

Friction technique to treat a case of knee pain.

Anterior tibial compartment syndrome
The anterior tibial originates in the anterior periosteum of the tibia. This muscle cushions the foot when the heel impacts the ground (upon landing). Repetitive pulls on the periosteum produce inflammation of this area, which cannot expand due to the robust covering of the muscle compartment. The vessels and nerves that circulate in the front of the leg are compressed and the leg goes limp. Myofascial techniques allow the therapist to create space in order to decompress the leg.

Jumper's knee
The quadriceps tendon or patellar ligament is overloaded when the athlete frequently repeats the jumping movement (the ankle-knee-hip triple extensor mechanism). Tendinitis is usually located in the upper section of the patella or in the anterior tibial tuberosity and is manifested by pain, swelling and functional impairment. These are repetitive injuries that become chronic if the underlying problem is not treated. They respond very well to deep transverse massage.

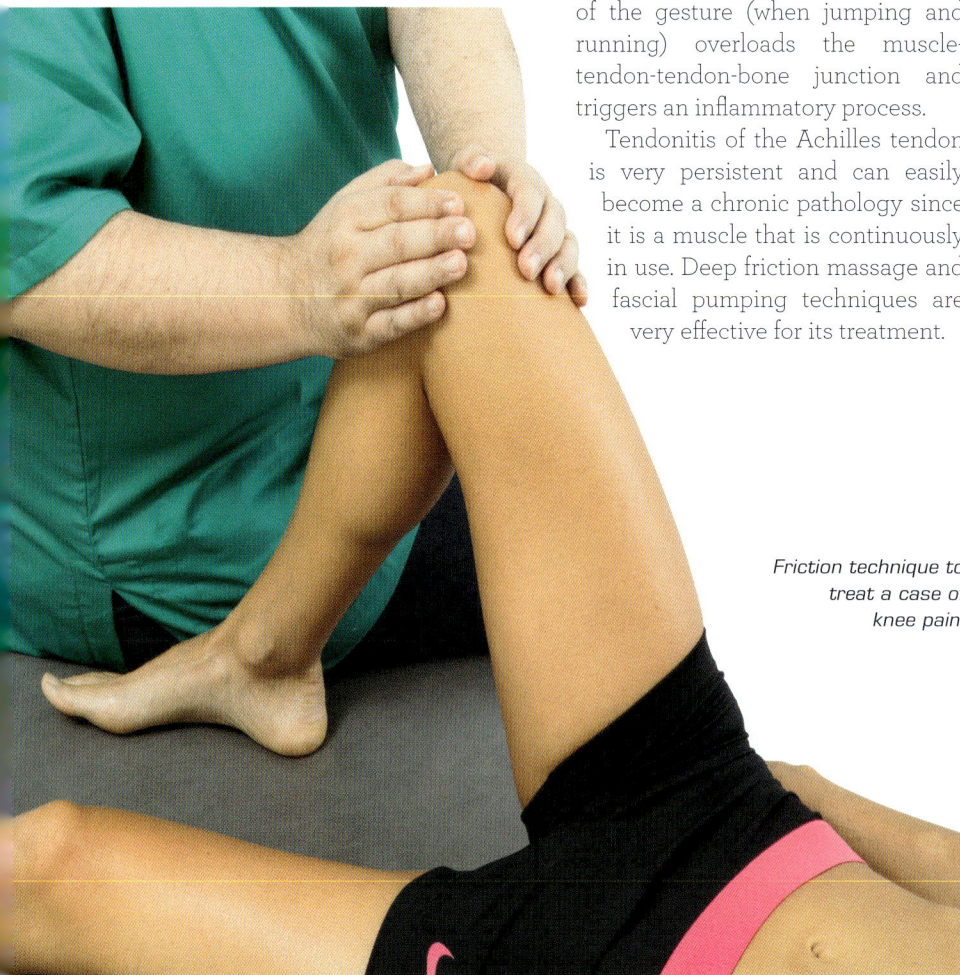

Chondromalacia patella

The term refers to wear of the articular cartilage by excessive compression of the patella against the femur when pulled by the quadriceps. The femur and tibia form the "valgus" anatomical angle (physiological), and the quadriceps moves the patella outwards. This normal procedure regresses into overload as the knee extension is constantly repeated and is aggravated when we lift large weights (squat), if the body is overweight or the knees have a pathological valgus. Treatment of chondromalacia indicates kinesiotherapy, postural reeducation and collagen supplements. Massage is indicated in the periphery to improve nutrition of the articular cartilage.

Lower back pain

Lower back pain (lumbago) can originate in a diverse number of tissues. Contractures, paravertebral fibrillation fractures or vertebral ligament sprains may present as lumbago. Fissured or ruptured intervertebral discs and disc herniation (the nucleus pulposus slips from its ring) produce lumbar and even lumbosciatic pain. Herniated discs are well known as a "cause" of lower back pain, and may be attributed to be the cause of the pain. However, osteoarthritis, rheumatoid arthritis and certain diseases such as tuberculosis or cancer are other possible causes of lower back pain. Treatment will depend on the specific cause of lower back pain.

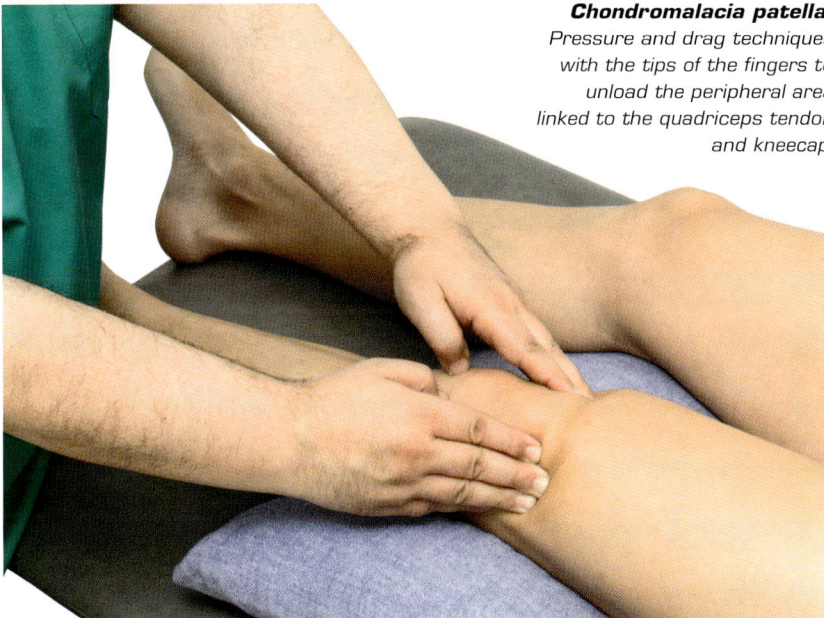

Chondromalacia patella.
Pressure and drag techniques with the tips of the fingers to unload the peripheral area linked to the quadriceps tendon and kneecap.

Lower back pain.
Characteristic response to back pain after physical activity.

Precautions

◆ When a sports injury occurs the necessary measures must be taken to ease the recovery process. The contemporary approach to rehabilitation moves away from the traditional indication of rest and opts for a therapy based on active recovery (movement plus exercise), adjusted according to what the injured party can tolerate, boosting recovery and improving function in the damaged area. It is also advisable to review how sports equipment is used and how sport movements are executed.

Movement patterns correspond to the sequence of activated myofascial tissue and to the joints that turn when making a certain movement; in other words, the momentary organization of the muscular contractions that occur when we move. If a movement moves the body, we speak of locomotive patterns (jumping or racing). Manipulative patterns are generated by hitting, kicking, picking up, etc., to move or collect objects; body segments move but the body as a whole is not displaced.

Common injuries to manipulators

Carpal tunnel syndrome

This injury consists of entrapment of the median nerve. The flexor retinaculum of the carpus is located at the front of the wrist and serves to contain and direct the tendons of the forearm muscles. With heavy, hard striking with the hand (volleyball or basketball), grasping the racket handle, or leaning on the wrists heavily when seated at a desk (computer mouse), the tendons and the retinaculum become fibrous and thick, compromising the nerve passage. This compression produces a tingling sensation, muscle weakness and pain.

To prevent this problem, hygienic measures such as wrist discharge support, localized rest and stretching are used. The most suitable manual treatment involves mobilization of the wrist bones, soft tissues, median nerve and wrist flexor tendons.

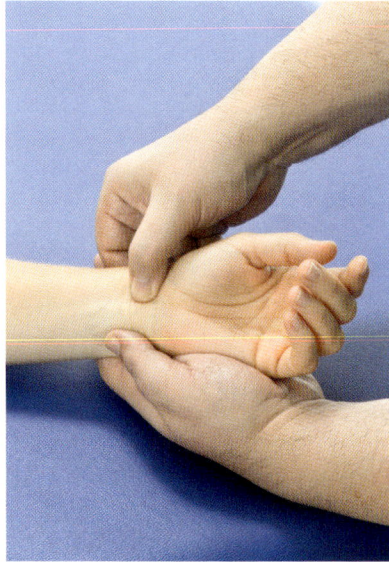

Friction technique in the wrist to treat carpal tunnel syndrome.

Palmar pump for forearm discharge of the forearm in an epicondilitis case.

Tendonitis of the epicondyles

The anterior muscles of the forearm originate in the anterior epicondyle and the posterior muscles in the posterior epicondyle, both located near the elbow. These are flexor and extensor muscles, respectively, involved in swinging the arm forward (anterior epicondyles) and backwards (posterior epicondyles). Repetitive impacts when hitting a ball with a racket or golf club transmit the stress of an impact and its vibrations to the tendons where the muscles originate. This overload produces inflammation and pain that can reach the palm of the hand. Therefore medial epicondilitis is often called golfer's elbow, while lateral epicondilitis is called tennis elbow. These chronic inflammations respond well to the transverse friction technique and myofascial stretching.

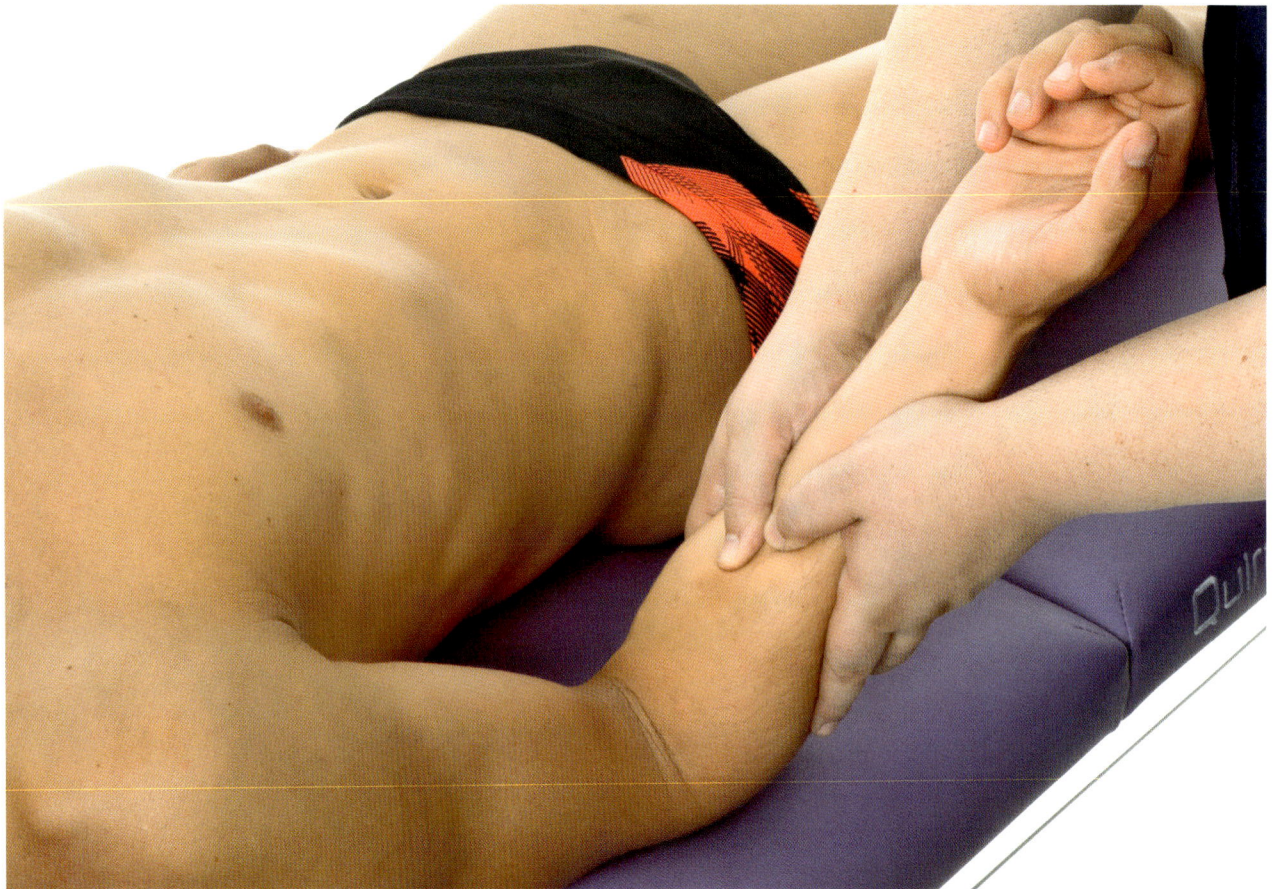

Soccer player.
Kicking the ball.

Rotator cuff syndrome

The shoulder muscles surrounding the head of the humerus form a lining called the rotator cuff. The tendons of these muscles pass through narrow boney and muscular channels, and when they overload they become inflamed. The sport movements that produce these overloads are fist blows, volleyball shots, tennis serves and the judo floor pinning techniques.

The most frequently affected sites are the supraspinatus tendon and the long biceps section (tendonitis) and inflammation of the subacromial bursa (bursitis). The most frequent symptoms are pain in the injured tissue, limitation in range of movement and loss of functional capacity. A painful shoulder should be clinically examined and is treated by joint mobilizations and transverse deep friction at the inflamed areas.

Osteopathy of the pubis

It is an insertion inflammation (tendon-periosteum) of the tendons of the thigh adductors. This adductor and flexor muscle of the hip intervenes during running, in changes of direction and in soccer passes. Inflammation is caused by instability in the forces that converge in the pelvis: abdominals, adductors and obliques. By frequently repeating this movement the athlete overloads the area where the adductor tendons are inserted, and they become inflamed. Local and irradiated pain appears in the pelvis and thigh, and the hip becomes inflamed and limited in range of movement, meaning that the athlete is unable repeat the mechanics of the movement. Treatment includes joint mobilization of the hip, stretching, and transverse friction of the tendons.

Judo. *Floor pinning techniques.*

Glossary

A

Activation (nervous). Application of low-intensity stimulus with high demand, inciting the central nervous system.

Active washing. Soft and low-intensity cardiovascular exercises (40-60% of VO_2 max) that can be combined with static stretches and used after exertion, normalizing circulation, metabolism and tone.

Analgesic. Relief or reduction of painful sensations.

B

Biochemistry. The scientific study of the chemical composition of living things.

Biomechanics. Study of the application of the laws of mechanics to living beings to investigate biological functions.

C

Contracture. High resistance and rigidity or permanent, involuntary and lasting contraction of muscle fibers.

Cortisol. Steroid hormone produced by the adrenal gland released in response to stress.

D

Dermatome. Area of the skin that receives sensory fibers from a single spinal nerve.

E

Eversion. External rotation of the foot with elevation of its outer edge.

Extracellular matrix. A collection of extracellular materials that are part of a tissue in which cells are "immersed."

F

Fibrosis. Thickening and retraction of the connective tissue that most often results from inflammation or trauma.

Fractal. Geometric object whose basic structure, fragmented or apparently irregular is repeated at different scales.

G

Ganglion or lymph node. Nodular structure that forms part of the lymphatic system and generates in clusters, especially in the neck, armpits and groin.

H

Holism. Methodological and epistemological outlook that postulates how the systems and their properties must be analyzed in their entirety and not only through the parts that they compose.

I

Iatrogenic. Refers to a condition or pathology that occurs as a side effect of treatment prescribed by a physician.

Inhibition. Decrease or cessation of the normal functions of a part of an organism mentally, physically or chemically.

Immobilization. A technique that limits the movement of a bone or injured joint by using an orthotic (splints, cast, bandages, etc.).

J

Joint play (movements). Involuntary movements in response to external forces.

K

Kibler (skinfold test). Test involving folding the skin to diagnose areas of hyperalgesia. Exploration is performed by moving the fold obliquely to the direction of the dermatomes.

Kinesthesia. The sensation of movement or the capacity of spatial location of the individual. It also refers to the branch of science that studies human movement.

L

Ligament. A fibrous, resistant cord that projects into the periosteum through which it joins the joints of the bones.

Lymphatic drainage (manual). Massage technique that acts on the superficial lymphatic system to improve the removal of retained fluid (edema).

M

Mechanical transduction. Transduction process of cellular signals in response to mechanical stimuli. Converts the mechanical stimulus into a chemical sequence due to distortion of the membranes, leading to a search for membrane components that can mediate the appropriate mechanochemical conversion.

Muscle hypertonia. Exaggerated and permanent muscle tension presented in a muscle at rest.

N

Neuropeptides. Small molecules formed by the union of two or more amino acids that originate by cerebral synaptic transduction.

Nociceptive. A common form of pain that appears as a consequence of the application of stimuli that produce damage or injury of somatic or visceral organs. Appears in the context of acute or chronic pain.

O

Osteophytes. Pathological neoformations of bone tissue prominent in the form of hard, well-circumscribed protuberances on the outer surface of a bone.

Osteopathy. Science that studies and applies a global concept of manual therapy, based on the idea that all systems of the organism are interrelated with each other, so that any restriction or discomfort in one area affects all other areas.

Overcompensation. The body creates organic adaptations in response to stress as a stimulus.

P

Pattern of movement. Initial movements from which all technical movements are derived.

Proprioception. Informs the body of the position of all muscles and encompasses the ability to feel the relative position of the body.

R

Reflex. Automatic and involuntary response performed by a living being at the presentation of a given stimulus. The reflex response usually involves a movement, although it may also consist of the activation of secretion in a gland.

Relapse. Repetition of a disease shortly after convalescence.

S

Spasm. Abrupt, involuntary and persistent contraction of muscle fibers.

Systemic. Of the totality of a system, as opposed to local or related to it.

T

Tendinosis. (Commonly called tendonitis.) An accumulation of small tendon lesions at the cellular level that involves a chronic degenerative pathology without inflammation.

Tensegrity (systems of). Refers to integrated tension or tension integration. A structural principle based on the use of compressed, isolated components within a continuous tensioned network. Compressed areas do not touch each other and are bonded only by means of tensile components, which spatially delimit the system.

Threshold. In physiology this refers to a certain minimum limit value below which a certain phenomenon cannot occur.

V

Vascular. Relative to the vessels or conduits through which blood or other liquids circulate in animals or plants.

Vasodilation. Increased caliber of a vessel caused by relaxation of muscle fibers.

Vasoconstriction. Decreased caliber of a vessel caused by contraction of muscle fibers.

Z

Zone (hypothenar). Leading part of the motor muscles of the little finger in the inner part of the palm of the hand.

Zone (tenar). A muscle mass located in the human hand and shaped like a drop of water which forms the base of the thumb.

Bibliography

Andrade, C-K. *Masaje basado en resultados*. Editorial Paidotribo, 2004.

Benjamín, B. *Listen to your pain*. Group Penguin Books, 1984.

Biel, D. *Guía topográfica del cuerpo humano*. Editorial Paidotribo, 2016.

Bosco, J. *Danza y medicina*. Librería deportiva Esteban Sanz S.L., 2001.

Bienfait, M. *Bases elementales técnicas de la terapia manual*. Editorial Paidotribo, 2008.

La reeducación postural por medio de las terapias manuales. Editorial Paidotribo, 2005.

Biriukov, A. *Masaje deportivo*. 4.ª edición. Editorial Paidotribo, 2003.

Bossy, J. *Bases neurobiológicas de las reflexoterapias*. Editorial Masson, 1985.

Busquet, L. *Las cadenas musculares* (tomos I, II, III y IV). Editorial Paidotribo, 2006-2007.

Cardinali, D. P. *Manual de neurofisiología*. Ediciones Díaz de Santos, 1991.

Curtis-Barnes. *Invitación a la biología*. 5.ª edición. Editorial Médica Panamericana, 2006.

Chaitow, L. *Terapia manual: valoración y diagnóstico*. McGraw-Hill Interamericana, 2001.

Técnica neuromuscular. Ediciones Bellaterra S.A., 1981.

Dolto, B. *Cinesiterapia práctica*. Editorial Paidotribo, 1995.

Franscoo, P. *Examen clínico del paciente con lumbalgia, compendio práctico de reeducación*. Editorial Paidotribo, 2003.

Fritz, S. *Fundamentos del masaje terapéutico*. Elsevier, 2005.

García Vilanova, N. *La tonificación muscular, teoría y practica*. Editorial Paidotribo, 2016.

Geneser, F. *Histología*. 3.ª edición. Editorial Médica Panamericana, 2000.

Gladman, G. *El masaje en el deporte*. 4.ª edición. Editorial Síntesis, 1961.

Guirao, M. *Anatomía de la consciencia, neuropsicoanatomía*. 2.ª edición. Editorial Masson, 1997.

Heiman, F. *Compendio de terapia manual*. Editorial Paidotribo, 2006.

Hilde, S. *Fisioterapia, teoría y registro de hallazgos*. Editorial Paidotribo, 2003.

Hoffa, Gocht, Storck, Lüdke. *Técnica del Masaje*. 1985 Editorial JIMS. Barcelona.

Howse, J. *Técnica de la danza y prevención de lesiones*. Editorial Paidotribo, 2002.

Ingberg, D. *The architecture of the life*. *Scientific american*. Jan 1998.

Kaltenborn. *Fisioterapia manual. Columna vertebral*. 2.ª edicion en español. McGraw-Hill Interamericana, 2004.

Kapandji. A.I. Fisiología articular. *Tomo 2. Miembro inferior*. 6ª/2012 Editorial médica Panamericana.

Fisioterapia manual. Extremidades. 2.ª edicion en español. McGraw-Hill Interamericana, 2004.

Kuprian. *Sport et physiothérapie*. Editorial Masson, 1987.

Le Boulch, Jean. *Hacia una ciencia del movimiento humano*. Paidós, 1984.

Lloret, M. *Anatomía aplicada a la actividad física y deportiva*. Editorial Paidotribo, 2000.

Llusá, M. - Meri, A. *Manual y atlas fotográfico de anatomía del aparato locomotor*. Editorial Médica Panamericana, 2004.

McConail y Basmagiani. 1977. *Muscles and movements*. Williams & Wilkinks.

Montagu, A. *El sentido del tacto, comunicación humana a través de la piel*. Colección Aurion, Aguilar, 1981.

Myers, T. *Vías anatómicas. Meridianos miofasciales para terapeutas manuales y del movimiento*. Elsevier Churchill Livingstone, 2014.

Orozco, L. *Tecnopatías del músico. Introducción a la medicina de la danza*. Editorial Aritza, 1996.

Paoletti, S. *Las fascias. El papel de los tejidos en la mecánica humana*. Editorial Paidotribo, 2004, 2013.

Pedersen BK, Febbraio MA. 2008. *Muscle as an endocrine organ: focus on muscle-derived interleukin-6*. Physiological Reviews.

Pérez-Caballer. *Patología del aparato locomotor en ciencias de la salud*. Editorial Médica Panamericana, 2004.

Pilat, A. *Terapias miofasciales: Inducción miofascial. Aspectos teóricos y aplicaciones prácticas*. Mcgraw-Hill Interamericana, 2003.

Piret y Beziers. *La coordinación motriz*. Masson, París, 1971.

Rasch-Burke. *Kinesiología y anatomía aplicada*. 6.ª edición. Editorial El Ateneo, 1986.

Riggs, A. *Masaje de los tejidos profundos. Guía visual de las técnicas*. 2.ª edición. Editorial Paidotribo, 2015.

Schoenfeld BJ & Contreras B, et al. 2013. *Influence of Resistance Training Frequency on Muscular Adaptations in Well-Trained Men*. Physiological Reviews.

Solé y Forn, J. *Terapéutica física. Masaje terapéutico*. Tobella y Costa Impresiones, Barcelona, 1906.

Souchard, Ph. *Principios de la reeducación postural global*. Editorial Paidotribo, 2012.

Reeducación postural global. Monográfico n.º 4. Edita I.T.G. Bilbao, 2003.

Stretching global activo (tomos I y II). Editorial Paidotribo, 2016.

S. Butler, D. *Movilización del sistema nervioso*. Editorial Paidotribo, 2009.

Stecco, L. *Atlas of physiology of muscular fascia*. Piccin Nuova Libraria S.p.A., 2016.

Viel, E. *Diagnóstico fisioterápico, concepción, realización y aplicación en la práctica libre y hospitalaria*. Editorial Masson, 1999.

Viladot Voegeli, A. *Lecciones básicas de biomecánica del aparato locomotor*. Editorial Masson, 2004.

Lecciones sobre patología del pie. Ediciones Mayo, 2011.